Food Composition and Dedicated Databases: Key Tools for Human Health and Public Nutrition

Food Composition and Dedicated Databases: Key Tools for Human Health and Public Nutrition

Editors

Alessandra Durazzo
Massimo Lucarini

MDPI • Basel • Beijing • Wuhan • Barcelona • Belgrade • Manchester • Tokyo • Cluj • Tianjin

Editors
Alessandra Durazzo
CREA-Research Centre for
Food and Nutrition
Italy

Massimo Lucarini
CREA-Research Centre for
Food and Nutrition
Italy

Editorial Office
MDPI
St. Alban-Anlage 66
4052 Basel, Switzerland

This is a reprint of articles from the Special Issue published online in the open access journal *Nutrients* (ISSN 2072-6643) (available at: https://www.mdpi.com/journal/nutrients/special_issues/Food_Composition_Dedicated_Databases).

For citation purposes, cite each article independently as indicated on the article page online and as indicated below:

LastName, A.A.; LastName, B.B.; LastName, C.C. Article Title. *Journal Name* **Year**, *Volume Number*, Page Range.

ISBN 978-3-0365-6850-8 (Hbk)
ISBN 978-3-0365-6851-5 (PDF)

Contents

About the Editors

Alessandra Durazzo

Alessandra Durazzo was awarded a Master's degree in Chemistry and Pharmaceutical Technology cum laude in 2003 and a PhD in Horticulture in 2010. Senior researcher at the CREA Research Centre for Food and Nutrition. The core of her research is the study of chemical, nutritional, and bioactive components of food, with regard to the wide spectrum of substances classes and their nutraceutical features. For several years, she was involved in national and international research projects on the evaluation of several factors (agronomic practices, processing, etc.) that affect food quality, the levels of bioactive molecules and the total antioxidant properties, as well as on their possible impact on the biological role played by bioactive components in human physiology. Particular attention is given to the study of alternative sources of nutraceutical compounds such as agro-food waste and the application of nanotechnologies on pharmaceutical and nutraceutical compounds. Her research activities are also addressed towards the development, management, and updating of databases of bioactive compounds, nutraceuticals, and dietary supplements; particular attention is given towards the harmonization of analytical procedures and classification and codification of dietary supplements.

Massimo Lucarini

Massimo Lucarini was awarded a master's degree in Industrial Chemistry "cum laude" at the University of Rome "La Sapienza", Italy (1992), and a PhD in Chemistry (University of Rome "La Sapienza"). Senior researcher at the CREA Research Centre for Food and Nutrition. His research activity is mainly aimed at the evaluation of nutrient content, molecules with biological and anti-nutrient activity in foods and diets, and studies of the stability and technological treatments of food products using specific process markers. Particular interest is given to the evaluation of the nutritional quality of foods, the bioavailability of nutrients and bioactive components and their interaction with the food matrix (using in vitro models and cellular models), and to applications in the nutraceutical field; recent attention has been paid to the exploitation of waste from the agri-food industry, with a view to sustainable agri-food production. An integral part of Dr. Lucarini's research carried is linked to institutional activity, including: Food Composition Tables, Dietary Guidelines for Healthy Eating, and the evaluation of fraud risk in the agri-food system. In relation to the food production system, the effects of technological treatments on molecules of nutritional interest are also evaluated; he is also interested in using natural substances with strong antioxidant properties to improve the shelf-life of food products. His research activity is also aimed at the development of new analytical methods, the exchange of scientific information, and the acquisition of new skills both at national and international level, through training courses and participation in congresses and seminars. His work is disseminated through the production of scientific articles, interviews released in national journals and broadcasting systems, the creation of web pages, and participation in congresses and educational and informative activities.

nutrients

MDPI

Editorial

Food Composition and Dedicated Databases: Key Tools for Human Health and Public Nutrition

Alessandra Durazzo * and Massimo Lucarini *

CREA—Research Centre for Food and Nutrition, Via Ardeatina 546, 01178 Rome, Italy
* Correspondence: alessandra.durazzo@crea.gov.it (A.D.); massimo.lucarini@crea.gov.it (M.L.)

To better understand nutrition, food chemistry, and medicine, it is important to investigate biologically active constituents, which requires a detailed knowledge and coverage of the composition of compounds of nutritional and nutraceutical character. The categorization of substances and thus the implementation of specific and dedicated databases have now emerged, based on both analytical data and collected data derived from the literature through a standardized and harmonized approach [1].

Food composition databases aim to produce, collect, and present data in a standardized format to "speak a common language", which allows the comparison of data from different national databases to foster an exchange and collaboration between countries [2,3]. Simultaneously, research is focused on the development of databases and models on metabolites in humans and novel dietary biomarkers [4–6].

The development of databases of nutrients, bioactive compounds, metabolites and dietary sypplements are key tools for human health and public nutrition and represent resources for a broad range of applications in different fields, i.e. food, nutraceutical, pharmaceutical, epidemiology and medicinal areas [7–12].

The initial construction of a dataset of specific nutrients, bioactive compounds, or bioactive compounds' class and their inclusion in a specified and standardized database should be monitored. Moreover, an update and expansion of the database for a more comprehensive source of data and information is encouraged. Databases dedicated to particular and characteristic categories of foods are also promoted: traditional, certified, and recipe databases [13–16].

Hoteit el al. [17], aiming at studying non-conjugated-industrially-produced-trans fatty in Lebanese foods, especially regarding Elaidic acid and Linolelaidic acid, monitored 145 food samples consisting of 3 categories: traditional dishes, Arabic sweets, and market food products. The results showed that approximately 93% of the products tested in Lebanon, between 2019 and 2021, met the World Health Organization recommendations, while approximately 7% exceeded the limit [17].

Beltrá et al. [18] studied sodium content of foods sold in the Spanish market, as results of the BADALI Project. Balakrishna et al. [19] identified the nutrient patterns in South African foods to support the National Nutrition Guidelines and Policies. Marcotrigiano et al. [20] reported the results obtained from a field investigation on nutritional and hygienic features in the Apulia region (Southern Italy) as an integrated control plan in primary schools.

First and foremost, the design and construction of food databases require the exact identification of foods from an adequate food nomenclature and a precise description of the foods. There is a general consensus on the importance of the nomenclature, description, and classification of foods and food groups. A coherent food description system is essential for comparing and/or exchanging data from different databases, and the data of the same nature from different organizations and countries. Moreover, matching procedures for linking different databases should be encouraged [21].

Food composition and other dedicated databases, as well as metabolomic databases and biomarker repositories, represent a unique data resource for nutritionists, dietitians,

Citation: Durazzo, A.; Lucarini, M. Food Composition and Dedicated Databases: Key Tools for Human Health and Public Nutrition. *Nutrients* **2021**, *13*, 4003. https://doi.org/10.3390/nu13114003

Received: 20 October 2021
Accepted: 3 November 2021
Published: 10 November 2021

and researchers for several applications, i.e., dietary assessments, exposure studies, food labeling, epidemiological studies, and clinical trials. Concerning dietary assessment, Witkowska et al. [22] reported the assessment of plant sterols in the diet of adult polish population with the use of a newly developed database. Regarding food labeling, Castro et al. [23] reported the comparison of healthiness, labeling, and price between private and branded label packaged foods in New Zealand (2015–2019).

Applications and the utilization of databases from nutrition- and medicine-related fields in other contexts are explored, and current research trends are defined. Delgado et al. [24] described the usefulness and limitations of food databases with particular emphasis what concerns sustainable diets, the food 'matrix effect', missing compounds, safe processing, and in guiding innovation in foods, as well as in shaping consumers' perceptions and food choices.

Author Contributions: A.D. and M.L. have made a substantial, direct, and intellectual contribution to the work, and approved it for publication. All authors have read and agreed to the published version of the manuscript.

Funding: This research received no external funding.

Institutional Review Board Statement: Not applicable.

Informed Consent Statement: Not applicable.

Data Availability Statement: Not applicable.

Acknowledgments: We would like to thank the authors and the reviewers of the publications in this Special Issue for their invaluable contributions and effort. We are also grateful to the editorial board members and support staff of the journal for their kind support during the preparation of this Special Issue.

Conflicts of Interest: The authors declare no conflict of interest.

References

1. Durazzo, A.; D'Addezio, L.; Camilli, E.; Piccinelli, R.; Turrini, A.; Marletta, L.; Marconi, S.; Lucarini, M.; Lisciani, S.; Gabrielli, P.; et al. From Plant Compounds to Botanicals and Back: A Current Snapshot. *Molecules* **2018**, *23*, 1844. [CrossRef]
2. Traka, M.H.; Plumb, J.; Berry, R.; Pinchen, H.; Finglas, P.M. Maintaining and updating food composition datasets for multiple users and novel technologies: Current challenges from a UK perspective. *Nutr. Bull.* **2020**, *45*, 230–240. [CrossRef]
3. Kapsokefalou, M.; Roe, M.; Turrini, A.; Costa, H.S.; Martinez-Victoria, E.; Marletta, L.; Berry, R.; Finglas, P. Food Composition at Present: New Challenges. *Nutrients* **2019**, *11*, 1714. [CrossRef]
4. Tsugawa, H. Advances in computational metabolomics and databases deepen the understanding of metabolisms. *Curr. Opin. Biotechnol.* **2018**, *54*, 10–17. [CrossRef]
5. Lai, Z.; Tsugawa, H.; Wohlgemuth, G.; Mehta, S.; Mueller, M.; Zheng, Y.; Ogiwara, A.; Meissen, J.; Showalter, M.; Takeuchi, K.; et al. Identifying metabolites by integrating metabolome databases with mass spectrometry cheminformatics. *Nat. Methods* **2018**, *15*, 53–56. [CrossRef]
6. Maruvada, P.; Lampe, J.W.; Wishart, D.S.; Barupal, D.; Chester, D.N.; Dodd, D.; Feunang, Y.D.; Dorrestein, P.C.; Dragsted, L.O.; Draper, J.; et al. Perspective: Dietary Biomarkers of Intake and Exposure—Exploration with Omics Approaches. *Adv. Nutr.* **2019**, *11*, 200–215. [CrossRef]
7. Durazzo, A.; Lucarini, M.; Camilli, E.; Marconi, S.; Gabrielli, P.; Lisciani, S.; Gambelli, L.; Aguzzi, A.; Novellino, E.; Santini, A.; et al. Dietary Lignans: Definition, Description and Research Trends in Databases Development. *Molecules* **2018**, *23*, 3251. [CrossRef] [PubMed]
8. Durazzo, A.; Zaccaria, M.; Polito, A.; Maiani, G.; Carcea, M. Lignan Content in Cereals, Buckwheat and Derived Foods. *Foods* **2013**, *2*, 53–63. [CrossRef] [PubMed]
9. Durazzo, A.; Camilli, E.; D'Addezio, L.; Piccinelli, R.; Mantur-Vierendeel, A.; Marletta, L.; Finglas, P.; Turrini, A.; Sette, S. Development of Dietary Supplement Label Database in Italy: Focus of FoodEx2 Coding. *Nutrients* **2019**, *12*, 89. [CrossRef] [PubMed]
10. Plumb, J.; Durazzo, A.; Lucarini, M.; Camilli, E.; Turrini, A.; Marletta, L.; Finglas, P. Extractable and Non-Extractable Antioxidants Composition in the eBASIS Database: A Key Tool for Dietary Assessment in Human Health and Disease Research. *Nutrients* **2020**, *12*, 3405. [CrossRef]
11. Saldanha, L.G.; Dwyer, J.T.; Bailen, R.A. Modernization of the national institutes of health dietary supplement label database. *J. Food Compos. Anal.* **2021**, *102*, 104058. [CrossRef]

12. Dwyer, J.; Saldanha, L.; Bailen, R.; Durazzo, A.; Le Donne, C.; Piccinelli, R.; Andrews, K.; Pehrsson, P.; Gusev, P.; Calvillo, A.; et al. Commentary: An impossible dream? Integrating dietary supplement label databases: Needs, challenges, next steps. *J. Food Compos. Anal.* **2021**, *102*, 103882. [CrossRef]
13. Durazzo, A.; Lisciani, S.; Camilli, E.; Gabrielli, P.; Marconi, S.; Gambelli, L.; Aguzzi, A.; Lucarini, M.; Maiani, G.; Casale, G.; et al. Nutritional composition and antioxidant properties of traditional Italian dishes. *Food Chem.* **2017**, *218*, 70–77. [CrossRef] [PubMed]
14. Durazzo, A.; Camilli, E.; D'Addezio, L.; Sette, S.; Marconi, S.; Piccinelli, R.; Le Donne, C.; Turrini, A.; Marletta, L. Italian composite dishes: Description and classification by LanguaL™ and FoodEx2. *Eur. Food Res. Technol.* **2020**, *246*, 287–295. [CrossRef]
15. Durazzo, A.; Lucarini, M.; Santini, A.; Camilli, E.; Gabrielli, P.; Marconi, S.; Lisciani, S.; Aguzzi, A.; Gambelli, L.; Novellino, E.; et al. Antioxidant Properties of Four Commonly Consumed Popular Italian Dishes. *Molecules* **2019**, *24*, 1543. [CrossRef]
16. Durazzo, A.; Camilli, E.; Marconi, S.; Lisciani, S.; Gabrielli, P.; Gambelli, L.; Aguzzi, A.; Lucarini, M.; Kiefer, J.; Marletta, L. Nutritional composition and dietary intake of composite dishes traditionally consumed in Italy. *J. Food Compos. Anal.* **2019**, *77*, 115–124. [CrossRef]
17. Hoteit, M.; Zoghbi, E.; Rady, A.; Shankiti, I.; Ibrahim, C.; Al-Jawaldeh, A. Non-Conjugated-Industrially-Produced-Trans Fatty in Lebanese Foods: The Case of Elaidic and Linolelaidic Acids. *Nutrients* **2021**, *13*, 3664. [CrossRef] [PubMed]
18. Beltrá, M.; Borrás, F.; Ropero, A.B. Sodium Content of Foods Sold in the Spanish Market. Results from the BADALI Project. *Nutrients* **2021**, *13*, 3410. [CrossRef]
19. Balakrishna, Y.; Manda, S.; Mwambi, H.; van Graan, A. Identifying Nutrient Patterns in South African Foods to Support National Nutrition Guidelines and Policies. *Nutrients* **2021**, *13*, 3194. [CrossRef]
20. Marcotrigiano, V.; Stingi, G.D.; Fregnan, S.; Magarelli, P.; Pasquale, P.; Russo, S.; Orsi, G.B.; Montagna, M.T.; Napoli, C.; Napoli, C. An Integrated Control Plan in Primary Schools: Results of a Field Investigation on Nutritional and Hygienic Features in the Apulia Region (Southern Italy). *Nutrients* **2021**, *13*, 3006. [CrossRef]
21. Durazzo, A.; Camilli, E.; D'Addezio, L.; Le Donne, C.; Ferrari, M.; Marconi, S.; Marletta, L.; Mistura, L.; Piccinelli, R.; Scalvedi, M.L.; et al. Food Groups and Individual Foods: Nutritional Attributes and Dietary Importance. In *Reference Module in Food Science*; Elsevier: Amsterdam, The Netherlands, 2018; ISBN 9780081005965. [CrossRef]
22. Witkowska, A.; Waśkiewicz, A.; Zujko, M.; Mirończuk-Chodakowska, I.; Cicha-Mikołajczyk, A.; Drygas, W. Assessment of Plant Sterols in the Diet of Adult Polish Population with the Use of a Newly Developed Database. *Nutrients* **2021**, *13*, 2722. [CrossRef] [PubMed]
23. Castro, T.; Mackay, S.; Young, L.; Ni Mhurchu, C.; Shaw, G.; Tawfiq, E.; Eyles, H. Comparison of Healthiness, Labelling, and Price between Private and Branded Label Packaged Foods in New Zealand (2015–2019). *Nutrients* **2021**, *13*, 2731. [CrossRef] [PubMed]
24. Delgado, A.; Issaoui, M.; Vieira, M.C.; de Carvalho, I.S.; Fardet, A. Food Composition Databases: Does It Matter to Human Health? *Nutrients* **2021**, *13*, 2816. [CrossRef] [PubMed]

nutrients

Article

Assessment of Plant Sterols in the Diet of Adult Polish Population with the Use of a Newly Developed Database

Anna Maria Witkowska [1,*,†], Anna Waśkiewicz [2,†], Małgorzata Elżbieta Zujko [1], Iwona Mirończuk-Chodakowska [1], Alicja Cicha-Mikołajczyk [2] and Wojciech Drygas [2,3]

1 Department of Food Biotechnology, Faculty of Health Sciences, Medical University of Bialystok, Szpitalna 37, 15-295 Bialystok, Poland; malgorzata.zujko@umb.edu.pl (M.E.Z.); iwona.mironczuk-chodakowska@umb.edu.pl (I.M.-C.)

2 Department of Epidemiology, Cardiovascular Disease Prevention and Health Promotion, National Institute of Cardiology, Alpejska 42, 04-628 Warsaw, Poland; awaskiewicz@ikard.pl (A.W.); acicha@ikard.pl (A.C.-M.); wdrygas@ikard.pl (W.D.)

3 Department of Social and Preventive Medicine, Faculty of Health Sciences, Medical University of Lodz, Hallera 1, 90-001 Lodz, Poland

* Correspondence: anna.witkowska@umb.edu.pl; Tel.: +48-85-6865088; Fax: +48-85-6865089

† A.M.W. and A.W. contributed equally to this work.

Abstract: Plant sterols are compounds with multiple biological functions, mainly cholesterol-reducing. There are no comprehensive databases on plant sterols, which makes it difficult to estimate their intake in the Polish population. This work attempted to use international food databases, additionally supplemented by scientific data from the literature, to create a database of plant sterols, which would cover various kinds of foods and dishes consumed in Poland. The aim was to assess the size and sources of dietary plant sterols in the adult population of Poland. The literature search was conducted using PubMed, Web of Science, Scopus, and Google Scholar to identify possible sources of published food composition data for plant sterols. The study group consisted of 5690 participants of the WOBASZ II survey. We identified 361 dietary sources of plant sterols based on the consumption of foods and dishes reported by participants. Cereals and fats provided 61% of the total plant sterols, and together with vegetables and fruits, this totaled 80%. The median intake of plant sterols in the Polish population was 255.96 mg/day, and for men and women 291.76 and 230.61 mg/day, respectively. Canola oil provided the most plant sterols at 16.92%, followed by white bread at 16.65% and soft margarine at 8.33%. The study found that plant sterol intake in Poland is comparable to other populations, and women's diets are more dense in plant sterols. Due to the lack of literature sources on plant sterol content in some foods, future studies should expand and complete the databases on plant sterol content in foods.

Keywords: plant sterols; database; Polish population

Citation: Witkowska, A.M.; Waśkiewicz, A.; Zujko, M.E.; Mirończuk-Chodakowska, I.; Cicha-Mikołajczyk, A.; Drygas, W. Assessment of Plant Sterols in the Diet of Adult Polish Population with the Use of a Newly Developed Database. *Nutrients* **2021**, *13*, 2722. https://doi.org/10.3390/nu13082722

Academic Editors: Massimo Lucarini and Alessandra Durazzo

Received: 28 May 2021
Accepted: 4 August 2021
Published: 7 August 2021

1. Introduction

Plant sterols are bioactive phytocompounds with a molecular structure similar to cholesterol [1]. The absorption of dietary cholesterol from diets rich in phytosterols is reduced by various mechanisms, mainly associated with the displacement of cholesterol from lipid micelles [2]. To date, more than 250 phytosterols have been identified, which include plant sterols and their saturated forms, stanols [3,4]. In various food sources, β-sitosterol is predominant and accounts for approximately 80% of the phytosterol intake in the diet [5]. Clinical evidence shows that phytosterols have a moderate LDL- and triglyceride-lowering effect [6,7]. Phytosterols are also considered moderately active antioxidants [8] and have immunomodulatory properties [9]. Sitosterol may suppress obesity-related chronic inflammation by reducing circulating interleukin-6 and TNF-α [10]. A growing body of evidence suggests that phytosterols may be an alternative and/or complementary therapy for patients with obesity and diabetes [3]. The consumption of naturally occurring plant sterols has been found to be associated with a lower risk of first myocardial infarction in men [11]. In addition, high doses of plant sterols in the diet, especially β-sitosterol, have been found to prevent the development of cancer [12,13].

Food sources with the highest plant sterol content include vegetable oils, mainly corn oil (746 mg/100 g), and sesame seeds (714 mg/100 g) [14]. A good source of phytosterols is nuts, which provide 30–220 mg/100 g of phytosterols, and cereals that contain phytosterols in the amount of 35–198 mg/100 g [15]. Vegetables contain smaller amounts of phytosterols, with 4–40 mg/100 g, and fruits contain 4–24 mg/100 g [15]. Consumption studies have shown that due to the frequency and volume of consumption, the suppliers of plant sterols are mainly bread, cereals, fats, and vegetables [3,5]. As studies show, population intakes of plant sterols are variable [5,11,16–21].

There is a need to develop databases of biologically active compounds to calculate population intakes [22]. Unlike the various databases on food composition, there are no comprehensive databases on plant sterols, which makes it difficult to estimate the intake of plant sterols in populations, as well as their further calculations in epidemiological studies. Earlier population-based studies used different databases prepared for individual studies with different methodologies [5,11,16–21]. Some studies used plant sterol databases [16,18,20], but others prepared individual databases based on experimental data [5,11,17,19,21]. There is currently no evaluation of plant sterols at the Polish population level, but an attempt has been made in a pilot study on a sample of students [23].

This work attempted to use international food databases, additionally supplemented by scientific data from the literature, to create a database of plant sterols, which would cover various kinds of foods and dishes consumed in Poland. The aim was to assess the size and sources of dietary plant sterols in the adult population of Poland.

2. Materials and Methods

2.1. Plant Sterol Database and Calculation of Dietary Intake

Since there is no plant sterol database in Poland, its establishment for the purpose of this study was based on international databases, which were published in English and are publicly available [14,24]. A literature review was conducted to search for reliable data sources that would supplement the data taken from international databases. The literature search was conducted using PubMed, Web of Science, Scopus, and Google Scholar to identify possible sources of published food composition data for plant sterols. The search terms included phytosterols, plant sterols, β-sitosterol, campesterol, and stigmasterol combined with food, cereals, vegetables, fruit, berries, nuts, seeds, legumes, beverages, coffee, tea, wine, soda, chocolate, pastry, and cookies.

The plan was to select data sources that were as complete as possible in terms of individual plant sterols (β-sitosterol, campesterol, and stigmasterol). For the total plant sterol content, the full data reported by databases or scientific sources were used or, in the absence of relevant data, the available data for plant sterol content were aggregated. The quality of the data was assessed according to the procedure described by Rand et al. [25], which takes into account the analytical method used, the number of samples, the sample handling procedures, the sampling plan for the selection of foods, and the analytical and quality assurance. The currently available techniques for sterol analysis are gas chromatography (GC), high-pressure liquid chromatography (HPLC), and supercritical fluid chromatography (SFC). GC/FID (flame-ionization detection) or GC/MS (mass spectrometry) can be considered the methods of choice for the determination of phytosterols in foods and diets [26]. For most of the studies, all of the quality criteria were met. For some food products, the number of studies was limited to only one publication; although they did not meet all quality criteria, they were included in the developed database due to lack of other publication sources. Finally, data from 13 data sources were included in the database, with 11 studies meeting the Rand criteria and 2 not meeting these criteria.

In this study, data for fats and oils were extracted from the British database of Food Composition [24], the USDA Database [14], and Normen et al. [27]. Data on plant sterols in cereals were extracted from the British database of Food Composition [24] and Normen et al. [28]. Most of the data for vegetables and potatoes were taken from Normen et al. [29]. Data gaps in the vegetables group were filled in from the publications by Han et al. [30], Piironen et al. [31], Ryan et al. [32], the British database of Food Composition [24], and the USDA Nutrient Database [14]. The plant sterol contents in fruits and berries were compiled from the USDA Database [14], Piironen et al. [31], Normen et al. [29], and Han et al. [30]. The plant sterol contents in nuts and seeds were taken from the USDA Database [14], the British database of Food Composition [24], and Normen et al. [27]. The plant sterols for legumes were compiled from Li et al. [33], Han et al. [30], the USDA Database [14], Ryan et al. [32], and Yamaya et al. [34]. Data for fruit and vegetable juices, sodas, tea, and beer were taken from Decloedt et al. [35]. Data for the plant sterols in wines were taken from Ruggiero et al. [36]. The plant sterol content in the sterolic fraction of coffee was taken from Čížková et al. [37] and recalculated per 100 g of coffee. For pastry and cookies, data were extracted from the British database of Food Composition [24], the USDA Database [14], and Piironen et al. [31]. For chocolate and chocolate candies, data were compiled from Normen et al. [27]. Data on plant sterols in foods are available in Supplementary Table S1.

For the dishes, the individual ingredients were extracted according to recipes of the National Institute of Food and Nutrition of Poland, taking into account the yield factors of the dishes. Data on plant sterols in dishes are available in Supplementary Table S2.

Finally, foods were grouped into 10 categories: cereals (flour, bread, breakfast cereals, bran, groats, and pasta), fruit (processed and non-processed), vegetables (processed and non-processed), potatoes, legumes, fats and oils (oils, margarine, and mayonnaise), coffee (instant and infusion), cookies and cakes, chocolate (chocolate and chocolate candies and bars), and other foods (tea, beer, wine, sodas, mustard, nuts, and seeds). Foods enriched with phytosterols were not included in these calculations because not all manufacturers were willing to disclose their formulations regarding individual phytosterols.

The process used to estimate plant sterols in foods is given in Figure 1.

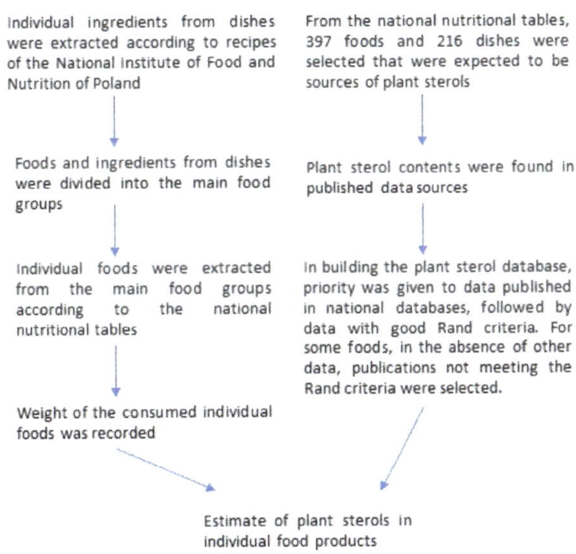

Figure 1. The process used to estimate plant sterol intake.

2.2. Study Group and Data Collection

The study group consisted of 5690 participants (2554 men and 3136 women) of the National Multicenter Health Survey II (the Polish acronym is WOBASZ II). WOBASZ II is a cross-sectional study representative of the Polish adult population aged 20 years and over, which was carried out by the National Institute of Cardiology (formerly the Institute of Cardiology), Warsaw, Poland, in the years 2013–2014, in collaboration with five national medical universities. The design and methods of the WOBASZ II survey have been described in detail elsewhere [38]. Daily food consumption data were collected by trained interviewers using a single 24-h dietary recall method. The overall evaluation included a sample of 6170 participants, 480 of whom were excluded due to missing or unreliable dietary recalls. A flowchart of the participants is shown in Figure 2. The WOBASZ II study was approved by the Bioethics Committee of the National Institute of Cardiology (no. 1344), as was the current study (no. 1837). Written informed consent was obtained from all participants.

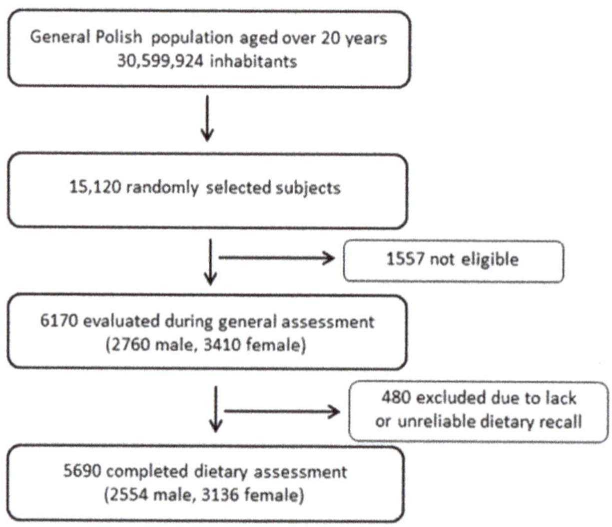

Figure 2. Flow chart of the study participants.

Data on the demographic status, diseases, leisure-time physical activity, tobacco use, community size, marital status, and education level of the participants were collected using a standardized questionnaire developed for the WOBASZ II survey. Height and weight measurements were taken by personnel trained in standard procedures. Body mass index (BMI) was calculated from body weight in kilograms divided by the square of the height in meters. Blood pressure (BP) was measured three times on the right arm after 5 min of rest in a sitting position at 1 min intervals, and final BP was reported as the mean of the second and third measurements. The general characteristics of the study group are shown in Table 1.

The present study identified 361 dietary sources of plant sterols based on the consumption of foods and dishes reported by participants in the WOBASZ II survey. A small proportion of subjects who consumed phytosterol-enriched products was found (Table 1). Plant sterol daily intake was determined by multiplying the daily consumption of individual food items by the respective total plant sterols, such as the β-sitosterol, campesterol, and stigmasterol contents, in these food items and then summed up.

Table 1. General description of the studied population.

Trait	Men and Women N = 5690	Men N = 2554	Women N = 3136	*p* *
Age (year), mean ± SD	49.58 ± 16.43	48.79 ± 16.27	50.23 ± 16.54	0.0023
median (IQR)	50.00 (36.00–62.00)	49.00 (35.00–61.00)	51.00 (37.00–62.00)	
BMI (kg/m^2), mean ± SD	27.17 ± 5.19	27.42 ± 4.55	26.96 ± 5.65	<0.0001
median (IQR)	26.63 (23.54–30.15)	27.07 (24.34–30.02)	26.12 (22.87–30.39)	
Systolic BP (mmHg), mean ± SD	130.67 ± 19.34	134.44 ± 18.19	127.6 ± 19.71	<0.0001
median (IQR)	127.5 (117.5–141.0)	131.5 (122.0–144.5)	124.0 (113.5–138.0)	
Diastolic BP (mmHg), mean ± SD	80.23 ± 10.81	81.51 ± 10.91	79.19 ± 10.62	<0.0001
median (IQR)	80.0 (72.5–87.0)	81.0 (74.0–88.0)	78.5 (72.0–85.5)	
Fasting glucose (mmol/L), mean ± SD	5.50 ± 1.46	5.65 ± 1.6	5.38 ± 1.32	<0.0001
median (IQR)	5.21 (4.84–5.72)	5.35 (4.96–5.84)	5.12 (4.77–5.58)	
Total cholesterol (mmol/L), mean ± SD	5.20 ± 1.27	5.21 ± 1.33	5.19 ± 1.22	0.7223
median (IQR)	5.14 (4.38–5.93)	5.15 (4.36–5.97)	5.14 (4.41–5.90)	
LDL-cholesterol (mmol/L), mean ± SD	3.15 ± 1.03	3.19 ± 1.04	3.11 ± 1.02	0.0002
median (IQR)	3.07 (2.42–3.78)	3.15 (2.46–3.86)	3.01 (2.39–3.72)	
Diseases (%)				
Hypertension [1]	45.22	49.56	41.69	<0.0001
Hypercholesterolemia [2]	67.30	68.86	66.03	0.0262
Diabetes [3]	10.82	11.86	9.96	0.0249
Age groups (%)				
20–40 years	33.46	34.92	32.27	0.0045
41–60 years	38.60	38.32	38.83	
61–74 years	20.42	20.52	20.34	
>74 years	7.52	6.24	8.56	
Commune size (%)				
<8.000 inhabitants	35.20	33.83	36.32	0.0849
8.000–40.000 inhabitants	30.67	30.70	30.64	
>40.000 inhabitants	34.13	35.47	33.04	
Marital status (%)				
married	66.71	70.19	63.87	<0.0001
single [4]	33.29	29.81	36.13	
Level of education [5] (%)				
under middle	17.12	14.74	19.06	<0.0001
middle	38.89	36.89	40.52	
academic	19.85	17.09	22.09	
vocational	24.14	31.28	18.33	
Smoking status (%)				
current smokers	23.28	28.95	18.66	<0.0001
past smokers	25.46	33.62	18.82	
never smokers	51.26	37.43	62.52	
Leisure-time physical activity [6] (%)				
low level	54.25	54.98	53.67	0.2856
middle level	15.29	14.81	15.68	
high level	28.08	27.50	28.54	
seasonally	2.38	2.71	2.11	
BMI (kg/m^2) (%)				
underweight (BMI < 18.5)	1.61	0.90	2.20	<0.0001
normal (BMI 18.5–24.99)	34.91	30.07	38.88	
overweight (BMI 25–29.99)	37.25	43.93	31.76	
obesity (BMI ≥ 30)	26.23	25.10	27.16	
Use of phytosterol-enriched margarines (%)	1.90	1.96	1.85	0.7660

* *p* calculated for differences between men and women. [1] Hypertension: systolic blood pressure SBP ≥140 mmHg or diastolic blood pressure DBP ≥90 mmHg, or use of antihypertensive drugs. [2] Hypercholesterolemia: TC ≥5 mmol/L or LDL-C ≥3 mmol/L or the participant was taking lipid-lowering medication. [3] Diabetes: blood glucose level was ≥7.0 mmol/L or diabetes was declared in an interview. [4] Singles: widows/widowers, unmarried, divorced, in separation. [5] Education level: under middle—no education, partial or completed education for primary level, partial secondary education; middle—secondary education, partial academic education; academic—tertiary education; vocational—vocational based on primary or on middle school. [6] Physical activity at leisure (for example, jogging, cycling, swimming, gardening for at least 30 min a day): low level—no such physical activity, once a week or less; middle level—every second or third day; high level—everyday, almost every day; seasonally (e.g., skiing in winter or on the plot in summer).

2.3. *Data Analysis*

Total phytosterol intake, including β-sitosterol, campesterol, and stigmasterol, was calculated by multiplying the daily consumption of individual food items by the respective phytosterol contents in these products. Additionally, the contribution of individual groups of food products and their ingredients to the consumption of different phytosterols was studied. Descriptive statistics were applied to describe the continuous variables (means and standard deviations, as well as median and interquartile range), and the percentages of the respective values were used for categorized variables. The contributions of food categories and individual food items to the intake of particular total and individual phytosterols are presented as percentages. To investigate the differences between men and women, a non-parametric Wilcoxon test or Chi-square test was used, respectively, for quantitative and qualitative variables. The level of significance was considered $p < 0.05$. Data analyses were processed using Statistical Analysis System (SAS; version 9.4, SAS Institute Inc., Cary, NC, USA).

3. Results

This study identified the top 10 food categories that provided plant sterols for the Polish population, which were cereals, vegetable fats and oils, vegetables, fruits, coffee, cookies and cakes, chocolate products, potatoes, and legumes. The other food products providing lower amounts of plant sterols were classified into the category of "other food products". Among all of these categories, cereals and fats provided 61% of the total plant sterols, and together with vegetables and fruits, this totaled 80%. Median total plant sterol intake in this study was 255.96 mg/day, and divided by men and women was 291.76 and 230.61 mg/day, respectively (Table 2). Considering individual foods (mg/day), canola oil provided the most plant sterols at 16.92%, followed by white bread at 16.65% and soft margarine at 8.33%. Among vegetables and fruits, there was no single significant source of plant sterols, but raw fruits and vegetables provided the predominant amounts of plant sterols (9.78% and 7.27%, respectively). This pattern of plant sterol sources was reflected in men, while among women, the main contributor was canola oil, followed by white bread, raw fruits, raw vegetables and soft margarine. Gender differences were found for most sources of plant sterol intake.

Figure 3 shows the intake of plant sterols in the Polish population (total, men, women) compared to other populations. With a plant sterol intake of 255.96 mg/day, the data for Poles are within the range for other populations.

Tables 3–5 show the contribution of food categories to the consumption of individual plant sterols such as β-sitosterol, campesterol, and stigmasterol. The median β-sitosterol consumption was 160.85 mg/day, while the intake of campesterol and stigmasterol was 47.45 mg/day and 22.10 mg/day, respectively.

The main food categories providing β-sitosterol were cereals (29.19%), fats (28.86%), fruits (14.20%), and vegetables (8.70%), with a total share of 80.95% of the β-sitosterol supply (Table 3). Among the food products, β-sitosterol was supplied by canola oil (15.88%), followed by wheat bread (14.88%) and soft margarine (9.02%). Women had a lower β-sitosterol intake compared to men at 146.28 mg/day vs. 180.84 mg/day, respectively.

The main sources of campesterol were fats (44.95%) and cereal products (31.81%), which together accounted for 76.76% of the campesterol intake (Table 4). For individual products, campesterol was supplied by canola oil (33.43%), white bread (16.30%), and soft margarines (8.11%). Men consumed more campesterol compared to women (56.71 vs. 40.88 mg/day, respectively).

As for stigmasterol, its main sources were the following product groups: coffee (25.10%), vegetables (23.22%), fats (16.85%), and cereal products (12.93%). The foods supplying the highest amounts of stigmasterol included coffee (as a food product; 25.10%), soft margarine (11.82%), and white bread (6.48%). The median intake of stigmasterol was higher in men at 23.49 mg/day compared to women at 21.11 mg/day.

Table 2. Contributions of food categories and individual food products to total plant sterol intake (PS), listed according to diminishing order of contribution.

Food Categories		**All** N = 5690	**Men** N = 2554	**Women** N = 3136	p *
Cereals	mg/day (mean ± SD), median (IQR)	90.65 ± 56.38 79.15 (53.94–114.87)	112.51 ± 63.28 102.44 (69.27–143.02)	72.85 ± 42.42 66.77 (45.90–91.39)	<0.0001
	Contribution to PS (%)	32.04	35.08	28.88	<0.0001
	Major sources (% contribution) **	wheat bread (16.65), rolls (6.64), rye bread (5.38)	wheat bread (20.59), rolls (6.85), rye bread (4.82)	wheat bread (12.56), rolls (6.43), rye bread (5.96)	-
Fats	mg/day (mean ± SD), median (IQR)	81.94 ± 92.30 51.65 (19.05–114.67)	98.34 ± 107.10 64.75 (24.30–138.68)	68.58 ± 75.63 44.47 (15.76–97.22)	<0.0001
	Contribution to PS (%)	28.95	30.66	27.20	0.0042
	Major sources (% contribution) **	oils (19.11) *including: canola oil (16.92), sunflower oil (2.06), olive oil (0.04)*, soft margarines (8.33), mayonnaise (1.05)	oils (20.02) *including: canola oil (18.03), sunflower oil (1.88), olive oil (0.06)*, soft margarines (9.08), mayonnaise (1.05)	oils (18.17) *including: canola oil (15.77), sunflower oil (2.25), olive oil (0.03)*, soft margarines (7.56), mayonnaise (1.05)	-
Fruits	mg/day (mean ± SD), median (IQR)	27.76 ± 31.23 20.19 (0–40.38)	25.69 ± 31.46 17.50 (0–39.37)	29.44 ± 30.95 21.62 (3.62–42.39)	<0.0001
	Contribution to PS (%)	9.81	8.01	11.67	<0.0001
	Major sources (% contribution) **	raw fruits (9.78) *including: apples (4.47), bananas (1.04), grapes (0.78), pears (0.52), plums (0.48), strawberries (0.37)*	raw fruits (7.98) *including: apples (4.10), bananas (0.87), grapes (0.57), pears (0.44), plums (0.40), strawberries (0.27)*	raw fruits (11.65) *including: apples (5.42), bananas (1.23), grapes (1.00), pears (0.60), plums (0.55), strawberries (0.48)*	-
Vegetables	mg/day (mean ± SD), median (IQR)	25.37 ± 24.22 20.05 (10.12–33.45)	26.04 ± 24.16 21.10 (10.53–34.73)	24.83± 24.26 19.22 (9.92–32.34)	0.0028
	Contribution to PS (%)	8.97	8.12	9.85	0.0224
	Major sources (% contribution) **	raw vegetables (7.27), *including: tomatoes (1.11), carrots (0.90), cabbage (0.84), cauliflowers (0.77), peppers (0.45), beetroot (0.49), lettuce (0.47), cucumbers (0.42)*, vegetable preserves (1.32)	raw vegetables (6.45) *including: tomatoes (1.02), carrots (0.77), cabbage (0.77), cauliflowers (0.64), beetroot (0.50), peppers (0.39), lettuce (0.38), cucumbers (0.38)*, vegetable preserves (1.37)	raw vegetables (8.11) *including: tomatoes (1.21), carrots (1.03), cabbage (0.91), cauliflowers (0.91), peppers (0.52), beetroot (0.49), lettuce (0.47), cucumbers (0.42)*, vegetable preserves (1.26)	-

Table 2. *Cont.*

Food Categories		**All** N = 5690	**Men** N = 2554	**Women** N = 3136	*p* *
Coffee	mg/day (mean ± SD), median (IQR)	19.24 ± 20.39 21.47 (0–26.84)	17.56 ± 20.84 21.47 (0–26.84)	20.61 ± 19.91 21.47 (0–26.84)	<0.0001
	Contribution to PS (%)	6.80	5.48	8.17	<0.0001
Cookies, cakes	mg/day (mean ± SD), median (IQR)	11.57± 23.84 0 (0–16.50)	11.36 ± 24.40 0 (0–15.00)	11.72 ± 23.40 0 (0–17.60)	0.0055
	Contribution to PS (%)	4.08	3.54	4.65	0.0332
Chocolate products	mg/day (mean ± SD), median (IQR)	6.46 ± 22.86 0 (0–0)	6.78 ± 7.20 0 (0–0)	6.19 ± 21.39 0 (0–0)	0.0477
	Contribution to PS (%)	2.28	2.11	2.46	0.3936
Potatoes	mg/day (mean ± SD), median (IQR)	6.12 ± 6.34 6.05 (0–11.01)	7.30 ± 7.20 6.05 (0–12.10)	5.16 ± 5.35 4.15 (0–8.07)	<0.0001
	Contribution to PS (%)	2.16	2.27	2.05	0.5511
Legumes	mg/day (mean ± SD), median (IQR)	3.84 ± 18.56 0 (0–0)	4.70 ± 21.97 0 (0–0)	3.13 ± 15.19 0 (0–0)	0.9949
	Contribution to PS (%)	1.36	1.47	1.24	0.4277
Other food products	mg/day (mean ± SD),	10.02	10.49	9.68	-
	Contribution to PS (%)	3.55	3.26	3.83	0.2434
Total plant sterol intake	mg/day (mean ± SD), median (IQR)	282.97 ± 144.50 255.96 (184.98–347.98)	320.77 ± 160.93 291.76 (209.96–399.07)	252.19 ± 121.20 230.61 (167.73–308.2)	<0.0001
	Contribution to PS (%)	100	100	100	-

* *p* calculated for differences between men and women. ** In the total and each food category, only individual food products with the strongest impact on the total plant sterol intakes were listed.

Table 3. Contributions of food categories and individual food products to β-sitosterol intake (β-SIT), listed according to diminishing order of contribution.

Food Categories		**All** **N = 5690**	**Men** **N = 2554**	**Women** **N = 3136**	*p* *
Cereals	mg/day (mean ± SD), median (IQR)	51.37 ± 31.69 44.98 (30.27–64.94)	63.57 ± 35.62 58.45 (39.20–81.10)	41.44 ± 23.84 37.87 (26.28–51.81)	<0.0001
	Contribution to β-SIT (%)	29.19	32.13	26.20	<0.0001
	Major sources (% contribution) **	wheat bread (14.88), rolls (6.04), rye bread (4.75)	wheat bread (18.56), rolls (6.29), rye bread (4.28)	wheat bread (11.14), rolls (5.79), rye bread (5.22)	-
Fats	mg/day (mean ± SD), median (IQR)	50.78 ± 54.93 33.86 (12.70–71.60)	60.81 ± 63.51 42.15 (16.50–88.00)	42.61 ± 45.17 28.40 (11.00–61.14)	<0.0001
	Contribution to β-SIT (%)	28.86	30.73	26.95	0.0017
	Major sources (% contribution) **	oils (18.43) *including: canola oil (15.88), sunflower oil (2.40), olive oil (0.07), soybean oil (0.07),* soft margarines (9.02), mayonnaise (0.87)	oils (19.40) *including: canola oil (17.06), sunflower oil (2.20), olive oil (0.09), soybean oil (0.05),* soft margarines (9.87), mayonnaise (0.88)	oils (17.44) *including: canola oil (14.68), sunflower oil (2.60), olive oil (0.05), soybean oil (0.09),* soft margarines (8.16), mayonnaise (0.86)	-
Fruits	mg/day (mean ± SD), median (IQR)	25.00 ± 27.70 19.50 (0–38.13)	23.21 ± 27.96 15.90 (0–36.36)	26.45 ± 27.41 19.50 (3.29–39.00)	<0.0001
	Contribution to β-SIT (%)	14.20	11.73	16.73	<0.0001
	Major sources (% contribution) **	raw fruits (14.17) *including:* *apples (7.37), bananas (1.29), grapes (1.05), pears (0.81), plums (0.62), strawberries (0.55)*	raw fruits (11.68) *including:* *apples (6.41), bananas (1.08), grapes (0.77), pears (0.69), plums (0.53), strawberries (0.40)*	raw fruits (16.70) *including:* *apples (8.34), bananas (1.51), grapes (1.33), pears (0.93), plums (0.72), strawberries (0.70), peaches (0,56)*	-
Vegetables	mg/day (mean ± SD), median (IQR)	15.31 ± 14.51 11.96 (5.88–20.15)	15.68 ± 14.69 12.60 (6.16–20.84)	15.01 ± 14.36 11.56 (5.73–19.60)	0.0037
	Contribution to β-SIT (%)	8.70	7.92	9.49	0.0347
	Major sources (% contribution) **	raw vegetables (7.11), *including:* *cabbage (0.98), carrots (0.99), tomatoes (0.91), cauliflowers (0.81), peppers (0.54), beetroot (0.43), onion (0.36), cucumbers (0.33),* vegetable preserves (1.23)	raw vegetables (6.36) *including:* *cabbage (0.92), carrots (0.86), tomatoes (0.84), cauliflowers (0.68), peppers (0.47), beetroot (0.43), onion (0.38), cucumbers (0.32),* vegetable preserves (1.28)	raw vegetables (7.88) *including:* *cabbage (1.06), carrots (1.13), tomatoes (0.99), cauliflowers (0.94), peppers (0.61), beetroot (0.42), onion (0.34), cucumbers (0.35),* vegetable preserves (1.18)	-

Table 3. *Cont.*

Food Categories		All N = 5690	Men N = 2554	Women N = 3136	p *
Coffee	mg/day (mean ± SD), median (IQR)	9.91 ± 10.50 11.06 (0–13.83)	9.05 ± 10.74 11.06 (0–13.83)	10.62 ± 10.26 11.06 (0–13.83)	<0.0001
	Contribution to β-SIT (%)	5.64	4.57	6.72	0.0005
Cookies, cakes	mg/day (mean ± SD), median (IQR)	7.04 ± 14.04 0 (0–10.20)	6.97 ± 14.69 0 (0–10.00)	7.10 ± 13.49 0 (0–10.40)	0.0058
	Contribution to β-SIT (%)	4.00	3.52	4.49	0.0645
Chocolate products	mg/day (mean ± SD), median (IQR)	3.89 ± 13.77 0 (0–0)	4.10 ± 14.76 0 (0–0)	3.73 ± 12.89 0 (0–0)	0.0485
	Contribution to β-SIT (%)	2.22	2.07	2.36	0.4699
Potatoes	mg/day (mean ± SD), median (IQR)	4.35 ± 4.50 4.30 (0–7.82)	5.18 ± 5.12 4.30 (0–8.60)	3.67 ± 3.80 2.95 (0–5.73)	<0.0001
	Contribution to β-SIT (%)	2.47	2.62	2.32	0.4742
Legumes	mg/day (mean ± SD), median (IQR)	2.21 ± 11.91 0 (0–0)	2.71 ± 14.03 0 (0–0)	1.80 ± 9.83 0 (0–0)	0.9852
	Contribution to β-SIT (%)	1.25	1.37	1.14	0.4522
Other food products	mg/day (mean ± SD),	6.12	6.61	5.71	-
	Contribution to β-SIT (%)	3.47	3.34	3.60	0.5732
Total β-sitosterol intake	mg/day (mean ± SD), median (IQR)	175.98 ± 88.00 160.85 (115.80–218.15)	197.89 ± 98.28 180.84 (131.20–246.86)	158.14 ± 74.02 146.28 (105.89–196.13)	<0.0001
	Contribution to β-SIT (%)	100	100	100	-

* p calculated for differences between men and women. ** In the total and each food category, only individual food products with the strongest impact on the total plant sterol intakes were listed.

Table 4. Contributions of food categories and individual food products to campesterol intake (CAMP), listed according to diminishing order of contribution.

Food Categories		All N = 5690	Men N = 2554	Women N = 3136	*p* *
Fats	mg/day (mean ± SD), median (IQR)	26.55 ± 35.53 12.52 (3.75–37.85)	32.02 ± 41.34 16.70 (4.65–45.65)	22.10 ± 29.23 10.10 (3.00–31.11)	<0.0001
	Contribution to CAMP (%)	44.95	46.30	43.47	0.0311
	Major sources (% contribution) **	oils (34.60) *including:* *canola oil (33.43), sunflower oil (1.07), soybean oil (0.08), olive oil (0.01),* soft margarines (8.11), mayonnaise (1.87)	oils (35.50) *including:* *canola oil (34.48), sunflower oil (0.94), soybean oil (0.06), olive oil (0.01),* soft margarines (8.58), mayonnaise (1.81)	oils (33.61) *including:* *canola oil (32.26), sunflower oil (1.21), soybean oil (0.11), olive oil (0.01),* soft margarines (7.60), mayonnaise (1.94)	-
Cereals	mg/day (mean ± SD), median (IQR)	18.79 ± 11.57 16.50 (11.11–23.84)	23.34 ± 13.00 21.46 (14.48–30.02)	15.09 ± 8.65 13.79 (9.37–19.06)	<0.0001
	Contribution to CAMP (%)	31.81	33.74	29.67	0.0009
	Major sources (% contribution) **	wheat bread (16.30), rolls (6.91), rye bread (5.49)	wheat bread (19.52), rolls (6.92), rye bread (4.76)	wheat bread (12.74), rolls (6.91), rye bread (6.29)	-
Vegetables	mg/day (mean ± SD), median (IQR)	3.04 ± 3.81 2.00 (0.72–3.99)	3.06 ± 3.90 2.09 (0.74–4.08)	3.04 ± 3.75 1.95 (0.69–3.91)	0.2297
	Contribution to CAMP (%)	5.16	4.43	5.97	0.0098
	Major sources (% contribution) **	fresh vegetables (4.29) *including:* *cauliflowers (1.07), cabbage (0.93), carrots (0.70), peppers (0.51), tomatoes (0.36),* vegetable preserves (0.68)	fresh vegetables (3.62) *including:* *cabbage (0.75), cauliflowers (0.71), carrots (0.49), peppers (0.36), tomatoes (0.28),* vegetable preserves (0.67)	fresh vegetables (5.04) *including:* *cauliflowers (1.07), cabbage (0.93), carrots (0.70), peppers (0.51), tomatoes (0.36),* vegetable preserves (0.69)	-
Cookies, cakes	mg/day (mean ± SD), median (IQR)	2.99 ± 6.30 0 (0–3.90)	2.99 ± 6.69 0 (0–3.60)	3.00 ± 5.95 0 (0–4.16)	0.0049
	Contribution to CAMP (%)	5.07	4.32	5.90	0.0071
Coffee	mg/day (mean ± SD), median (IQR)	3.15 ± 3.33 3.51 (0–4.39)	2.87 ± 3.41 3.51 (0–4.39)	3.37 ± 3.26 3.51 (0–4.39)	<0.0001
	Contribution to CAMP (%)	5.33	4.15	6.63	<0.0001

Table 4. *Cont.*

Food Categories		All N = 5690	Men N = 2554	Women N = 3136	p *
Fruits	mg/day (mean ± SD), median (IQR)	1.52 ± 2.47 0.60 (0–1.91)	1.38 ± 2.53 0.54 (0–1.64)	1.63 ± 2.42 0.72 (0.15–2.23)	<0.0001
	Contribution to CAMP (%)	2.57	1.99	3.21	0.0044
	Major sources (% contribution) **	raw fruits (2.56) *including: apples (0.61), bananas (0.53), grapes (0.30), mandarins (0.27), plums (0.19), oranges (0.17)*	raw fruits (1.98) *including: apples (0.51), bananas (0.42), grapes (0.22), mandarins (0.21), plums (0.16), oranges (0.11)*	raw fruits (3.20) *including: apples (0.72), bananas (0.64), grapes (0.40), mandarins (0.34), oranges (0.24), plums (0.23)*	-
Chocolate products	mg/day (mean ± SD), median (IQR)	0.68 ± 2.41 0 (0–0)	0.71 ± 2.59 0 (0–0)	0.65 ± 2.25 0 (0–0)	0.0462
	Contribution to CAMP (%)	1.15	1.03	1.28	0.3669
Potatoes	mg/day (mean ± SD), median (IQR)	0.37 ± 0.38 0.37 (0–0.67)	0.44 ± 0.44 0.37 (0–0.73)	0.31 ± 0.32 0.25 (0–0.49)	<0.0001
	Contribution to CAMP (%)	0.63	0.64	0.61	0.9213
Legumes	mg/day (mean ± SD), median (IQR)	0.35 ± 1.75 0 (0–0)	0.42 ± 2.06 0 (0–0)	0.28 ± 1.44 0 (0–0)	0.9950
	Contribution to CAMP (%)	0.59	0.61	0.55	0.6768
Other food products	mg/day (mean ± SD),	1.62	1.93	1.36	-
	Contribution to CAMP (%)	2.74	2.79	2.67	0.8152
Total campesterol intake	mg/day (mean ± SD), median (IQR)	59.06 ± 41.44 47.45 (31.53–74.39)	69.16 ± 47.68 56.71 (37.17–86.39)	50.83 ± 33.38 40.88 (27.80–64.90)	<0.0001
	Contribution to CAMP (%)	100	100	100	-

* p calculated for differences between men and women. ** In the total and each food category, only individual food products with the strongest impact on the total plant sterol intakes were listed.

Table 5. Contributions of food categories and individual food products to stigmasterol intake (STIG), listed according to diminishing order of contribution.

Food Categories		All N = 5690	Men N = 2554	Women N = 3136	*p* *
Coffee	mg/day (mean ± SD), median (IQR)	6.18 ± 6.55 6.90 (0–8.63)	5.64 ± 6.70 6.90 (0–8.63)	6.62 ± 6.40 6.90 (0–8.63)	<0.0001
	Contribution to STIG (%)	25.10	21.41	28.53	<0.0001
Vegetables	mg/day (mean ± SD), median (IQR)	5.72 ± 5.48 4.34 (1.90–7.96)	6.00 ± 5.63 4.55 (1.95–8.48)	5.49 ± 5.34 4.12 (1.85–7.45)	0.0007
	Contribution to STIG (%)	23.22	22.76	23.66	0.4179
	Major sources (% contribution) **	raw vegetables (17.90) *including: tomatoes (4.63), beets (1.90), cucumbers (1.82), carrots (1.80), parsley (1.88), green beans (1.88), lettuce (1.24), celery (1.08),* vegetable preserves (4.10)	raw vegetables (17.15) *including: tomatoes (4.48), beets (2.02), cucumbers (1.83), carrots (1.64), parsley (1.89), green beans (1.52), lettuce (1.17), celery (1.12),* vegetable preserves (4.62)	raw vegetables (18.60) *including: tomatoes (4.76), beets (1.79), cucumbers (1.81), carrots (1.96), parsley (1.88), green beans (2.21), lettuce (1.31), celery (1.04),* vegetable preserves (3.62)	-
Fats	mg/day (mean ± SD), median (IQR)	4.15 ± 4.93 2.59 (0.81–5.77)	5.02 ± 5.79 3.23 (1.05–7.10)	3.44 ± 3.97 2.28 (0.66–4.97)	<0.0001
	Contribution to STIG (%)	16.85	19.05	14.84	<0.0001
	Major sources (% contribution) **	soft margarines (11.82), oils (3.75), mixed fats (0.65), mayonnaise (0.60)	soft margarines (13.90), oils (3.69), mixed fats (0.79), mayonnaise (0.63)	soft margarines (9.91), oils (3.80), mixed fats (0.52), mayonnaise (0.56)	-
Cereals	mg/day (mean ± SD), median (IQR)	3.19 ± 2.26 2.72 (1.82–3.98)	3.91 ± 2.43 3.53 (2.36–5.02)	2.60 ± 1.93 2.33 (1.57–3.19)	<0.0001
	Contribution to STIG (%)	12.93	14.83	11.18	<0.0001
	Major sources (% contribution) **	wheat bread (6.48), rolls (2.57), rye bread (2.48), cereals (0,66)	wheat bread (8.49), rolls (2.69), rye bread (2.44), cereals (0,47)	wheat bread (4.62), rolls (2.29), rye bread (2.69), cereals (0,84)	-
Chocolate products	mg/day (mean ± SD), median (IQR)	1.58 ± 5.59 0 (0–0)	1.66 ± 6.00 0 (0–0)	1.51 ± 5.22 0 (0–0)	0.0474
	Contribution to STIG (%)	6.40	6.29	6.52	0.7579
Legumes	mg/day (mean ± SD), median (IQR)	1.21 ± 6.04 0 (0–0)	1.47 ± 7.23 0 (0–0)	0.99 ± 4.86 0 (0–0)	0.9904
	Contribution to STIG (%)	4.91	5.58	4.28	0.0208

Table 5. *Cont.*

Food Categories		**All** **N = 5690**	**Men** **N = 2554**	**Women** **N = 3136**	p *
Fruits	mg/day (mean ± SD), median (IQR)	0.92 ± 1.81 0.18 (0–0.88)	0.82 ± 1.71 0.15 (0–0.60)	1.00 ± 1.88 0.22 (0.02–1.13)	<0.0001
	Contribution to STIG (%)	3.73	3.13	4.29	0.0211
	Major sources (% contribution) **	raw fruits (3.72) *including:* *bananas (1.68), apples (0.40), nectarines (0.30), plums (0.29), peaches (0.40)*	raw fruits (3.12) *including:* *bananas (1.33), apples (0.37), nectarines (0.27), plums (0.26), peaches (0.25)*	raw fruits (4.28) *including:* *bananas (1.68), apples (0.44), nectarines (0.33), plums (0.32), peaches (0.53)*	-
Potatoes	mg/day (mean ± SD), median (IQR)	0.61 ± 0.63 0.60 (0–1.10)	0.73 ± 0.72 0.60 (0–1.21)	0.52 ± 0.53 0.41 (0–0.81)	<0.0001
	Contribution to STIG (%)	2.48	2.77	2.22	0.1861
Cookies, cakes	mg/day (mean ± SD), median (IQR)	0.59 ± 1.52 0 (0–0.40)	0.57 ± 1.62 0 (0–0.25)	0.61 ± 1.44 0 (0–0.50)	0.0004
	Contribution to STIG (%)	2.40	2.17	2.61	0.2589
Other food products	mg/day (mean ± SD),	0.48	0.54	0.44	-
	Contribution to STIG (%)	1.98	2.01	1.88	0.7530
Total stigmasterol intake	mg/day (mean ± SD), median (IQR)	24.63 ± 14.49 22.10 (14.53–30.92)	26.36 ± 16.02 23.49 (15.14–32.91)	23.22 ± 12.94 21.11 (14.16–29.19)	<0.0001
	Contribution to STIG (%)	100	100	100	-

* p calculated for differences between men and women. ** In the total and each food category, only individual food products with the strongest impact on the total plant sterol intakes were listed.

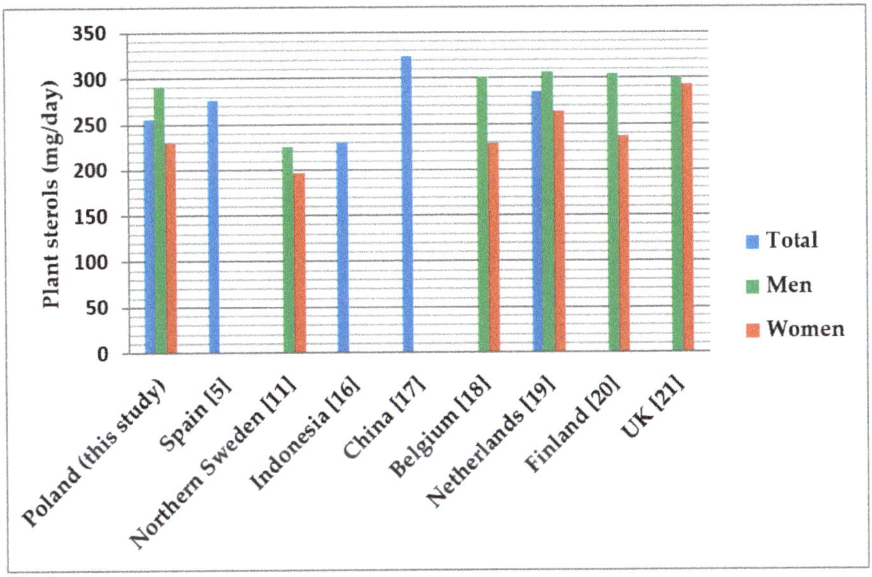

Figure 3. Intake of plant sterols in the Polish population and in other countries.

On a per milligram basis, men consumed more total and individual plant sterols (Table 6). However, per 1000 kcal, significantly more plant sterols as total and individual sterols were consumed by women ($p < 0.0001$), except for campesterol, for which the difference was not statistically significant.

Table 6. Comparison of total and individual sterol intakes (in mg and in mg/1000 kcal) by men and women.

Plant Sterols (mg)	Men	Women	*p*-Value
Total plant sterols	320.8	252.2	<0.0001
Total plant sterols/1000 kcal	141.0	154.2	<0.0001
β-sitosterol	197.8	158.1	<0.0001
β-sitosterol/1000 kcal	87.1	96.8	<0.0001
Campesterol	69.2	50.8	<0.0001
Campesterol/1000 kcal	29.8	30.4	0.2279
Stigmasterol	26.4	23.2	<0.0001
Stigmasterol/1000 kcal	12.0	14.8	<0.0001

4. Discussion

This is the first report on dietary plant sterol intake and its dietary sources in the Polish population. Due to the lack of plant sterols in Polish food composition tables, the database used for this study included international databases available in English supplemented with data from research papers on plant sterol contents in food products. In our study, the consumption of plant sterols from enriched food products was not taken into consideration, since the percentage of consumers of phytosterol-enriched products was low (2%). In comparison, it has been estimated that regular consumers of products with added plant sterols represent approximately 10–15% of the EU population [39].

Typical contemporary Western diets provide much lower amounts of phytosterols [40] than estimated for distant human ancestors, whose diet provided 1 g/day of phytosterols [41]. The dietary phytosterol intake in population studies is usually between 200 and 400 mg/day [21,42], even in those populations with more beneficial dietary habits [43], and this amount is too low to show significant LDL cholesterol-lowering effects demonstrated for 1 g of phytosterols [44]. Contrary to this, the PREDIMED study found that even small amounts of plant sterols from natural foods may exert a cholesterol-lowering effect [45]. A recent meta-analysis of 124 clinical studies demonstrated that a phytosterol intake between 0.6 and 3.3 g/day is associated with a gradual decrease in the concentration of LDL-cholesterol from 6% to 12% [46]. Scientific evidence indicates that even moderate doses of phytosterols delivered via a normal diet can provide a protective effect on the lipid profile by reducing cholesterol absorption [47,48], but a lipid-lowering effect may depend on the inter-individual variation in response to phytosterols [49].

The daily intake of total plant sterols in our study (255.96 mg/day) is similar to that of the Spanish population, where it was estimated to be 276 mg, with the largest contribution of beta-sitosterol (79.7%) [5]. In different populations, plant sterol intake ranged from 230 to 324 mg/day. Among other things, these differences may be due to the dietary habits of different populations or the availability of different food products on the market. Some differences may also be due to the food intake methodology. Some studies were based on a 24-h interview or dietary records, and others on a frequency of intake. Our results confirm earlier findings that β-sitosterol is the most important contributor (67.8%) to the intake of total dietary plant sterols. Regarding gender differences in plant sterol intake, in our study, the intake was 291.76 mg/day for men and 230.61 mg/day for women. These results are similar to most other populations where gender differences in plant sterol intake were observed among men and women [20]. These differences may be due to differences in food intake between the two sexes. Women tend to consume smaller portions of foods, which translates into fewer ingredients including plant sterols.

As per our study, the consumption pattern of total plant sterols from major food groups such as cereal products, vegetable oils and fats, vegetables, and fruits is similar to the intervention group in the PREDIMED study and to the U.K. population [21,45]. Of these, cereal products and oils provided nearly 61% of plant sterols, and when combined with vegetables and fruits, nearly 80%. However, unlike the PREDIMED study, where legumes were the fifth contributor to total plant sterols, in our study, the additional sources of plant sterols included coffee, cookies and cakes, chocolate products, and potatoes, while legumes were only ninth in providing plant sterols. Together, these minor sources of plant sterols accounted for 16.68% of plant sterol intake. The other sources of plant sterols accounted for 3.55%; these included, among others, nuts and seeds, which are normally a good source of plant sterols, but because of their low intake [50], they were not a significant source of plant sterols for the Polish population. The PREDIMED intervention study indicated an important role for the Mediterranean diet, in combination with nuts, in providing plant sterols in the diet and providing a cholesterol-lowering effect [45]. Considering this, Poles should be encouraged to increase their nut consumption and improve their dietary habits, which are far from the recommended for the prevention of cardiovascular diseases [51,52]. Regarding individual dietary sources of total plant sterols, canola oil and white bread predominated, followed by soft margarine. Similar to a Chinese study, canola oil was the main provider of plant sterols among vegetable fats and oils [53].

As in the study of EPIC-Norfolk population [21], women in the WOBASZ survey had a higher plant sterol density than men. Interestingly, when converted per 1000 kcal, the total plant sterol content did not differ from the values obtained in the EPIC-Norfolk study. For men and women in our study, the amount of plant sterols was 141.0 mg and 154.2 mg, respectively, and in the EPIC-Norfolk study, for men it was 137.33 mg and for women it was 152.4 mg/day.

Limitations

Some plant sterol values in this study may have been underestimated because only three major sterols (sitosterol, campesterol, and stigmasterol) are typically included in the totals, despite the contribution of other sterols. Although the compiled database facilitated the calculation of plant sterols, there are some shortcomings due to the lack of data for individual plant sterols. This is mainly due to the fact that the literature data do not provide information on the content of plant sterols in certain food products. For some foods, the values of plant sterols (total and individual) were not found, e.g., no studies were found for chard. No data were found for campesterol or stigmasterol in radishes, wines, and mushrooms. For foods such as chives, blueberries, cherries, pears, raspberries, blackcurrants, walnuts, and pumpkin seeds, no value was found for stigmasterol. Therefore, the values obtained for the sum of individual plant sterols could be lower than the total plant sterol content. Moreover, there are no specific data on the composition of plant sterols in enriched margarine, which is related, among other things, to proprietary manufacturing technologies. In addition, since a small percentage of study participants consumed phytosterol-enriched margarine, they were not included in the calculation of dietary plant sterols.

Furthermore, a limitation of the study is the inclusion in the plant sterol database of results from several less rigorous literature sources than those given in the Rand criteria.

This study used single 24-h recall as a tool to measure food intake, which is an appropriate method for large-scale studies. However, 24-h recall does not account for variability in food intake and may not describe a typical diet.

5. Conclusions

This is the first study to evaluate the intake of plant sterols in the Polish population. Since no plant sterols are listed in Polish food composition tables, a database was developed using published data sources. The study found that plant sterol intake in Poland is 255.96 mg/day, which is comparable to other populations, despite different methodologies of nutritional assessment and slightly different databases. The main dietary sources of plant sterols in this study were cereals, fats, vegetables, and fruits, which is consistent with data for other populations. This study found that women's diets are more dense in plant sterols, which is in agreement with other studies. Due to the lack of literature sources on plant sterol content in some foods, future studies should expand and complete the databases on plant sterol content in foods.

Supplementary Materials: The following are available online at https://www.mdpi.com/article/10.3390/nu13082722/s1, Table S1: Content of selected plant sterols (stigmasterol, campesterol, beta-sitosterol) and total plant sterols in food products (mg/100 g of product), Table S2: Content of selected plant sterols (stigmasterol, campesterol, beta-sitosterol) and total plant sterols in dishes (mg/100 g of dish).

Author Contributions: Conceptualization, A.W. and A.M.W.; methodology, A.M.W., A.W. and M.E.Z.; software, A.W., A.M.W. and A.C.-M.; validation, A.W., A.M.W. and I.M.-C.; formal analysis, A.C.-M. and A.W.; investigation, A.W. and A.M.W.; resources, A.M.W., A.W., M.E.Z., I.M.-C. and W.D.; data curation, A.W. and A.M.W.; writing—original draft preparation, A.M.W.; writing—review and editing, A.W., M.E.Z., I.M.-C., A.C.-M. and W.D.; visualization, A.W. and A.M.W.; supervision, A.W., A.M.W. and W.D.; project administration, A.M.W. and A.W.; funding acquisition, A.W. and A.M.W. All authors have read and agreed to the published version of the manuscript.

Funding: This research was funded by the National Institute of Cardiology (Grant no. 2.20/I/20) and Medical University of Bialystok (Grant no. SUB/3/DN/21/001/3317).

Institutional Review Board Statement: The study was conducted according to the guidelines of the Declaration of Helsinki, and approved by the Bioethics Committee of the National Institute of Cardiology (protocol code 1344, date of approval 5 November 2012, and protocol code 1837, date of approval 14 January 2020).

Informed Consent Statement: Informed consent was obtained from all subjects involved in the study.

Data Availability Statement: All data in this study are made available upon request to the authors at the following e-mail address: anna.witkowska@umb.edu.pl or awaskiewicz@ikard.pl.

Conflicts of Interest: The authors declare no conflict of interest.

References

1. Dufourc, E.J. Sterols and membrane dynamics. *J. Chem. Biol.* **2008**, *1*, 63–77. [CrossRef] [PubMed]
2. De Smet, E.; Mensink, R.P.; Plat, J. Effects of plant sterols and stanols on intestinal cholesterol metabolism: Suggested mechanisms from past to present. *Mol. Nutr. Food Res.* **2012**, *56*, 1058–1072. [CrossRef]
3. Vezza, T.; Canet, F.; de Marañón, A.M.; Bañuls, C.; Rocha, M.; Víctor, V.M. Phytosterols: Nutritional Health Players in the Management of Obesity and Its Related Disorders. *Antioxidants* **2020**, *9*, 1266. [CrossRef] [PubMed]
4. Feng, S.; Belwal, T.; Li, L.; Limwachiranon, J.; Liu, X.; Luo, Z. Phytosterols and their derivatives: Potential health-promoting uses against lipid metabolism and associated diseases, mechanism, and safety issues. *Compr. Rev. Food Sci. Food Saf.* **2020**, *19*, 1243–1267. [CrossRef]
5. Jiménez-Escrig, A.; Santos-Hidalgo, A.B.; Saura-Calixto, F. Common sources and estimated intake of plant sterols in the Spanish diet. *J. Agric. Food Chem.* **2006**, *54*, 3462–3471. [CrossRef] [PubMed]
6. Shaghaghi, M.A.; Harding, S.V.; Jones, P.J.H. Water dispersible plant sterol formulation shows improved effect on lipid profile compared to plant sterol esters. *J. Funct. Foods* **2014**, *6*, 280–289. [CrossRef]
7. Plat, J.; Brufau, G.; Dallinga-Thie, G.M.; Dasselaar, M.; Mensink, R.P. A plant stanol yogurt drink alone or combined with a low-dose statin lowers serum triacylglycerol and non-HDL cholesterol in metabolic syndrome patients. *J. Nutr.* **2009**, *139*, 1143–1149. [CrossRef]
8. Yoshida, Y.; Niki, E. Antioxidant effects of phytosterol and its components. *J. Nutr. Sci. Vitaminol.* **2003**, *49*, 277–280. [CrossRef]
9. Brüll, F.; De Smet, E.; Mensink, R.P.; Vreugdenhil, A.; Kerksiek, A.; Lütjohann, D.; Wesseling, G.; Plat, J. Dietary plant stanol ester consumption improves immune function in asthma patients: Results of a randomized, double-blind clinical trial. *Am. J. Clin. Nutr.* **2016**, *103*, 444–453. [CrossRef] [PubMed]
10. Kurano, M.; Hasegawa, K.; Kunimi, M.; Hara, M.; Yatomi, Y.; Teramoto, T.; Tsukamoto, K. Sitosterol prevents obesity-related chronic inflammation. *Biochim. Biophys. Acta Mol. Cell Biol. Lipids* **2018**, *1863*, 191–198. [CrossRef]
11. Klingberg, S.; Ellegård, L.; Johansson, I.; Jansson, J.H.; Hallmans, G.; Winkvist, A. Dietary intake of naturally occurring plant sterols is related to a lower risk of a first myocardial infarction in men but not in women in northern Sweden. *J. Nutr.* **2013**, *143*, 1630–1635. [CrossRef]
12. Ramprasath, V.R.; Awad, A.B. Role of phytosterols in cancer prevention and treatment. *J. AOAC Int.* **2015**, *98*, 735–738. [CrossRef]
13. Jiang, L.; Zhao, X.; Xu, J.; Li, C.; Yu, Y.; Wang, W.; Zhu, L. The Protective Effect of Dietary Phytosterols on Cancer Risk: A Systematic Meta-Analysis. *J. Oncol.* **2019**, *2019*, 7479518. [CrossRef]
14. Composition of Foods, Raw, Processed, Prepared. National Nutrient Database for Standard Reference Release 27. USDA, 2015. Modified in 2019. Available online: https://data.nal.usda.gov/dataset/composition-foods-raw-processed-prepared-usda-national-nutrient-database-standard-reference-release-27 (accessed on 25 May 2019).
15. Tolve, R.; Cela, N.; Condelli, N.; Di Cairano, M.; Caruso, M.C.; Galgano, F. Microencapsulation as a Tool for the Formulation of Functional Foods: The Phytosterols' Case Study. *Foods* **2020**, *9*, 470. [CrossRef]
16. Martianto, D.; Bararah, A.; Andarwulan, N.; Średnicka-Tober, D. Cross-Sectional Study of Plant Sterols Intake as a Basis for Designing Appropriate Plant Sterol-Enriched Food in Indonesia. *Nutrients* **2021**, *13*, 452. [CrossRef]
17. Wang, M.; Huang, W.; Hu, Y.; Zhang, L.; Shao, Y.; Wang, M.; Zhang, F.; Zhao, Z.; Mei, X.; Li, T.; et al. Phytosterol profiles of common foods and estimated natural intake of different structures and forms in China. *J. Agric. Food Chem.* **2018**, *66*, 2669–2676. [CrossRef]
18. Sioen, I.; Matthys, C.; Huybrechts, I.; Van Camp, J.; De Henauw, S. Consumption of plant sterols in Belgium: Estimated intakes and sources of naturally occurring plant sterols and β-carotene. *Br. J. Nutr.* **2011**, *105*, 960–966. [CrossRef] [PubMed]
19. Normen, A.L.; Brants, H.A.M.; Voorrips, L.E.; Anderson, H.A.; van den Brandt, P.A.; Goldbohm, R.A. Plant sterol intakes and colorectal cancer risk in The Netherlands Cohort Study on diet and cancer. *Am. J. Clin. Nutr.* **2001**, *74*, 141–148. [CrossRef] [PubMed]
20. Valsta, L.M.; Lemström, A.; Ovaskainen, M.L.; Lampi, A.M.; Toivo, J.; Korhonen, T.; Piironen, V. Estimation of plant sterol and cholesterol intake in Finland: Quality of new values and their effect on intake. *Br. J. Nutr.* **2004**, *92*, 671–678. [CrossRef] [PubMed]
21. Klingberg, S.; Andersson, H.; Mulligan, A.; Bhaniani, A.; Welch, A.; Bingham, S.; Khaw, K.T.; Andersson, S.; Ellegård, L. Food sources of plant sterols in the EPIC Norfolk population. *Eur. J. Clin. Nutr.* **2008**, *62*, 695–703. [CrossRef]
22. Amirabdollahian, F.; Ash, R. An estimate of phytate intake and molar ratio of phytate to zinc in the diet of the people in the United Kingdom. *Public Health Nutr.* **2010**, *13*, 1380–1388. [CrossRef]
23. Przysławski, J.; Stelmach, M.; Grygiel-Górniak, M.; Dubec, A. An assessment of dietary habits in the group of studying adolescents especially taking into consideration plant sterols intake—Pilot study. *Nowiny Lek.* **2009**, *77*, 299–304.
24. Food Standard Agency, Public Health England. McCance and Widdowson's the Composition of Foods: 2019. Available online: https://www.gov.uk/government/publications/composition-of-foods-integrated-dataset-cofid (accessed on 25 May 2019).
25. Rand, W.M.; Pennington, J.A.T.; Murphy, S.P.; Klensin, J.C. *Compiling Data for Food Composition Data Bases*; United Nations University Press: Tokyo, Japan, 1991. Available online: http://www.unu.edu/unupress/unupbooks/80772e/80772E00.htm (accessed on 10 January 2019).
26. Lagarda, M.J.; García-Llatas, G.; Farré, R. Analysis of phytosterols in foods. *J. Pharm. Biomed. Anal.* **2006**, *41*, 1486–1496. [CrossRef] [PubMed]

27. Normen, L.; Ellegard, L.; Brants, H.; Dutta, P.; Andersson, H. A phytosterol database: Fatty foods consumed in Sweden and the Netherlands. *J. Food Compos. Anal.* **2007**, *20*, 193–201. [CrossRef]
28. Normén, L.; Bryngelsson, S.; Johnsson, M.; Evheden, P.; Ellegård, L.; Brants, H.; Andersson, H.; Dutta, P. The Phytosterol Content of Some Cereal Foods Commonly Consumed in Sweden and in the Netherlands. *J. Food Compos. Anal.* **2002**, *15*, 693–704. [CrossRef]
29. Normén, L.; Johnsson, M.; Andersson, H.; van Gameren, Y.; Dutta, P. Plant sterols in vegetables and fruits commonly consumed in Sweden. *Eur. J. Nutr.* **1999**, *38*, 84–89. [CrossRef] [PubMed]
30. Han, J.H.; Yang, Y.X.; Feng, M.Y. Contents of phytosterols in vegetables and fruits commonly consumed in China. *Biomed. Environ. Sci.* **2008**, *21*, 449–453. [CrossRef]
31. Piironen, V.; Toivo, J.; Puupponen-Pimia, R.; Lampi, A.M. Plant sterols in vegetables, fruits and berries. *J. Sci. Food Agric.* **2003**, *83*, 330–337. [CrossRef]
32. Ryan, E.; Galvin, K.; O'Connor, T.P.; Maguire, A.R.; O'Brien, N.M. Phytosterol, squalene, tocopherol content and fatty acid profile of selected seeds, grains, and legumes. *Plant Foods Hum. Nutr.* **2007**, *62*, 85–91. [CrossRef]
33. Li, Y.C.; Li, C.L.; Li, R.; Chen, Y.; Zhang, M.; Guo, P.P.; Shi, D.; Ji, X.N.; Feng, R.N.; Sun, C.H. Associations of dietary phytosterols with blood lipid profiles and prevalence of obesity in Chinese adults, a cross-sectional study. *Lipids Health Dis.* **2018**, *17*, 54. [CrossRef]
34. Yamaya, A.; Endo, Y.; Fujimoto, K.; Kitamura, K. Effects of genetic variability and planting location on the phytosterol content and composition in soybean seeds. *Food Chem.* **2007**, *102*, 1071–1075. [CrossRef]
35. Decloedt, A.I.; Van Landschoot, A.; Watson, H.; Vanderputten, D.; Vanhaecke, L. Plant-Based Beverages as Good Sources of Free and Glycosidic Plant Sterols. *Nutrients* **2017**, *10*, 21. [CrossRef]
36. Ruggiero, A.; Vitalini, S.; Burlini, N.; Bernasconi, S.; Iriti, M. Phytosterols in grapes and wine, and effects of agrochemicals on their levels. *Food Chem.* **2013**, *141*, 3473–3479. [CrossRef]
37. Čížková, H.; Soukupová, V.; Voldřich, M.; Ševčík, R. Differentiation of coffee varieties according to their sterolic profile. *J. Food Nutr. Res.* **2007**, *46*, 28–34.
38. Drygas, W.; Niklas, A.A.; Piwońska, A.; Piotrowski, W.; Flotyńska, A.; Kwaśniewska, M.; Nadrowski, P.; Puch-Walczak, A.; Szafraniec, K.; Bielecki, W.; et al. Multi-center National Population Health Examination Survey (WOBASZ II study): Assumptions, methods and implementation. *Kardiol. Pol.* **2016**, *74*, 681–690. [CrossRef] [PubMed]
39. European Food Safety Authority (EFSA). Consumption of Food and Beverages with Added Plant Sterols. *EFSA J.* **2008**, *133*, 1–21. [CrossRef]
40. Ras, R.T.; van der Schouw, Y.T.; Trautwein, E.A.; Sioen, I.; Dalmeijer, G.W.; Zock, P.L.; Beulens, J.W.J. Intake of phytosterols from natural sources and risk of cardiovascular disease in the European Prospective Investigation into Cancer and Nutrition-the Netherlands (EPIC-NL) population. *Eur. J. Prev. Cardiol.* **2015**, *22*, 1067–1075. [CrossRef] [PubMed]
41. Jew, S.; AbuMweis, S.S.; Jones, P.J.H. Evolution of the human diet: Linking our ancestral diet to modern functional foods as a means of chronic disease prevention. *J. Med. Food* **2009**, *12*, 925–934. [CrossRef]
42. Sirirat, R.; Heskey, C.; Haddad, E.; Tantamango-Bartley, Y.; Fraser, G.; Mashchak, A.; Jaceldo-Siegl, K. Comparison of phytosterol intake from FFQ with repeated 24-h dietary recalls of the Adventist Health Study-2 calibration sub-study. *Br. J. Nutr.* **2019**, *121*, 1424–1430. [CrossRef]
43. Jaceldo-Siegl, K.; Lütjohann, D.; Sirirat, R.; Mashchak, A.; Fraser, G.E.; Haddad, E. Variations in dietary intake and plasma concentrations of plant sterols across plant-based diets among North American adults. *Mol. Nutr. Food Res.* **2017**, *61*. [CrossRef]
44. Moreau, R.A. Composition of Plant Sterols and Stanols in Supplemented Food Products. *J. AOAC Int.* **2015**, *98*, 685–690. [CrossRef] [PubMed]
45. Escurriol, V.; Cofán, M.; Serra, M.; Bulló, M.; Basora, J.; Salas-Salvadó, J.; Corella, D.; Zazpe, I.; Martínez-González, M.A.; Ruiz-Gutiérrez, V.; et al. Serum sterol responses to increasing plant sterol intake from natural foods in the Mediterranean diet. *Eur. J. Nutr.* **2009**, *48*, 373–382. [CrossRef]
46. Ras, R.T.; Geleijnse, J.M.; Trautwein, E.A. LDL cholesterol-lowering effect of plant sterols and stanols across different dose ranges: A meta-analysis of randomised controlled studies. *Br. J. Nutr.* **2014**, *112*, 214–219. [CrossRef]
47. Sanclemente, T.; Marques-Lopes, I.; Fajo-Pascual, M.; Cofán, M.; Jarauta, E.; Ros, E.; Puzo, J.; García-Otín, A.L. Naturally-occurring phytosterols in the usual diet influence cholesterol metabolism in healthy subjects. *Nutr. Metab. Cardiovasc. Dis.* **2012**, *22*, 849–855. [CrossRef]
48. Racette, S.B.; Lin, X.; Lefevre, M.; Spearie, C.A.; Most, M.M.; Ma, L.; Ostlund, R.E., Jr. Dose effects of dietary phytosterols on cholesterol metabolism: A controlled feeding study. *Am. J. Clin. Nutr.* **2010**, *91*, 32–38. [CrossRef] [PubMed]
49. Trautwein, E.A.; Vermeer, M.A.; Hiemstra, H.; Ras, R.T. LDL-Cholesterol Lowering of Plant Sterols and Stanols-Which Factors Influence Their Efficacy? *Nutrients* **2018**, *10*, 1262. [CrossRef]
50. Witkowska, A.M.; Waśkiewicz, A.; Zujko, M.E.; Szcześniewska, D.; Śmigielski, W.; Stepaniak, U.; Pająk, A.; Drygas, W. The Consumption of Nuts is Associated with Better Dietary and Lifestyle Patterns in Polish Adults: Results of WOBASZ and WOBASZ II Surveys. *Nutrients* **2019**, *11*, 1410. [CrossRef] [PubMed]
51. Waśkiewicz, A.; Szcześniewska, D.; Szostak-Węgierek, D.; Kwaśniewska, M.; Pająk, A.; Stepaniak, U.; Kozakiewicz, K.; Tykarski, A.; Zdrojewski, T.; Zujko, M.E.; et al. Are dietary habits of the Polish population consistent with the recommendations for prevention of cardiovascular disease?—WOBASZ II project. *Kardiol. Pol.* **2016**, *74*, 969–977. [CrossRef] [PubMed]

52. Drywień, M.E.; Hamulka, J.; Jezewska-Zychowicz, M. Perceived Nutrition and Health Concerns: Do They Protect against Unhealthy Dietary Patterns in Polish Adults? *Nutrients* **2021**, *13*, 170. [CrossRef]
53. Yang, R.; Xue, L.; Zhang, L.; Wang, X.; Qi, X.; Jiang, J.; Yu, L.; Wang, X.; Zhang, W.; Zhang, Q.; et al. Phytosterol Contents of Edible Oils and Their Contributions to Estimated Phytosterol Intake in the Chinese Diet. *Foods* **2019**, *8*, 334. [CrossRef]

nutrients

MDPI

Article

Comparison of Healthiness, Labelling, and Price between Private and Branded Label Packaged Foods in New Zealand (2015–2019)

Teresa Castro [1,2,*], Sally Mackay [1], Leanne Young [3], Cliona Ni Mhurchu [3,4], Grace Shaw [1], Essa Tawfiq [3] and Helen Eyles [1,3]

1 Department of Epidemiology and Biostatistics, The University of Auckland, Auckland 1010, New Zealand; sally.mackay@auckland.ac.nz (S.M.); shagr003@student.otago.ac.nz (G.S.); h.eyles@auckland.ac.nz (H.E.)
2 Nutrition Section, The University of Auckland, Auckland 1010, New Zealand
3 National Institute for Health Innovation, The University of Auckland, Auckland 1010, New Zealand; leanne.young@auckland.ac.nz (L.Y.); c.nimhurchu@auckland.ac.nz (C.N.M.); e.tawfiq@auckland.ac.nz (E.T.)
4 The George Institute for Global Health, Newtown, NSW 2042, Australia
* Correspondence: t.castro@auckland.ac.nz

Abstract: We aimed to compare New Zealand private label (PL) and branded label (BL) packaged food products in relation to their current (2019) healthiness (sodium and sugar contents, and estimated Health Star Rating (HSR) score), display of the voluntary HSR nutrition label on the package, and price. Healthiness and HSR display of products were also explored over time (2015 to 2019). Data were obtained from Nutritrack, a brand-specific food composition database. Means and proportions were compared using Student *t*-tests and Pearson chi-square tests, respectively. Changes over time were assessed using linear regression and chi-square tests for trends (Mantel–Haenzel tests). Altogether, 4266 PL and 19,318 BL products across 21 food categories were included. Overall, PL products in 2019 had a significantly lower mean sodium content and price, a higher proportion of products with estimated HSR $\geq$ 3.5/5 (48.9% vs. 38.5%) and were more likely to display the HSR on the pack compared with BL products (92.4% vs. 17.2%, respectively). However, for most food categories, no significant difference was found in mean sodium or sugar content between PL and BL products. In the period 2015–2019, there were no consistent changes in estimated HSR score, sodium or sugar contents of PL or BL products, but there was an increase in the proportion of both PL and BL products displaying HSR labels. In most food categories, there were PL options available which were similar in nutritional composition, more likely to be labelled with the HSR, and lower in cost than their branded counterparts.

Keywords: supermarket packaged foods; private labels; generic labels; branded labels; health star rating; sugar; sodium; healthiness; price; public health policy

Citation: Castro, T.; Mackay, S.; Young, L.; Ni Mhurchu, C.; Shaw, G.; Tawfiq, E.; Eyles, H. Comparison of Healthiness, Labelling, and Price between Private and Branded Label Packaged Foods in New Zealand (2015–2019). *Nutrients* **2021**, *13*, 2731. https://doi.org/10.3390/nu13082731

Academic Editor: Joerg Koenigstorfer

Received: 12 July 2021
Accepted: 5 August 2021
Published: 9 August 2021

1. Introduction

New Zealand (NZ) has a high prevalence of nutrition-related disease (NCD) [1], with poor diets characterized by energy-dense, nutrient-poor foods and beverages, accounting for nearly 20% of illness and early death in 2017 [1]. Furthermore, there is an inequitable food environment in NZ which promotes the consumption of unhealthy foods [2–7], including the steady growth of the packaged food industry, which consequently has an important role to play in improving population diets and preventing NCDs [2]. In NZ, supermarkets account for ~75% of all purchases of packaged foods [8], and the supermarket food environment consists of a duopoly of two supermarket retailers: Foodstuffs and Woolworths [9]. These two supermarket retailers provide groceries to eight supermarket chains across the country [9].

The availability of private label (PL) and branded label (BL) products in supermarkets is important for generating price competitiveness and to offer consumers options in terms of quality and variety [10]. Both supermarket retailers in NZ offer PL options, also known as "own-brands", "generic brands", "store brands", or "economy lines". In 2020, PL sales accounted for 10.2% of packaged foods sold in NZ stores and online [8]. In 2020, Foodstuffs reported that they were on target to reach $1.3 billion in sales from PL products [11]. In 2019, Woolworths Group expanded their PL range and released 640 new PL products into the market [12]. Given the relevant presence of PL products in NZ supermarkets, it is important that these products are equitable from a health perspective.

Since 2016, both NZ supermarket retailers have made commitments to improve the healthiness of their PL products, specifically through reformulation focused on reducing sugar, sodium, and saturated fat content [13–15]. For example, one retailer has committed to all PL products being nutritionally on par, or better than, the average comparable BL products (by 2018) [13]. Both retailers have also committed to displaying the voluntary Health Star Rating (HSR) nutrition label on all applicable PL products [13–15]. The HSR system was introduced into New Zealand and Australia in mid-2014 as a front-of-pack labelling scheme endorsed by the government to enable consumers to easily compare the healthiness of similar types of products on a scale of 0.5 to 5.0, with a higher score indicating a healthier product [16]. Products with an HSR $\geq$ 3.5 stars are generally considered a healthier choice [17].

Two previous NZ studies have examined differences in the healthiness of PL and BL products, with both examining sodium content [18,19], and one examining price [18]. In 2002, across 11 of 15 food categories assessed, mean sodium content was found to be lower for most PL compared with BL options in analyses involving unmatched products. This same study also found that, for 11 food categories, PL products were, on average, cheaper than BL options. This finding contrasts with that described by Monro et al. (2015) [19], where the mean sodium content of PL products was found to be higher than that of BL products. However, it is important to highlight the small number of food categories (n = 8) and variation in product types included in this latter study, which limits the generalizability of the findings.

To the best of our knowledge, there are no recent and comprehensive studies in NZ comparing the healthiness and price of PL and BL packaged food products, or changes in their healthiness over time. Consumers have the right to be informed when looking for healthier, cost-effective food options, and this is particularly important for addressing equity. Furthermore, NZ food retailers need to know how healthy their PL products are compared with branded options. Therefore, our aim was to compare the healthiness, display of HSR, and price of PL and BL packaged food products sold at major NZ supermarket chains. Specific questions were: (1) Do the healthiness, display of HSR and price differ between PL and BL food products on the market in 2019? (2) Has the healthiness and display of HSR on PL and BL packaged food products changed over time (five years from 2015 to 2019)?

2. Materials and Methods

2.1. Outcomes and Data Sources

In this study, the following indicators were described and compared between PL and BL packaged food categories (FCs):

(i) Healthiness: Data on the sodium and sugar contents (mg/100 g) and estimated Health Star Rating (HSR) were extracted from the Nutritrack database for the years 2015 to 2019. Nutritrack is a packaged food database managed by the National Institute for Health Innovation at the University of Auckland and includes information for packaged foods sold in four major NZ supermarket chains (New World, Four Square, Countdown and PAK'nSAVE). Annual cross-sectional supermarket surveys are undertaken using a systematic process at the same time each year (February to May) in Auckland, New Zealand. Photographs of packaged foods and beverages that display a nutrition information panel (NIP) are taken using a customized smartphone application and names, brands, labelling, ingredients and NIP information are entered into a secure online system [20]. Information is collected for ~75% of unique packaged foods and beverages purchased in NZ [8]. By 2018, only 21% of the NZ supermarket packaged products displayed the manufacturer-calculated HSR score on the pack [2]. Thus, for the purposes of this study, we estimated the HSR score for all products using the stepwise approach and the HSR Calculator 2018 provided by The New Zealand Ministry for Primary Industries; further details are available in Tawfiq et al. (2021) [5]. Estimated HSR scores were categorized as <3.5 stars (unhealthy) and ≥3.5 stars (healthy) for analyses [16].

(ii) Products displaying HSR on the pack: Information on whether products were displaying HSR on the pack was also sourced from the Nutritrack database for the years 2015 to 2019. This outcome was classified as "yes" or "no".

(iii) Price: Mean price for each product in Nutritrack 2019 (mean NZ$/product package) was calculated using price information from the Nielsen New Zealand Homescan® panel between October 2018 and October 2019. Nielsen market research data are one of the largest and most up-to-date datasets available to monitor household food purchases [21]. The Nielsen New Zealand Homescan® panel is a sample of approximately 2500 households, designed to be representative of NZ households in terms of geographic and demographic characteristics. Nielsen New Zealand Homescan® excludes data for households who scan items inconsistently or do not meet the minimum spending criteria. We used data for 1,800 NZ households who purchased food in stores, and this approach was consistent with that used for in-store data collection for Nutritrack 2019. Households are based in major and secondary urban sites (according to the definition of Statistics NZ [22]), which accounts for 92% of the country's population [5]. Price information was estimated from all product purchases made by panel members between October 2018 and October 2019. We excluded all pricing data from stores other than supermarkets, grocery stores, fruit and vegetable stores, convenience stores, fish and meat stores, and bakeries. Using the product barcodes as the key linking variable, information on the mean price of each product (NZ$) in NZ Nielsen Homescan® was linked to Nutritrack. After data merging, the mean price of product packs was converted to mean price (in NZ$) per 100 g of product, to enable comparison across different package sizes.

2.2. Selection of Products, Exclusion Criteria and Data Preparation

In Nutritrack, individual products are categorized into a standardized hierarchical structure of five levels (L1 to L5), the top three comprising 15 food groups, 59 categories and 177 smaller subcategories [23,24]. Information on PL or BL status was retrieved from company websites and manually added to each unique product in the Nutritrack data [20]. The selection of the food categories in Nutritrack for inclusion in the current analyses was first based on the rationale that reformulation of products within that category should be feasible, e.g., fresh dairy milk was excluded. In order to guarantee statistical power for the analyses, food categories also had to have at least 30 [25] PL products available in 2019 for the comparison of means, or at least 100 PL products available across all years from 2015 to 2019 for comparisons of mean changes over time. Food categories were initially selected using Nutritrack food group classification level 2 (L2), for example, fish/processed fish. However, when a food category at L2 contained a range of nutritionally diverse products, minor food categories at Nutritrack levels (L3, 4 and 5) were selected instead (for example, canned fish was used rather than the aggregated group fish and seafood products/processed fish). In total, 21 FCs were selected for inclusion, comprising 24,205 products in the period 2015–2019. If selected food categories at any level contained any minor food categories at L3, L4 or L5 with less than five PL products in the period 2015–2019, products within the minor (s) category at L3, L4 or L5 were excluded (n = 408; 1.7%). For example, anchovies were not included in the canned fish category because there were <5 anchovy products across the five years. Products where nutrient data were only available in reconstituted form, and products with multiple NIPs, such as meal kits, were also excluded (n = 209; 0.9% and n = 4; 0.02%, respectively). Thus, the total number of products included in the analyses from 2015 to 2019 was 23,584 (4266 PL and 19,318 BL products). Products included in analyses corresponded to 31.1%, 40.7% and 29.9% of all products, PL products, and BL products available in the Nutritrack database, respectively (2015–2019). Table S1 (Supplementary file) presents the selected food categories, their minor food categories and the number of products assessed in each category (in total, for PL and for BL).

Sugar content of PL and BL products was not compared within food categories that are not key sources of sugar, i.e., canned fish, canned vegetables, pickled vegetables, salted nuts, processed meats, and crispy and salty snacks. Similarly, sodium content was not compared for ice-cream and fruit in syrup/juice, as these are not major sources of sodium. In total, 170 (0.7%) products had missing information for sodium and sugar, so it was not possible to estimate and HSR for them. Estimations were not calculated for a further 152 (0.6%) products as there were errors in their sodium and/or sugar contents in Nutritrack. Table S2 (Supplementary file) shows the number and percentage of products with missing information for sugar content, sodium content, or estimated HSR, according to the food category. Information was available for all products and years on whether HSR was displayed on pack. Of the 4896 selected products in Nutritrack in 2019 (PL and BL), 431 (8.8%) were not included in the Nielsen Homescan data for 2019, and, thus, information on price for these products was missing.

2.3. Statistical Analyses

Descriptive statistics were performed to describe means and standard deviations (SD), value ranges and proportions. There were not enough PL products available to allow paired comparisons to assess product reformulation over time. Therefore, in this study, we compared means and proportions in 2019 and changes in means and proportions in the period 2015–2019 (overall and at the food category level).

Food categories with 30 or more products were considered sufficiently large for the central limit theorem to apply [25]. T-tests for independent samples were applied to compare statistically significant differences in means between PL and BL products in 2019. Pearson chi-square tests were performed to assess whether there were statistically significant differences in proportions of PL and BL products displaying HSR on the pack and with estimated HSR $\geq$ 3.5 in 2019.

Mean changes in sugar and sodium content in the period 2015–2019 were assessed separately within BL and PL products (overall and by FC). To estimate the average change in sodium (mg/100 g) or sugar (g/100 g) contents from 2015 to 2019, linear regression models were performed with sodium or sugar as the dependent variable. Year was included in the model as the independent variable—as a continuous variable, coded as: 2015 = 0, 2016 = 0.25, 2017 = 0.50, 2018 = 0.75, and 2019 = 1 [3,26]. Mean percentage change in sugar and sodium content across the five years was calculated by dividing the adjusted mean change in sugar or sodium from 2015 to 2019 by the mean value in 2015 (multiplied by 100%). Overall and within the food categories, five-year trends in the proportion of PL and BL products with HSR $\geq$ 3.5, and products displaying the HSR were examined using chi-square tests for trends (linear-by-linear associations using Mantel–Haenzel tests). Analyses of changes over time were performed for all PL and BL food categories as all contained at least 100 products with information available in the period 2015–2019. Analyses were performed using SPSS software (version 25, IBM SPSS Statistics), and all tests were two sided at the level of significance of 5%.

3. Results

3.1. Comparison of PL and BL Products in 2019

3.1.1. Healthiness: Mean Sugar and Sodium Content and Proportion of Products with Estimated HSR ≥ 3.5

Tables 1 and 2 describe and compare the indicators of healthiness between PL and BL products by food category in 2019. Overall, PL products had statistically significantly lower mean sodium content than BL products. However, there were significant differences in mean sodium content between PL and BL products for only two of the 19 food categories assessed (canned fish and canned vegetables, both with lower means for PL products). Overall, there were no differences in the mean sugar content of PL and BL products, and a significant difference in mean sugar content between PL and BL products for only one of the 14 food categories assessed, i.e., canned fruit, with a lower mean for PLs (Table 1).

Table 1. Mean (standard deviation) of sodium and sugar content of branded and private label supermarket products in 2019, by food category and overall.

Food Categories	Sodium (mg/100 g)		p Value *	Sugar (g/100 g)		p Value *
	N	Mean (SD)		n	Mean (SD)	
Savoury biscuits						
BL	295	624.6 (263.7)	0.132	295	3.0 (2.5)	0.512
PL	50	565.7 (200.1)		50	3.4 (4.3)	
Sweet biscuits						
BL	356	286.4 (135.7)	0.154	356	32.9 (11.2)	0.876
PL	73	261.2 (143.1)		73	33.1 (9.6)	
Everyday sliced breads						
BL	105	398.3 (73.3)	†	105	2.8 (1.0)	†
PL	9	373.3 (22.9)		9	3.2 (0.76)	
Other breads						
BL	223	421.0 (144.9)	0.100	225	2.8 (2.0)	0.625
PL	31	375.0 (146.0)		31	2.6 (1.5)	
Cakes/Muffins: ready-to-eat			†			†
BL	98	288.9 (138.9)		98	30.5 (12.3)	
PL	15	216.2 (174.1)		15	31.6 (13.9)	
Breakfast cereals: ready-to-eat			0.690			0.764
BL	280	165.9 (155.1)		275	17.2 (8.1)	
PL	41	176.5 (180.3)		41	17.6 (8.3)	
Cereal bars						
BL	156	152.0 (106.0)	0.212	156	25.4 (9.7)	0.357
PL	34	127.3 (95.5)		34	27.0 (4.9)	
Ice-cream						
BL	–	–	–	340	22.0 (6.1)	0.337
PL	–	–		25	23.1 (3.3)	
Canned fish						
BL	149	423.5 (153.8)	<0.001	–	–	–
PL	41	331.3 (96.0)		–	–	
Fruit—canned in syrup/juice			–			0.027
BL	–	–		71	13.4 (5.5)	
PL	–	–		67	11.7 (3.7)	
Nuts—salted						
BL	73	459.6 (324.2)	0.183	–	–	–
PL	31	375.6 (195.7)		–	–	
Vegetables—canned						
BL	181	221.9 (161.2)	<0.001	–	–	–
PL	60	135.3 (114.0)		–	–	

Table 1. *Cont.*

Food Categories	Sodium (mg/100 g)		*p* Value *	Sugar (g/100 g)		*p* Value *
	N	Mean (SD)		n	Mean (SD)	
Vegetables—pickled						
BL	177	1057.9 (1023.2)	0.946	–	–	–
PL	29	1044.9 (484.8)		–	–	
Salamis, hams, bacon						
BL	263	1204.3 (438.8)	0.794	–	–	–
PL	44	1185.9 (392.1)		–	–	
Sausages, hotdogs						
BL	102	764.8 (187.0)	†	–	–	–
PL	20	616.0 (93.0)		–	–	
Raw or frozen meats with flavour/coated						
BL	153	495.3 (204.3)	0.232	–	–	–
PL	38	451.1 (200.4)		–	–	
Mayonnaise and salad dressing						
BL	156	686.9 (452.4)	0.963	154	11.8(12.9)	0.441
PL	34	690.6 (282.5)		34	10.0 (6.4)	
Pasta sauces						
BL	156	407.9 (242.7)	0.388	151	4.4 (2.1)	0.850
PL	39	371.9 (184.1)		39	4.9 (2.1)	
Savoury spreads and dips						
BL	305	492.7 (344.8)	0.526	302	12.3 (14.2)	0.196
PL	32	453.6 (157.6)		32	9.0 (9.0)	
Peanut butter and other nut-based spreads						
BL	80	163.8 (151.9)	†	80	5.9 (3.1)	0.662
PL	25	208.4 (160.7)		25	5.6 (3.1)	
Crisps and salty snacks						
BL	216	625.1 (365.4)	0.952	–	–	–
PL	35	621.2 (306.5)		–	–	
Overall (all products)						
BL	3524	506.3 (540.5)	0.001	2608	15.5 (13.7)	0.404
PL	681	443.3 (362.1)		475	14.9 (12.7)	

SD: standard deviation; BL: branded label; PL: private label. * *p*-values of Student *t*-tests for comparison of means of two independent samples. †: Comparisons between means were not performed when PL had n < 30 products with information on sodium or sugar contents. –Nutrient content not assessed as food category does not represent relevant source of the nutrient.

Overall, there was a statistically significantly higher proportion of PL products with an estimated HSR ≥ 3.5 (48.9%) compared to BL products (38.5%; Table 2, 2019 column). However, there were statistically significant differences in the proportion of products with an estimated HSR ≥ 3.5 between PL and BL for just three of the 21 food categories assessed, i.e., PL canned fruit, and savoury spread and dips had a higher proportion of products with HSR ≥ 3.5. Cereal bars had a lower proportion of products with HSR ≥ 3.5 (Table 2). In 2019, the proportion of PL products with an estimated HSR ≥ 3.5 ranged from 0% (for sweet biscuits, cakes/muffins, ice-cream and mayonnaise/salad dressings) to 100% (for everyday sliced breads, canned fish, canned fruit and peanut butter and other nut-based spreads). Within BL products, the proportion of products with estimated HSR ≥ 3.5 ranged from 0% (for cakes/muffins; mayonnaise/salad dressings) to 98.9% (for canned vegetables) (Table 2).

Table 2. Number and proportion of branded and private label products with an estimated HSR of > 3.5 by year (2015–2019), food category and overall.

Food Categories	Estimated HSR ≥ 3.5										p for Trend † Changes in Proportions In the Period 2015–2019
	2015		2016		2017		2018		2019		
	n	%	n	%	N	%	n	%	n	%	
Savoury biscuits											
BL	50	19.7	43	17.4	47	18.7	57	20.3	57	19.3	0.776
PL	5	10.6	11	20.8	10	16.4	6	9.5	7	14.0	0.676
Sweet biscuits											
BL	2	0.7	2	0.6	2	0.7	3	0.8	4	1.1	0.471
PL	0	0	0	0	0	0	0	0	0	0	‡
Everyday sliced breads											
BL	100	89.3	98	95.1	103	96.3	104	96.3	100	95.2	0.055
PL	26	81.3	24	80.0	15	100	16	100	9	100	0.011
Other breads											
BL	119	61.3	124	60.8	123	62.4	141	65.9	147	66.2	0.156
PLPL	28	45.2	32	42.1	28	59.6	22	59.5	23	74.2	0.002
Cakes/Muffins: ready-to-eat											
BL	1	1.4	0	0	0	0	0	0	0	0	0.130
PL	0	0	0	0	0	0	0	0	0	0	‡
Breakfast cereals: ready-to-eat											
BL	140	64.5	168	64.6	163	66.0	166	64.3	171	62.4	0.613
PL	23	52.3	30	56.6	27	58.7	23	65.7	30	73.2	0.033
Cereal bars											
BL	11	7.4	24	13.1	32	17.7	38	21.1	38	24.4 *	<0.001
PL	3	8.6	3	6.7	4	10.0	3	8.6	3	8.8 *	0.847
Ice-cream											
BL	3	1.2	5	1.8	6	2.2	7	2.2	17	5.0	0.004
PL	0	0	0	0	0	0	0	0	0	0	‡
Canned fish											
BL	157	96.3	127	93.4	137	92.6	130	90.3	141	94.6	0.285
PL	72	97.3	79	98.8	76	98.7	50	100	41	100	0.147
Fruit—canned in syrup/juice											
BL	85	93.4	65	86.7	64	86.5	57	85.1	62	87.3 **	0.197
PL	66	94.3	65	89.0	69	90.8	62	92.5	67	100 **	0.123
Nuts—salted											
BL	33	73.3	59	80.8	52	81.3	57	83.8	56	76.7	0.731
PL	22	78.6	24	77.4	24	82.8	30	90.9	27	90.0	0.083
Vegetables—canned											
BL	173	96.1	188	96.9	193	97.0	182	94.8	176	98.9	0.173
PL	68	98.6	72	98.6	75	98.7	52	100	58	96.7	0.606
Vegetables—pickled											
BL	53	40.8	48	33.8	79	45.1	77	42.1	63	37.5	0.913
PL	6	33.3	7	33.3	3	18.8	6	22.2	7	24.1	0.344
Salamis, hams, bacon											
BL	20	7.0	16	6.1	23	8.2	25	8.4	28	10.7	0.066
PL	1	2.6	1	2.5	1	2.5	2	4.9	1	2.3	0.858
Sausages, hotdogs											
BL	12	10.3	11	8.3	11	8.0	3	2.8	6	5.9	0.063
PL	0	0	3	6.5	0	0	0	0	1	5.0	0.982
Raw or frozen meats with flavour/coated											
BL	64	49.2	96	57.1	85	53.1	101	58.7	84	54.9	0.358
PL	30	75.0	27	69.2	14	58.3	20	58.8	23	62.2	0.138

Table 2. *Cont.*

Food Categories	Estimated HSR ≥ 3.5										*p* for Trend † Changes in Proportions In the Period 2015–2019
	2015		2016		2017		2018		2019		
	n	%	n	%	N	%	n	%	n	%	
Mayonnaise and salad dressings											
BL	1	0.7	2	1.4	0	0	0	0	0	0	0.086
PL	0	0	0	0	0	0	0	0	0	0	‡
Pasta sauces											
BL	95	63.3	96	64.0	106	63.1	91	65.0	99	65.6	0.661
PL	13	61.9	19	63.3	15	55.6	16	51.6	19	48.7	0.177
Savoury spreads and dips											
BL	47	20.2	60	22.2	86	30.6	96	30.1	95	31.6 **	<0.001
PL	20	52.6	19	47.5	15	51.7	16	51.6	18	56.3 **	0.679
Peanut butter and other nut-based spreads											
BL	43	91.5	57	89.1	58	93.5	68	89.5	74	92.5	0.792
PL	9	56.3	20	83.3	20	90.9	20	90.9	25	100	<0.001
Crisps and salty snacks											
BL	1	0.5	7	3.5	6	2.9	12	5.5	15	7.0	0.001
PL	0	0	2	4.3	2	4.7	2	5.7	3	8.6	0.087
Overall (all products)											
BL	1283	33.6	1377	33.7	1462	35.2	1501	34.5	1522	38.5 ***	0.082
PL	404	43.2	448	43.2	419	46.5	362	43.5	380	48.9 ***	0.342

HSR: Health Star Rating: BL: branded label; PL: private label. † *p*-values of chi-square tests for linear trend (linear-by-linear associations using Mantel–Haenzel tests). Comparisons of changes in proportions within private and branded labels in the period 2015–2019. ‡: Zero products with estimated HSR ≥ 3.5. * Pearson chi-square tests: $p < 0.05$; ** $p < 0.005$; *** $p < 0.001$. Comparisons of proportions between private and branded labels in 2019. Missing for estimated HSR-2015–2019 (n): savoury biscuits (19); sweet biscuits (101); everyday sliced breads (20); other breads (38); cakes/muffins: ready-to-eat (28); breakfast cereals: ready-to-eat (19); cereal bars (13); cheese: everyday cheeses (21); ice-cream (17); canned fish (38); fruit—canned in syrup/juice (13); nuts—salted (13); vegetables—canned (37); vegetables—pickled (67); processed meats-I (26); processed meats-II (18); processed meats-III (15); mayonnaise and salad dressings (56); pasta sauces (19); spreads I—savoury (27); spreads II—peanut butter and other nut-based spreads (4); crisps and snacks (18).

3.1.2. Display of HSR on the Pack

The proportions of PL and BL products (overall and for each food category) displaying HSR on the pack in 2019 are shown in Table 3. Overall, PL products had a substantially higher prevalence of HSR label display than BL products (92.4% vs. 17.2%). Within food categories, more PL products displayed HSR on the pack compared to BL counterparts (with the only exception being cereal bars, where there was no difference in HSR uptake between PL and BL products in 2019). The proportion of PL products with HSR displayed on the pack ranged from 41.2% (cereal bars) to 100% (everyday sliced breads and peanut butter and other nut-based spreads). Within BL food categories the proportions ranged from 1.0% (cakes/muffins) to 55.7% (breakfast cereals; Table 3).

Table 3. Number and proportion of branded and private label products displaying HSR on the pack by year (2015–2019), food category and overall.

Food Categories	HSR Displayed on Pack										*p* for Trend † Changes in Proportions In the Period 2015–2019
	2015		2016		2017		2018		2019		
	n	%	n	%	n	%	n	%	n	%	
Savoury biscuits											
BL	0	0.0	7	2.8	28	11.0	32	11.4	49	16.6 ***	<0.001
PL	0	0.0	26	49.1	34	55.7	45	71.4	40	80.0 ***	<0.001

Table 3. *Cont.*

Food Categories	HSR Displayed on Pack										*p* for Trend † Changes in Proportions In the Period 2015–2019
	2015		2016		2017		2018		2019		
	n	%	n	%	n	%	n	%	n	%	
Sweet biscuits											
BL	0	0.0	1	0.3	12	3.8	23	6.1	34	9.6 ***	<0.001
PL	0	0.0	18	20.2	33	50.8	56	90.3	66	90.4 ***	<0.001
Everyday sliced breads											
BL	0	0.0	0	0.0	0	0.0	2	1.8	16	15.2 ***	<0.001
PL	0	0.0	6	16.7	6	40.0	8	50.0	9	100.0 ***	<0.001
Other breads											
BL	0	0.0	2	0.9	8	4.0	11	5.0	22	9.7 ***	<0.001
PL	0	0.0	4	5.3	5	10.8	27	73.0	30	96.8 ***	<0.001
Cakes/Muffins: ready-to-eat											
BL	0	0.0	0	0.0	1	1.2	0	0.0	1	1.0 ***	0.360
PL	0	0.0	0	0.0	2	20.0	6	42.9	13	86.7 ***	<0.001
Breakfast cereals: ready-to-eat											
BL	5	2.3	106	39.8	119	48.2	136	52.7	156	55.7 ***	<0.001
PL	0	0.0	15	27.8	21	45.7	25	71.4	38	92.7 ***	<0.001
Cereal bars											
BL	0	0.0	14	7.6	34	18.8	39	21.7	55	35.3	<0.001
PL	1	2.8	13	28.9	14	35.0	13	37.1	14	41.2	<0.001
Ice-cream											
BL	0	0.0	1	0.3	4	1.4	5	1.6	11	3.2 ***	<0.001
PL	0	0.0	9	30.0	22	64.7	19	70.4	24	96.0 ***	<0.001
Canned fish											
BL	0	0.0	0	0.0	7	4.7	17	11.5	32	21.5 ***	<0.001
PL	0	0.0	20	25.0	34	44.2	39	78.0	40	95.2 ***	<0.001
Fruit—canned in syrup/juice											
BL	0	0.0	0	0.0	2	2.7	6	9.0	5	6.9 ***	<0.001
PL	0	0.0	14	19.2	25	32.9	42	62.7	62	89.9 ***	<0.001
Nuts—salted											
BL	0	0.0	7	9.1	26	40.0	26	38.2	24	32.0 ***	<0.001
PL	0	0.0	5	16.1	20	69.0	28	84.8	26	83.9 ***	<0.001
Vegetables—canned											
BL	3	1.5	6	4.5	24	11.9	37	19.3	44	24.2 ***	<0.001
PL	0	0.0	13	17.3	23	30.3	34	65.4	54	90.0 ***	<0.001
Vegetables—pickled											
BL	0	0.0	0	0.0	13	7.1	20	10.4	26	14.7 ***	<0.001
PL	0	0.0	1	4.5	2	11.8	14	50.0	18	60.0 ***	<0.001
Processed meats I: salamis, hams, bacon											
BL	0	0.0	0	0.0	17	6.0	18	6.1	17	6.5 ***	<0.001
PL	1	2.6	10	24.4	21	52.5	29	70.7	39	88.6 ***	<0.001
Processed meats II: sausages, hotdogs											
BL	0	0.0	0	0.0	3	2.2	11	10.2	6	5.8 ***	<0.001
PL	0	0.0	1	2.2	7	29.2	9	33.3	13	65.0 ***	<0.001
Processed meats III: raw or frozen meats with flavour/coated											
BL	0	0.0	1	0.6	46	28.4	51	29.7	46	30.1 ***	<0.001
PL	0	0.0	2	5.0	17	70.8	25	73.5	27	71.1 ***	<0.001

Table 3. *Cont.*

Food Categories	HSR Displayed on Pack										*p* for Trend † Changes in Proportions In the Period 2015–2019
	2015		2016		2017		2018		2019		
	n	%	n	%	n	%	n	%	n	%	
Mayonnaise and salad dressings											
BL	0	0.0	0	0.0	1	0.5	2	1.1	2	1.3 ***	0.054
PL	0	0.0	5	21.7	6	28.6	22	84.6	33	97.1 ***	<0.001
Pasta sauces											
BL	0	0.0	13	8.6	29	17.3	32	22.5	56	35.9 ***	<0.001
PL	0	0.0	9	30.0	15	55.6	21	67.7	36	92.3 ***	<0.001
Spreads I: savoury spreads and dips											
BL	0	0.0	12	4.4	33	11.7	35	11.0	39	12.7 ***	<0.001
PL	0	0.0	5	12.5	12	41.4	16	51.6	32	100.0 ***	<0.001
Spreads II: peanut butter and other nut-based spreads											
BL	0	0.0	14	21.9	19	30.2	22	28.6	21	26.3 ***	0.002
PL	0	0.0	12	50.0	19	86.4	19	86.4	25	100.0 ***	<0.001
Crisps and salty snacks											
BL	0	0.0	1	0.5	5	2.4	10	4.6	14	6.5 ***	<0.001
PL	0	0.0	2	4.3	27	62.8	30	85.7	34	97.1 ***	<0.001
Overall (all products)											
BL	08	0.2	188	4.4	431	10.2	540	12.2	681	17.2 ***	<0.001
PL	02	0.2	195	18.5	398	44.1	569	68.2	718	92.4 ***	<0.001

HSR: Health Star Rating: BL: branded label; PL: private label. † *p*-values of chi-square tests for linear trend (linear-by-linear associations using Mantel–Haenzel tests). Comparisons of changes in proportions within private and branded labels in the period 2015–2019. *** $p < 0.001$. Comparisons of proportions between private and branded labels in 2019.

3.1.3. Price

In 2019, overall, the mean price of PL products was statistically significantly lower than the mean price of BL products. There were also statistically significant differences in mean price between PL and BL products for 11 of the 16 food categories where it was possible to compare price. Mean prices of savoury biscuits, sweet biscuits, other breads, breakfast cereals, cereal bars, canned fish, canned fruit, canned vegetables, savoury spreads and dips, peanut butter and other nut-based spreads and crisps and salty snacks were significantly lower for PL options than BL options. There were no significant differences in mean price between PL and BL products for the other food categories assessed (Table 4).

Table 4. Mean (SD) price (in New Zealand dollars) between branded label and private label products by food category and overall, 2019.

Food Categories	Mean Price (NZ$/100 g)		*p* †
	n	Mean (SD)	
Savoury biscuits			
BL	274	2.33 (1.24)	<0.001
PL	50	1.51 (1.37)	
Sweet biscuits			
BL	306	2.07 (1.76)	<0.001
PL	72	1.06 (0.84)	
Everyday sliced breads			
BL	92	0.66 (0.42)	‡
PL	9	0.57 (0.52)	

Table 4. *Cont.*

Food Categories	**Mean Price (NZ$/100 g)**		p †
	n	**Mean (SD)**	
Other breads			
BL	210	1.30 (0.61)	<0.001
PL	30	0.82 (0.26)	
Cakes/Muffins: ready-to-eat			
BL	83	1.84 (0.75)	‡
PL	15	1.53 (0.66)	
Breakfast cereals: ready-to-eat			
BL	244	1.59 (0.96)	<0.001
PL	40	0.84 (0.38)	
Cereal bars			
BL	139	2.07 (1.04)	<0.001
PL	33	1.15 (0.09)	
Ice-cream			
BL	299	1.59 (1.33)	‡
PL	25	0.66 (0.42)	
Canned fish			
BL	140	1.94 (0.82)	<0.001
PL	42	1.27 (0.44)	
Fruit—canned in syrup/juice			
BL	68	0.50 (0.14)	0.009
PL	68	0.41 (0.22)	
Nuts—salted			
BL	61	3.57 (3.07)	‡
PL	29	2.11 (0.95)	
Vegetables—canned			
BL	163	0.53 (0.20)	<0.001
PL	59	0.31 (0.14)	
Vegetables—pickled			
BL	142	1.94 (1.38)	0.068
PL	29	1.45 (0.92)	
Processed meats I: salamis, hams, bacon			
BL	236	3.07 (2.36)	0.178
PL	39	2.54 (1.76)	
Processed meats II: sausages, hotdogs			
BL	87	1.74 (0.97)	‡
PL	9	0.86 (0.43)	
Processed meats III: raw or frozen meats with flavour/coated			0.658
BL	141	1.69 (0.71)	
PL	30	1.76 (0.92)	
Mayonnaise and salad dressing			
BL	127	1.66 (0.97)	0.052
PL	32	1.29 (0.84)	
Pasta sauces			
BL	124	1.24 (1.31)	0.756
PL	39	1.31 (1.17)	
Spreads I: savoury spreads and dips			
BL	236	1.91 (1.00)	<0.001
PL	32	1.15 (0.59)	
Spreads II: peanut butter and other nut-based spreads			0.018
BL	65	2.25 (1.59)	
PL	24	1.40 (1.15)	
Crisps and salty snacks			
BL	203	2.10 (1.12)	<0.001
PL	35	1.23 (0.32)	

Table 4. *Cont.*

Food Categories	Mean Price (NZ$/100 g)		p †
	n	Mean (SD)	
Overall (all products)			
BL	3440	1.83 (1.44)	<0.001
PL	741	1.16 (0.98)	

NZ$: New Zealand dollars; SD: standard deviation; BL: branded label; PL: private label. † Student t-tests for comparison of means of two independent samples. ‡: Comparisons between means were not performed when PL had n < 30 products.

3.2. Changes in Healthiness and HSR Display from 2015 to 2019

3.2.1. Healthiness Changes in Mean Sodium Content

Information on the mean sodium content across the five years for all products and for PL and BL products separately, as well as in their minimum and maximum values, is available in Table S3 (Supplementary file). Figure 1 shows the mean change in sodium content from 2015 to 2019.

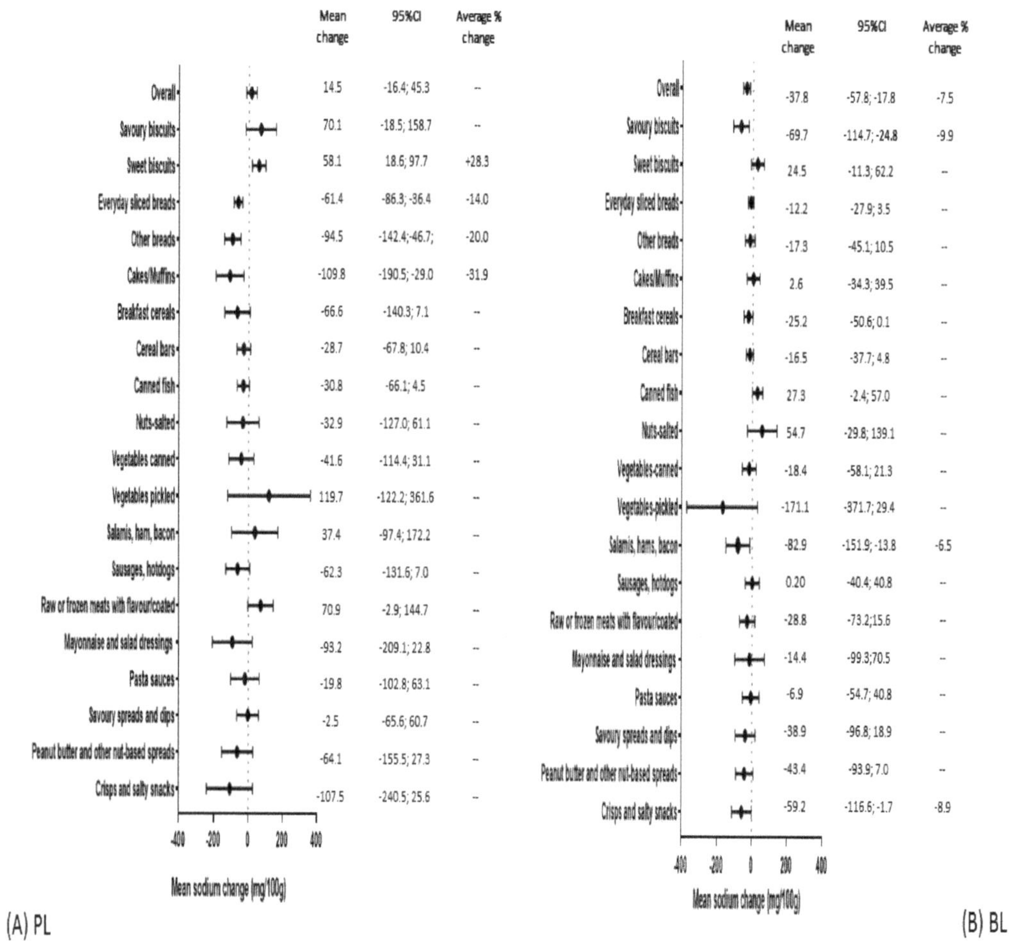

Figure 1. Mean sodium changes in the period 2015–2019 according to brand (overall and by food categories): supermarket private labels (**A**) and branded labels (**B**).

Overall, there were no significant changes in mean sodium content within all PL products and there was a significant mean sodium [mg/100 g (95% CI)] reduction of −37.8 (−57.8; −17.8) within all BL products (average percentage change of −7.5%). At the food category level, three PL food categories significantly reduced mean sodium content over time (everyday sliced breads, other breads, and cakes/muffins), with the mean percentage change >10% and the respective mean [g/100 g (95% CI)] reductions of −61.4 (−86.3; −36.4); −94.5 (−142.4; −46.7) and −109.8 (−190.5; −29.0). The mean sodium content of PL sweet biscuits increased over time by 58.1 mg/100 g (95%CI: 18.6; 97.7), corresponding to a percentage increase of 28.3% (Figure 1A). Similarly, three BL food categories significantly reduced mean sodium content over time (savoury biscuits, salamis, hams and bacon and crisps and salty snacks), all with an average percentage reduction of <10% and with the respective mean (95% CI) reductions of: −69.7 (−114.7; −24.8); −82.9 (−151.9; −13.8) and −59.2 (−116.6; −1.7) mg/100 g (Figure 1B).

3.2.2. Healthiness: Changes in Mean Sugar Content

Information on the mean sugar content across the five years for all products and for PL and BL products separately, as well as in their value ranges, is available in Table S4 (Supplementary file). Figure 2 shows the change in sugar content from 2015 to 2019.

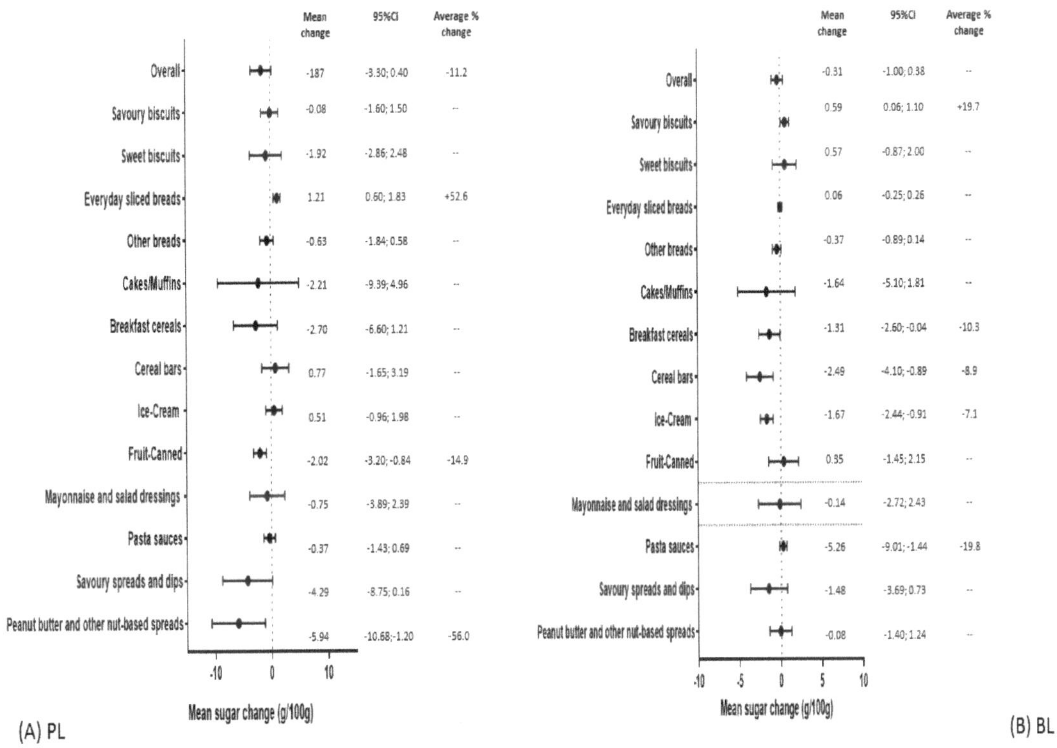

Figure 2. Mean sugar changes in the period 2015–2019 according to brand (overall and by food categories): supermarket private labels (**A**) and branded labels (**B**).

Overall, there were no significant changes in mean sugar content within all BL products and there was a significant mean sugar [g/100 g (95% CI)] reduction of −1.87 (−3.3; −0.04) within all PL products (average percentage change of −11.2%). At the food category level, two PL categories significantly reduced sugar content over time, i.e., canned fruit, and peanut butter and other nut-based spreads reduced mean sugar content by, respectively [g/100 g (95%CI)]: −2.02 (−3.20; −0.84) and −5.94 (−10.68; −1.20), corresponding to an average percentage drop of >10%. The mean sugar content of everyday sliced breads increased over time by 1.21 g/100 g (95%CI: 0.60; 1.83), (Figure 2A). Four BL food categories significantly reduced sugar content over time, i.e., breakfast cereals and pasta sauces reduced mean sugar content by [g/100 g (95%CI)]: −1.31 (−2.60; −0.04) and −5.26 (−9.01; −1.44), respectively, corresponding to an average percentage drop of >10%. Branded label cereal bars and ice-creams showed a mean sugar reduction of [g/100 g (95%CI)]: −2.49 (−4.10; −0.89) and −1.67 (−2.44; −0.91), respectively, corresponding to an average percentage drop of <10%. Across the five years, BL savoury biscuits increased mean sugar content by 0.59 g/100 g (0.06; 1.10) (Figure 2B).

3.2.3. Healthiness: Changes in Proportions of Products with an Estimated HSR ≥ 3.5

Table 2 describes the changes in the proportion of PL and BL products with an HSR ≥ 3.5 in the period 2015–2019 (overall and by food category). Across the five years, overall, there were no significant changes in the proportion of PL or BL products with an HSR ≥ 3.5. However, analyses within food categories indicated statistically significant increases in the proportion of products with estimated HSR ≥ 3.5 over time for four PL food categories (everyday sliced breads, other breads, breakfast cereals and spreads II) and four BL food categories (spreads I, crisps and salty snacks, cereal bars and ice-creams) (Table 2).

3.2.4. Changes in the Proportion of Products Displaying HSR

There was a statistically significant increase in the proportion of products displaying HSR over time within all PL and BL food categories, the only exception being BL cakes/muffins and BL mayonnaise and salad dressings (Table 3).

4. Discussion

4.1. Statement of Principal Findings

In 2019, PL products had, overall, a lower mean sodium content in relation to all BL products. Overall, PL and BL products had similar mean sugar contents, and the mean sodium and sugar content of most food categories was not significantly different between PL and BL products. Overall, a higher proportion of PL products had an estimated HSR ≥ 3.5 (48.9%), compared to BL products (38.5%). Considerably more PL products displayed the HSR on the pack than BL products (92.4% vs. 17.2%), and PL products were overall lower in price than BL options. There were no consistent changes over time (2015–2019) in any of the healthiness outcomes (sodium, sugar, or estimated HSR) of PL and BL products, but an increase in display of HSR on the pack was observed over time for all PL and BL food categories.

4.2. Findings in Relation to Other Studies

4.2.1. Healthiness

Results of our study showing a lower mean sodium content of PL products overall differ to those reported by a previous NZ study that compared the sodium content of PL and BL products in supermarkets between 2003 and 2013 [19]. Note, however, that the previous study compared matched means of PL and BL products available in both years and in only eight categories, rather than comparing means of PL and BL overall [19]. These aspects limit direct comparisons to our findings. Results of our study, however, align with studies conducted in Australia [27–29] and other countries [30–33], which showed that, despite differences in healthiness for a small number of FCs between PL and BL products (in both directions), overall, there were no systematic differences in healthiness between PL and BL products [27–33]. These studies used various methods. Ahuja et al. (2017) [30] undertook chemical analysis of 1,706 samples of PL and national brand products between 2010 and 2014 in the United States (US). In 2010 and 2012, a study in the United Kingdom assessed and compared the nutritional quality of 32 own brands and market brands processed foods most frequently consumed in the country. Products were sourced from supermarkets and their nutritional quality scoring was calculated according to the Food Standards Agency's Traffic Light System [31]. A Swiss study compared the nutritional quality of over 4000 processed foods distributed across 26 food categories. No differences were found between PL and BL products for total energy, protein, fat, and carbohydrates for most food categories. However, PL products had a lower fat, saturated fatty acid, and sodium content [33] in some food categories. In Australia in 2017, a study conducted in four major supermarket chains assessed 6269 products and found no differences in mean HSR in matched comparisons of PL and BL for any of the 10 food categories assessed [27]. Another Australian study also conducted in four major supermarket chains (in Sydney) but assessing a larger number of products (15,680 products, distributed in 15 food categories) found in 2013 that new supermarket PL products were 11% lower in sodium in relation to their BL counterparts [28]. An older study (2006–2008) involving 10 Australian supermarkets and 3204 products from 15 food categories identified that the contents of total and saturated fat were significantly greater for five and seven PL food categories, respectively, in relation to BL options. For sodium content, there were significant differences between PL and BL for seven food categories, but with no consistency in direction [29].

4.2.2. Display of HSR

Concern has been expressed by public health experts that voluntary uptake of the HSR label is slow and therefore it should be made mandatory [33]. Front-of-pack labelling provides visual information on product nutritional contents and studies have shown that it influences consumer's knowledge [34,35] and products reformulation [34]. Recent systematic review and meta-analyses including controlled experimental/intervention and interrupted time series found that findings about influence of front-of-pack labelling on consumers' consumption were limited and inconsistent. However, evidence from experimental and 'real-life' studies shows that front-of-pack labelling encouraged healthier purchasing [35]. An online randomized-controlled study of a large representative British sample found that front-of-pack labelling improved participants' ability to correctly rank products according to their healthiness [36]. A non-experimental prospective study reported that food reformulation occurred after the first phase of the Chilean Food Labelling and Advertising Law, with significant decreases in the amount of sugars and sodium in several groups of packaged foods and beverages between 2015 and 2017 [34].

A previous NZ study describing the state of the packaged food supply in 2018 indicated that products (PL and BL aggregated) displaying the HSR on the package had a higher mean HSR than products not displaying HSR values (mean $\pm$ SD, 3.2 $\pm$ 1.3 versus 2.5 $\pm$ 1.4, $p = 0.000$) [2]. Among the products examined in the period 2015–2019, our study indicated much greater uptake of HSR by PL products (92.4% in 2019) than BL products (17.2% in 2019). An Australian study in 2017 also reported a significantly higher proportion of supermarket PL products displaying HSR (57%) than BL products (28%) [27].

4.2.3. Price

The lower cost of PLs in relation to BLs reported in the current study corroborates with the 2003 NZ study that found lower mean price for 11 of 15 supermarket PL food categories examined (in relation to BL) [18]. A study looking at the cost of healthy and usual diets in NZ in 2015 found considerable savings (5.5%) if households purchased PL versions of brands compared to branded items [37,38]. Findings of our study are also similar to those reported in several other countries internationally, where, overall, supermarket PL products were lower priced in relation to BL options [31–33,39,40].

4.3. Findings in Relation to the Commitments Made by NZ Supermarkets

In our study, we did not evaluate separately how each of the two supermarket retailers met their commitments made in 2016, because there were insufficient PL products for most food categories assessed to provide robust comparisons. Thus, comparisons made include

PL FCs of both NZ supermarket retailers combined. As previously described, both NZ supermarket retailers committed to displaying HSR on almost all PL products by 2018–2020 [13–15]. The findings of our study confirm that this commitment is, overall, on track, as among the food categories examined, the majority of PL products (92.4%) were displaying HSR on the package in 2019. However, further effort to increase HSR uptake in some PL food categories is still required, e.g., cereal bars, sausages and hotdogs and raw or frozen meats with flavour/coating.

Both supermarket retailers committed to improving the nutrition of their PL products Our study found that in 2019, most PL food categories were of a similar nutritional quality to BL categories. In 2019, overall, a higher proportion of PL products had an HSR $\geq$ 3.5 in relation BL products (43.5% vs. 38.5%), and for 10 PL food categories the proportion of products with an HSR $\geq$ 3.5 was >50%. We found from 2015 to 2019 that only three PL food categories and two PL food categories changed, respectively, the mean sodium and sugar contents (with average reduction > 10%). Together, these findings indicate that the commitments of supermarkets retailers have been partially met, but more work is needed to increase the proportion of products with HSR $\geq$ 3.5 across all types of foods.

There are no public commitments made by the NZ supermarket retailers on price of PL products [11,13–15]. We believe that this is probably because PL products are usually considered lower cost options than branded products, and price is generally considered commercially sensitive. In our study, overall, PL products had a lower mean price than BL products, which indicates that, on average, PL products represent better value for money than BL options.

4.4. Strengths and Limitations of This Study

The strengths of our study include the fact that we assessed information on a large number of packaged foods and included data over five years to assess changes in healthiness and display of HSR over time. In total, 1/3 of packaged foods in the Nutritrack database from 2015 to 2019 were included in the analyses, corresponding to 40.7% of the PL products and 29.9% of the BL products. We also compared similar types of products and to improve the robustness of analyses, mean sodium and sugar contents were compared only for food categories that contained at least 30 PL products. Similarly, we only assessed mean changes in sodium and sugar contents over time if there were at least 100 PL products available for that period. Another strength is that, rather than using price information at a single point in time in 2019, we used information on product price over the whole year and calculated the mean price that consumers paid for products within this timeframe.

Findings of this study need to be interpreted taking limitations into account. A relevant limitation is that there were not enough PL products available within the food categories to allow for paired comparisons and to assess reformulation of individual products over time. Finally, the fact that the results of our study were not sales weighted or informed by product sales data represents another limitation. Sales data could provide valuable information on the most commonly purchased products and foods to better assess the public health impact of findings, including by sociodemographic group.

4.5. Implications of the Findings

In summary, PL products in major NZ supermarkets can be a good choice for consumers as they are usually lower in price, nutritionally similar to BL products, and more likely to display a HSR score. Retailers have made progress on their nutritional and labeling commitments regarding PL products. However, further positive movements can be made, including displaying the HSR on all products and establishing a systematic PL reformulation programme operating across all foods, but with an emphasis on categories with a high sales volume. These recommendations are important for public health given that PL products are driving the growth of sales in NZ supermarkets, and most NZ shoppers believe these supermarket own brands are 'just as good or better than' their branded counterparts [41].

To set a level playing field for all companies and retailers, and to help consumers make healthier choices, the government should make display of the HSR mandatory. While this study did not assess reformulation, providing targets for reformulation of common products would provide benchmarks for retailers and the wider food industry.

Supplementary Materials: The following are available online at https://www.mdpi.com/article/10.3390/nu13082731/s1, Table S1. Description and number of products assessed by food category from 2015 to 2019 (overall and by private and branded label). Table S2. Number and proportion of products with missing information for sugar content, sodium content, estimated HSR and HSR displayed on front-of-pack-labelling from 2015 to 2019, by food category (overall). Table S3. Sodium content within the selected food categories: mean (SD) from 2015 to 2019—in total, for branded and private labels. Table S4. Sugar content within the selected food categories: mean (SD) from 2015 to 2019—in total, for branded and private labels.

Author Contributions: Conceptualization, H.E., T.C., S.M., L.Y. and C.N.M.; methodology, H.E., T.C., S.M., L.Y. and C.N.M.; data preparation, H.E. and T.C.; formal analysis, T.C.; investigation, H.E., T.C., S.M., L.Y., C.N.M. and E.T.; writing—original draft preparation, T.C., S.M., H.E. and G.S.; writing—review and editing, L.Y., C.N.M., G.S. and E.T.; funding acquisition, C.N.M. All authors have read and agreed to the published version of the manuscript.

Funding: This research was funded by the National Health and Medical Research Council (NHMRC) funded Centre of Research Excellence in Reducing Salt Intake using Food Policy Interventions, grant number APP1117300.

Institutional Review Board Statement: Not applicable.

Informed Consent Statement: Not applicable.

Data Availability Statement: Because of the commercial and legal restrictions to the use of copyrighted material, it is not possible to share data openly, but unredacted versions of the dataset are available with a licensed agreement that they will be restricted to non-commercial use. For access to Nutritrack, please contact the National Institute for Health Innovation at the University of Auckland at enquiries@nihi.auckland.ac.nz.

Acknowledgments: We would like to thank John Faatui for data management and the fieldworkers for 2019 Nutritrack data.

Conflicts of Interest: The authors declare no conflict of interest. The funder had no role in the design of the study; in the collection, analyses, or interpretation of data; in the writing of the manuscript, or in the decision to publish the results.

References

1. Afshin, A.; Sur, P.J.; Fay, K.A.; Cornaby, L.; Ferrara, G.; Salama, J.S.; Mullany, E.C.; Abate, K.H.; Abbafati, C.; Abebe, Z.; et al. Health effects of dietary risks in 195 countries, 1990–2017: A systematic analysis for the Global Burden of Disease Study 2017. *Lancet* **2019**, *393*, 1958–1972. [CrossRef]
2. Mackay, S.; Eyles, H.; Gontijo de Castro, T.; Young, L.; Ni Mhurchu, C.; Swinburn, B. Which companies dominate the packaged food supply of New Zealand and how healthy are their products? *PLoS ONE* **2021**, *16*, e0245225. [CrossRef]
3. Gontijo de Castro, T.; Eyles, H.; Ni Mhurchu, C.; Young, L.; Mackay, S. Seven-year trends in the availability, sugar content and serve size of single-serve non-alcoholic beverages in New Zealand: 2013–2019. *Public Health Nutr.* **2021**, *24*, 1595–1607. [CrossRef] [PubMed]
4. Zorbas, C.; Eyles, H.; Orellana, L.; Peeters, A.; Mhurchu, C.N.; Riesenberg, D.; Backholer, K. Do purchases of price promoted and generic branded foods and beverages vary according to food category and income level? Evidence from a consumer research panel. *Appetite* **2020**, *144*, 104481. [CrossRef] [PubMed]
5. Tawfiq, E.; Bradbury, K.E.; Ni Mhurchu, C. Healthiness of foods and non-alcoholic beverages according to store type: A population-based study of household food and drink purchases in New Zealand. *SSM Popul. Health* **2021**, *14*, 100784. [CrossRef]
6. Ministry of Health. *Household Food Insecurity among Children in New Zealand*; Ministry of Health: Wellington, New Zealand, 2019. Available online: https://www.health.govt.nz/system/files/documents/publications/household-food-insecurity-among-children-new-zealand-health-survey-jun19.pdf (accessed on 2 March 2021).
7. Pearson, A.L.; Winter, P.R.; McBreen, B.; Stewart, G.; Roets, R.; Nutsford, D.; Bowie, C.; Donnellan, N.; Wilson, N. Obtaining Fruit and Vegetables for the Lowest Prices: Pricing Survey of Different Outlets and Geographical Analysis of Competition Effects. *PLoS ONE* **2014**, *9*, e89775. [CrossRef] [PubMed]

8. Euromonitor. Packaged Food in New Zealand—Country Report. 2020. Euromonitor Passport. Available online: http://www.portal.euromonitor.com.ezproxy.auckland.ac.nz/poral/analysis/tab (accessed on 23 June 2021).
9. Commerce Commission. *Market Study into the Retail Grocery Sector—Preliminary Issues Paper*; Commerce Commission: Wellington, New Zealand, 2020. Available online: https://comcom.govt.nz/__data/assets/pdf_file/0033/229857/Market-study-into-the-retail-grocery-sector-Preliminary-issues-paper-10-December-2020.pdf (accessed on 8 May 2021).
10. Volpe, R. *The Relationship between National Brand and Private Label Food Products: Prices, Promotions, Recessions, and Recoveries*; EER-129; U.S. Department of Agriculture, Economic Research Service: Washington, DC, USA. Available online: https://www.ers.usda.gov/webdocs/publications/44930/11070_err129_1_.pdf?v=41056 (accessed on 2 March 2021).
11. Foodstuffs South Island. Annual Report 2020. 2020. Available online: https://www.foodstuffs-si.co.nz/media/114485/food_j004067_annual_report_digital.pdf (accessed on 7 June 2021).
12. Wilson, J. Supermarket Power: Are Stores Playing Fair on Price? *Consumer*. 4 August 2020. Available online: https://www.consumer.org.nz/articles/supermarket-concentration (accessed on 21 February 2021).
13. Countdown. Corporate Responsibility Report 2016. 2016. Available online: https://s3-ap-southeast-2.amazonaws.com/countdownnz/Countdown+CR+Report_2016.pdf (accessed on 20 April 2021).
14. Foodstuffs South Island. Foodstuffs Announces Pledge to Help Reduce Childhood Obesity. 2016. Available online: https://www.foodstuffs-si.co.nz/about-us/news-announcements/2016/10/foodstuffs-announces-pledge-to-help-reduce-childhood-obesity (accessed on 10 May 2021).
15. Foodstuffs North Island. Looking after Our Patch: People, Community and Planet. 2018. Available online: https://www.foodstuffs.co.nz/news-room/looking-after-our-patch-people-community-and-planet (accessed on 10 May 2021).
16. Ministry for Primary Industries. *Health Star Rating: Monitoring Implementation for the Five Year Review. New Zealand Food Safety Technical Report No: 2018/08*; Ministry for Primary Industries: Wellington, New Zealand, 2018. Available online: https://www.mpi.govt.nz/dmsdocument/42478/direct (accessed on 25 March 2021).
17. Jones, A.; Rådholm, K.; Neal, B. Defining 'Unhealthy': A Systematic Analysis of Alignment between the Australian Dietary Guidelines and the Health Star Rating System. *Nutrients* **2018**, *10*, 501. [CrossRef]
18. Monro, D.; Young, L.; Wilson, J.; Chisholm, A. The sodium content of low cost and private label foods; implications for public health. *JNZ Diet Assoc.* **2004**, *58*, 4–10.
19. Monro, D.; Mhurchu, C.; Jiang, Y.; Gorton, D.; Eyles, H. Changes in the Sodium Content of New Zealand Processed Foods: 2003–2013. *Nutrients* **2015**, *7*, 4054–4067. [CrossRef]
20. Nutritrack Database. 2020. Nutritrack. Available online: https://diet.auckland.ac.nz/sites/default/files/201908/The_Nutritrack_Database.pdf (accessed on 15 March 2020).
21. Bandy, L.; Adhikari, V.; Jebb, S.; Rayner, M. The use of commercial food purchase data for public health nutrition research: A systematic review. *PLoS ONE* **2019**, *14*, e0210192. [CrossRef]
22. Stats NZ. Urban Accessibility-Methodology and Classification. 2020. Available online: https://www.stats.govt.nz/assets/Uploads/Methods/Urban-accessibility-methodology-and-classification/Download-document/Urban-accessibility-methodology-and-classification.pdf (accessed on 19 January 2021).
23. The Food Monitoring Group. Progress with a global branded food composition database. *Food Chem.* **2013**, *140*, 451–457. [CrossRef]
24. Dunford, E.; Webster, J.; Metzler, A.B.; Czernichow, S.; Mhurchu, C.N.; Wolmarans, P.; Snowdon, W.; L'Abbe, M.; Li, N.; Maulik, P.K.; et al. International collaborative project to compare and monitor the nutritional composition of processed foods. *Eur. J. Prev. Cardiol.* **2012**, *19*, 1326–1332. [CrossRef]
25. Boston University. Central Limit Theorem. 2016. Available online: http://sphweb.bumc.bu.edu/otlt/MPHModules/BS/BS704_Probability/BS704_Probability12.html (accessed on 9 February 2021).
26. Eyles, H.; Jiang, Y.; Blakely, T.; Neal, B.; Crowley, J.; Cleghorn, C.; Mhurchu, C.N. Five-year trends in the serve size, energy, and sodium contents of New Zealand fast foods: 2012 to 2016. *Nutr. J.* **2018**, *17*, 65. [CrossRef]
27. Kim, D.H.; Liu, W.G.A.; Rangan, A.; Gemming, L. A comparison of the Health Star Rating and nutrient profiles of branded and generic food products in Sydney supermarkets, Australia. *Public Health Nutr.* **2019**, *22*, 2132–2139. [CrossRef]
28. Trevena, H.; Neal, B.; Dunford, E.; Haskelberg, H.; Wu, J.H.Y. A Comparison of the Sodium Content of Supermarket Private-Label and Branded Foods in Australia. *Nutrients* **2015**, *7*, 7027–7041. [CrossRef]
29. Cleanthous, X.; Mackintosh, A.-M.; Anderson, S. Comparison of reported nutrients and serve size between private label products and branded products in Australian supermarkets. *Nutr. Diet.* **2011**, *68*, 120–126. [CrossRef]
30. Ahuja, J.K.C.; Pehrsson, P.R.; Cogswell, M. A Comparison of Concentrations of Sodium and Related Nutrients (Potassium, Total Dietary Fiber, Total and Saturated Fat, and Total Sugar) in Private-Label and National Brands of Popular, Sodium-Contributing, Commercially Packaged Foods in the United States. *J. Acad. Nutr. Diet.* **2017**, *117*, 770–777. [CrossRef]
31. Faulkner, G.P.; Livingstone, M.B.E.; McCaffrey, T.A.; Kerr, M.A. Supermarket own brand foods: Lower in energy cost but similar in nutritional quality to their market brand alternatives. *J. Hum. Nutr. Diet.* **2014**, *27*, 617–625. [CrossRef] [PubMed]
32. Darmon, N.; Caillavet, F.; Joly, C.; Maillot, M.; Drewnowski, A. Low-cost foods: How do they compare with their brand name equivalents? A French study. *Public Health Nutr.* **2009**, *12*, 808–815. [CrossRef]
33. Khalatbari-Soltani, S.; Marques-Vidal, P. Not as bad as you think: A comparison of the nutrient content of best price and brand name food products in Switzerland. *Prev. Med. Rep.* **2016**, *3*, 222–228. [CrossRef] [PubMed]

34. Reyes, M.; Tillie, L.S.; Popkin, B.; Kanter, R.; Vandevijvere, S.; Corval'an, C. Changes in the amount of nutrient of packaged foods and beverages after the initial implementationof the Chilean Law of Food Labelling and Advertising: A nonexperimental prospective study. *PLoS Med.* **2020**, *17*, e1003220. [CrossRef] [PubMed]
35. Croker, H.; Packer, J.; Russel, S.J.; Stansfield, C.; Viner, R.M. Front of pack nutritional labelling schemes: A systematic review and meta-analysis of recent evidence relating to objectively measured consumption and purchasing. *J. Hum. Nutr. Diet.* **2020**, *33*, 518–537. [CrossRef] [PubMed]
36. Packer, J.; Russel, S.J.; Rodout, D.; Hope, S.; Conolly, A.; Jessop, C.; Robinson, O.J.; Stoffel, S.T.; Viner, R.M.; Croker, H. Assessing the effectiveness of front of pack labels: Findings from an online randomized-controlled experiment in a representative British sample. *Nutrients* **2021**, *13*, 900. [CrossRef] [PubMed]
37. Mackay, S.; Sing, F.; Gerritsen, S.; Swinburn, B. *Food-EPI 2020 Benchmarking NZ Food Environments Full Report*; The University of Auckland: Auckland, New Zealand, 2020. [CrossRef]
38. Mackay, S.; Buch, T.; Vandevijvere, S.; Goodwin, R.; Korohina, E.; Funaki-Tahifote, M.; Lee, A.; Swinburn, B. Cost and Affordability of Diets Modelled on Current Eating Patterns and on Dietary Guidelines, for New Zealand Total Population, Māori and Pacific Households. *Int. J. Environ. Res. Public Health* **2018**, *15*, 1255. [CrossRef] [PubMed]
39. Chapman, K.; Innes-Hughes, C.; Goldsbury, D.; Kelly, B.; Bauman, A.; Allman-Farinelli, M. A comparison of the cost of generic and branded food products in Australian supermarkets. *Public Health Nutr.* **2012**, *16*, 1–7. [CrossRef]
40. Cooper, S.; Nelson, M. 'Economy' line foods from four supermarkets and brand name equivalents: A comparison of their nutrient contents and costs. *J. Hum. Nutr. Diet.* **2003**, *16*, 339–347. [CrossRef] [PubMed]
41. *Nielsen IQ Homescan Survey (March 2021)*; The State of the Nation Report; Total New Zealand: Auckland, New Zealand, 2021.

nutrients

MDPI

Review

Food Composition Databases: Does It Matter to Human Health?

Amélia Delgado [1,*], Manel Issaoui [2], Margarida C. Vieira [1,3], Isabel Saraiva de Carvalho [1,4] and Anthony Fardet [5]

1 Mediterranean Institute for Agriculture, Environment and Development, University of Algarve, 8005-139 Faro, Portugal; mvieira@ualg.pt (M.C.V.); icarva@ualg.pt (I.S.d.C.)
2 Department of Biotechnology, Faculty of Science and Technology of Sidi Bouzid, University of Kairouan, Sidi Bouzid 9100, Tunisia; manelissaoui80@gmail.com
3 Department of Food Engineering, Superior Institute of Engineering, University of Algarve, 8005-139 Faro, Portugal
4 FSLab—Food Science Laboratory, Faculty of Sciences and Technology, University of Algarve, 8005-139 Faro, Portugal
5 Unité de Nutrition Humaine, INRAE, Route de Theix, 63122 Saint-Genès-Champanelle, France; anthony.fardet@inrae.fr
* Correspondence: amdelgado@ualg.pt

Abstract: Food provides humans with more than just energy and nutrients, addressing both vital needs and pleasure. Food habits are determined by a wide range of factors, from sensorial stimuli to beliefs and, once commanded by local and seasonal availability, are nowadays driven by marketing campaigns promoting unhealthy and non-sustainable foodstuffs. Top-down and bottom-up changes are transforming food systems, driven by policies on SDGs and by consumer's concerns about environmental and health impacts. Food quality, in terms of taste, safety, and nutritional value, is determined by its composition, described in food composition databases (FDBs). FDBs are then useful resources to agronomists, food and mechanical engineers, nutritionists, marketers, and others in their efforts to address at maximum human nutrient needs. In this work, we analyse some relevant food composition databases (viz., purpose, type of data, ease of access, regularity of updates), inspecting information on the health and environmental nexus, such as food origin, production mode as well as nutritional quality. The usefulness and limitations of food databases are discussed regarding what concerns sustainable diets, the food 'matrix effect', missing compounds, safe processing, and in guiding innovation in foods, as well as in shaping consumers' perceptions and food choices.

Keywords: food data; natural substances; health promotion; sustainable foods; national food composition databases; one health

Citation: Delgado, A.; Issaoui, M.; Vieira, M.C.; Saraiva de Carvalho, I.; Fardet, A. Food Composition Databases: Does It Matter to Human Health?. *Nutrients* **2021**, *13*, 2816. https://doi.org/10.3390/nu13082816

Academic Editor: Joerg Koenigstorfer

Received: 15 July 2021
Accepted: 13 August 2021
Published: 17 August 2021

1. Introduction

Food databases (FDB), or more correctly food composition databases, contain detailed information on the nutritional composition of foods and on other relevant compounds (e.g., polyphenols, phytic acid). Food components primarily determine nutritional features and, in some cases, quality aspects. For example, polyphenols, which are abundant in plants, are often associated to bitter taste and astringency sensation of foods [1], while acting in favour of food safety by inhibiting foodborne pathogens and spoilage microbes. Polyphenols can be intentionally added to foods for their bioactive properties [2–4] or they can be key natural components, as happens in table olive fermentation [5,6]. During the spontaneous fermentation process, olive's polyphenols help to select the suitable microbial populations, resulting in taster and safer foodstuffs.

The applications of FDBs have been greatly evolving and, consequently, the awareness on some of their limitations. Firstly, FDBs consisted of printed tables listing the nutritional composition of selected foods, usually from a certain country and only available to a few specialists. Today, the most popular FDBs are open access online comprehensive datasets and resources, which may provide answers to simple queries or the download of large datasets; for this reason, the main FDBs are compatible among them and with many interface applications. Up to date food composition data are of capital importance for estimations in relation to nutrition and public health, for different purposes and calculations in food science and engineering, in managing agrobiodiversity and plant breeding, as well as in food regulatory aspects.

Today, food system sustainability is questioned to better address the SDGs, as are consumers' dietary shifts driven by environmental concerns [7]. The interconnection between public health and environmental issues is more and more acknowledged and translated into action [8], while FDBs' gaps have been noticed at the level of the environment-public health nexus [9]. Moreover, the strategic trend of using food by-products as ingredients in other foods (secondary raw materials) seems to be insufficiently addressed by existing FDB. The importance of FDBs is such that inaccurate food composition data can result in incorrect policies (regarding nutritional guidelines and the agri-food system), misleading food labelling, incorrect health claims, and inadequate food choices by the consumers, especially concerning industrially processed foods with added salt, fats, and/or sugars. Therefore, the awareness of relevant new trends and the adjustments to address them is as important as the frequency of FDB's data update.

A comprehensive review on the production, management, and use of food composition data was released by the FAO (United Nations Food and Agriculture Organization) in 2003 [10], dedicating one chapter to possible limitations of FDB. However, the nexus between food, health, and the environment was not considered because there was little or no awareness yet about it, since the agreement on 2030 agenda only took place in 2015 [11].

According to the FAO [10,12], the three pillars of FDBs should be: (a) the existence of international standards and guidelines for food composition data; (b) national and/or regional programs supporting the regular update of FDB; (c) professional training in aspects related to food composition. In order to ensure these foundations, InFoods (International Network of Food Data Systems) was established in 1984. This FDB is based in regional nodes, under a global coordination, and acts as a network of experts and as a taskforce to respond to users' needs, database content, organization, and operation, etc. InFoods keep standards in food nomenclature, terminology, and classification systems, in food component identifiers (tag names), in exchange of data between FDB, and in data quality [12]. In addition to its role in setting standards, the FAO/WHO Codex Alimentarius also keeps specific databases, notably on pesticides residues in food and on veterinary drug residues in food [13].

Whilst many countries maintain their own FDB, despite the broad variation of richness and adequacy, the majority of countries keep incomplete, outdated, and/or unreliable food composition datasets or none at all, as further detailed in Section 4.6, dedicated to national FDBs. In such cases, data need to be borrowed from other sources, and the international network of FDBs is, therefore, very important. A list of software tools to assist in nutrient intake estimations and in planning diets is provided in the InFoods webpage, in addition to specific software tools for labelling or for the calculation of food supply/availability [12].

Relevant information on food composition can be retrieved from the FAO [12], EuroFIR [14], USDA [15], and others. It is noteworthy that some national FDBs comply with international standards and are accessible online, in English. That is the case of ANSES-CIQUAL [16] and Frida Food Data [17], whose outstanding dimension, updates, and ease of use turn them into reference databases at the international level. Many other national databases are freely accessible online, in English. Even when their scope is limited, they can be valuable sources of information on specific/ethnical foods, following new trends on diets in compliance with the updated double pyramid model, which relates to the health and environmental impacts of diets [18]. The formats and variability of national FDBs are further discussed below.

The scope of this critical review is to provide new information on the most prominent FDBs freely available online and in English and to discuss their current and future uses, as well as their advantages and limitations in some current applications, e.g., their potential link with human health and their use for preventing chronic diseases.

The current work provides relevant information and links for prominent FDBs and discusses some of their gaps and trends. The need for environmental indicators linked to foods and the coverage of secondary raw materials are argued, and ways on how FDBs can offer better tools for action in the public-health, food, and environment nexus are discussed.

User recommendations and instructions as well as the cybersecurity aspects of FDBs are out of the scope of the current work.

2. Main Features and Historical Background

Originally, FDBs existed only in printed form, with the oldest ones dating back to the early 1800s. According to Church [19], the first food composition table dates from 1818, and it was elaborated in the form of a 'nutrition scale' aiming at managing food supply in prisons. Early in the 20th century, the USA pioneered standards and regulations aiming at controlling fraud and food safety, and as a result, the USDA's FDBs are among the most important and comprehensive in the world [15].

The FAO also established an important milestone in this regard when publishing 'Food Composition Tables for International Use' back in 1949, to assist in the assessment of food availability at the global level, on a per capita basis, a tool that evolved into today's food balance sheets, an interactive online tool compiling data on food availability worldwide [20]. The evolution of standards and definitions always have accompanied the pace of growing information, thus scouting and steering its usefulness, a basilar principle, which is more than valid when dealing with Big Data and machine learning algorithms. FDBs continue evolving, as does the knowledge on the chemical nature of food components and the mechanisms by which they exert influence on health and disease. FDBs remain central in nutritional research and guidance, despite the increasing awareness on the complexity and knowledge gaps of the role of food components and their interactions within food matrix [21], suggesting that a nutrient does not have the same health effects depending on the matrix in which it is embedded [22]. Because of that, FDBs are more and more comprehensive and interlinked, providing information on a growing list of features.

Besides whole food composition databases, some specialised ones, generally concerning one class of compounds, are accessible to researchers and other interested parties. In this scope, two classes of compounds have emerged recently: bioactive molecules (such as polyphenols) and microbial metabolites (e.g., butyric acid, accumulated during food fermentations and found to be beneficial in the gut). We open, herein, a parenthesis to categorize both types of compounds, because they have been increasingly noted in innovative foods that highlight health-related aspects.by. In the words of Biesalski et al. [23], a 'bioactive compound' is a 'compound that occurs in nature, part of the food chain, and that can interact with one or more compounds of the living tissue, by showing an effect on human health'. As a consequence, bioactive compounds in a food are chemically defined molecules with a proven function in the body and encompass vitamins, minerals, polyphenols, and others. Bioactive compounds are sometimes named as 'nutraceuticals', and there is some confusion around these concepts. According to Heinrich [24], the term nutraceutical is often misused as a synonym of 'functional food' and 'dietary supplement'. Still, according to the same author, 'functional foods' are foods that are part of a diet for which scientifically assessed health benefits are acknowledged, sometimes in the form of health claims. That is the case of the so-called 'function claims' in Article 13 of Reg. (EC) 1924/2006 and of 'risk reduction claims' in Article 14 of the same European regulation [25].

The designation 'dietary supplements' corresponds to ingestible preparations (whether synthetic or extracted from natural sources), which are consumed to supplement the diet, with the intention of conveying extra health benefits, or in balancing a (nutritionally poor) diet.

On the other hand, the 'Nutraceuticals' designation refers to substances with biological functions that are derived only from foods. Both dietary supplements and nutraceuticals may, thus, refer to products that are consumed in a form that resembles a medicine, and both are sold over-the-counter (OTC). Distinguishing these concepts can be further complicated by the fact that many substances fall within all three categories (functional food, nutraceutical, and dietary supplement). That is the case of beta-carotene, which occurs naturally in fruits, vegetables, and grains, but it can be also synthesised and, thus, also be sold as a dietary supplement and as a nutraceutical. Hence, the commonly found designation of 'superfoods' addresses such cases, although it is equally confusing and potentially misleading. Superfoods, functional foods, and nutraceuticals are commonly advertised as having remarkable health claims, such as being able to slow the aging process, having anti-tumoral properties, or in tackling obesity. Such claims are often problematic and difficult to substantiate. From a regulatory point of view, and still according to Heinrich [24], since foods themselves are not considered as therapeutic agents, therefore the claim that nutraceuticals or functional foods can treat disease cannot apply to a food substance.

The second food-related trend, the focus on microbial metabolites, is at an earlier research stage, and despite some penetration in the market (e.g., probiotics), the reach of related (mis)information is currently not significant.

3. Current Uses, State-of-the-Art, and Future Challenges of Food Composition Databases

The reference FDBs that were once tables on paper and later on physical digital supports are nowadays easily accessible online, holding and managing large quantities of data and metadata that can be inspected and downloaded. As previous versions, online FDBs mostly detail the composition of fresh produce as well as branded foodstuffs, discriminating energy sources and macronutrients into their components (e.g., amino acids, sugars, starch, fatty acids), as well as minerals (e.g., calcium, iron, sodium) and vitamins. Often, information on other features, as the content of dietary fibre and relevant bioactive constituents (e.g., carotenoids, polyphenols) is also included, and recently, more and more information has been made available, in pace with the development of convenient interfaces to access and use it.

FDBs have been evolving in adapting new ICT tools. A trend in establishing connections between different databases can be observed, thus expanding the available information while allowing the access either by specifically designed algorithms or by individual discrete users making simple searches.

Connections between FDBs complement information about a certain food or about the food sources for a certain compound; for example, bioactive compounds are included in the eBASIS database, in the US isoflavone database and in the French Phenol-Explorer database, all linked to EuroFIR and to FoodData central, as detailed below.

FDBs' interlinkage adheres to agreed international standards and guidelines, which are of the competence of InFOODs, the International Network of Food Data System from the FAO (UN, Food and Agriculture Organization). It acts as a network of regional datacentres with a central coordination, as well as a forum for the international harmonization and support for food composition activities. InFOODs aims at linking agriculture, biodiversity, food systems, health, and nutrition to achieve better nutrition worldwide. The network regularly issues publications on food composition and other food-related aspects, and its webpage provides access to searchable FDBs [12].

The standardization and harmonizing of food composition data from different countries with distinct metadata are essential to ensure efficient data linkage and the retrieval of information. Hence, tools and procedures have been developed aiming to guarantee interoperability between the databases. Langual is such a tool [26]. It is a food description thesaurus that stands for 'langua alimentaria' or 'language of food' and provides a standardised language for describing foods, specifically in classifying food products for information retrieval. Each of their over 40,000 foods is described by the means of numerical attributes on food composition (nutrients and contaminants), food consumption, and legislation. Langual establishes a correspondence between these food attributes (descriptors) and common language terms in different natural languages [26]. This important tool facilitates the linkage to many different food data banks from different countries, interpreting distinct designations and resolving ambiguities to ensure the correspondence between food and their attributes, thus contributing to coherent data exchange [27]. The food indexing system of Langual already considers food source (e.g., animal or plant species), food preservation (e.g., fresh, frozen), cooking, packaging, etc. However, the next generation of this European FDB thesaurus is even more complex and comprehensive. This global initiative under development—FoodOn—deals with a very comprehensive semantics encompassing descriptors for food safety, food security, agricultural practices, culinary, nutritional and chemical ingredients, and processes [26,27], as can be overviewed in Figure 1.

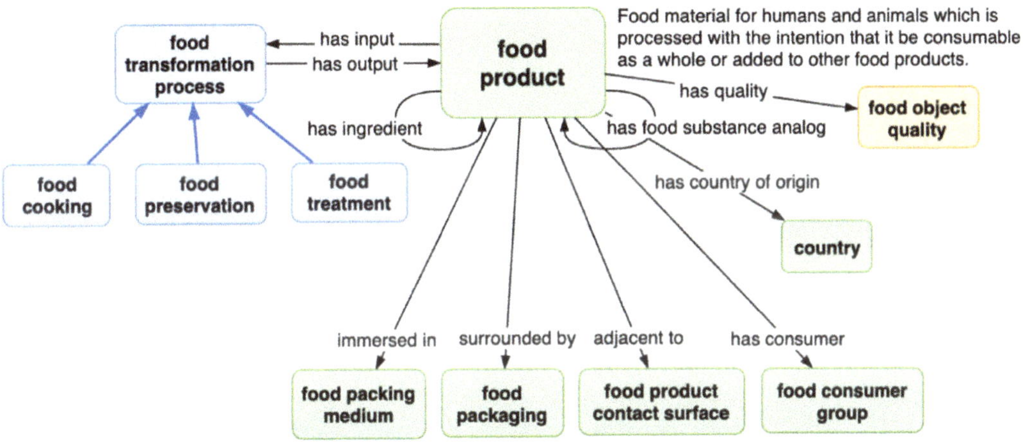

Figure 1. Some facets provided in FoodOn and their relations to a certain food product, of which primary objective is to provide the vocabulary to describe a given food. Reprinted with permission from ref. [26,27]. 2017. Roger A Smith (cc-by-sa/2.0).

The detail of such descriptions and relationships can be better understood by observing Figure 2, which refers to an apple. The degree of detail may increase, for example by adding information about ripeness at harvesting. Finally, it is worth mentioning that the Joint Food Ontology Workgroup GitHub (of FoodOn) is working to provide vocabulary for nutritional analysis, such as chemical food components relevant to the diet, as well as many aspects important to research. FoodOn relies on academic curators and some funding agencies' grants, mostly from Canada [28].

Since 2011, EFSA also maintains FoodEx2, a food classification and description system covering different food safety domains, notably including a description system for exposition assessment. The application range of FoodEx2 encompasses feed additives, food contact materials, food improvement agents, and pesticides [29].

Experimental science advances are based in data, including from FDB, and such figures are commonly fed into models, producing results from which conclusions are withdrawn. Nowadays, these processes can be easily automated by using a bot/API to download data from FDB, which can then be analysed with the assistance of an AI, allowing for instance rapid identification of patterns and trends. With more or less automation, the ability to provide reliable and significant results rely on the research's rigor and methodologies, as much as on the rigor and detail of the semantics and structure of the database from where the information was withdrawn. Specially developed apps may provide insights on more obvious relationships (e.g., between dietary intakes and health) or less obvious relationships (as between food composition and climate change). So, besides the traditional use in assessing nutrient intakes for diet planning, FDBs can have many more applications for different users in the food value chain, facilitated by IT tools that make it easier to manage and analyse large quantities of data and information. FDBs can, thus, be important tools for exploring the relationship between foods, diets, and nutrients' intake, regarding nutritional needs and micronutrient deficiencies; yet, a need to better categorise bioactive compounds in foods is emerging, as state-of-the-art knowledge has been disclosing more and more compounds from foods with important physiological roles. Another emerging trend relates to the environmental impact of foods and attempts in systematizing available information are mentioned below (see Section 4.7.3). The key nutritional components found in FDBs are only a few among the more than 26,000 distinct, definable biochemicals present in our food that remain unquantified [30].

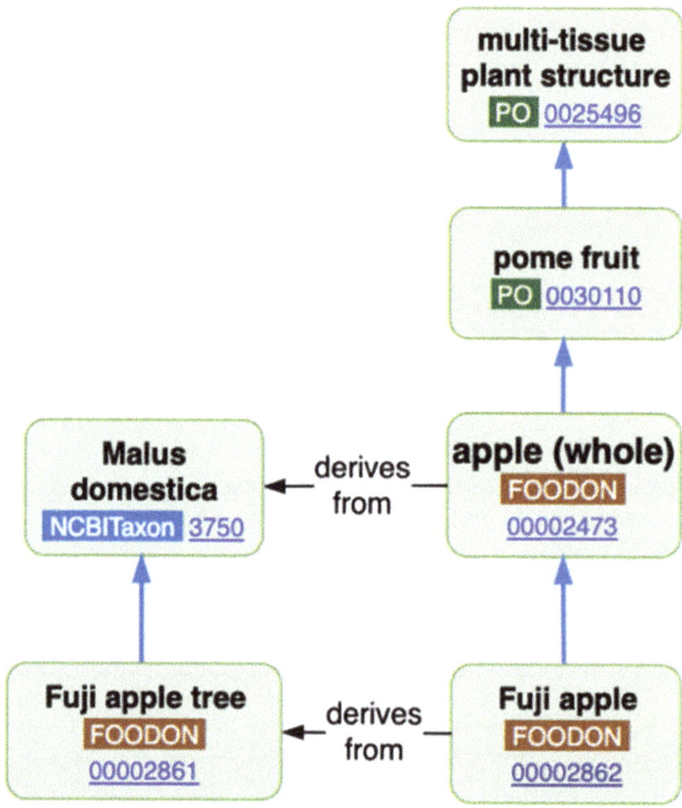

Figure 2. A basic food product, in this case an apple, can be a simple anatomical part, in this case a pome fruit, deriving from a particular plant species (*Malus domestica*) of a specific variety (Fuji). Reprinted with permission from ref. [26,27]. 2017. Roger A Smith (cc-by-sa/2.0).

Whole food databases are described below and summarised in Table 1. The inclusion criteria were 'freely accessible online', in 'English', and 'providing extensive datasets as well as corresponding metadata on food composition', while exclusion criteria were 'not in English' and/or 'absence of online access and/or information not easily accessible' and/or 'pay-per-use/subscription service' and/or 'not updated regularly'.

Table 1. Food composition databases publicly available, in English, allowing data searches and/or download.

Organization	Name of FDB	URL (Available at the Date of the Current Publication)	Discrimination of Food Composition	Source of Data	Ease of Access	Regularity of Updates	Citation/Site
USDA	FoodData Central	https://fdc.nal.usda.gov accessed on 17 August 2021;	Target important components that make sense in each food; highly discriminated	Laboratory analysis by state-of-the-art methods	Search by food name or by component + API for access with proprietary app; instructions and tips provided	Regularly updated (date is shown)	U.S. Department of Agriculture (USDA), Agricultural Research Service. FoodData Central: Version October 2020.:
TMIC	FoodB	www.foodb.ca accessed on 17 August 2021	Content range and average values for an extensive list of compounds	Literature and other FDB	Search by food name or browse foods by constituents	Frequency of updates not mentioned (last update in 2021)	www.foodb.ca accessed on 17 August 2021
DTU food (National food Institute (Denmark)	Fødevaredata (Frida Food Data)	http://frida.fooddata.dk/ accessed on 17 August 2021	DTU foods' database—Frida Food Data reflects the food supply in Denmark and targets professionals in food and nutrition	Laboratory analysis	Easily searched by food item (alphabetic order), food group or by parameters, which include waste and added sugar	Updated every few years (last update 29/10/19) and food composition referred to be quite stable over the past 50 years	Food data (frida.fooddata.dk accessed on 17 August 2021), version 4, 2019, National Food Institute, Technical University of Denmark
EuroFIR AISBL, International non-profit association	EuroFIR	https://www.eurofir.org/food-information/ accessed on 17 August 2021	The dataset presents energy, macronutrients, vitamins, and minerals as well as other bioactive compounds and daily recommended intakes for selected nutrients	Estimations from FDB by expert panels and targets food and nutrition professionals	Search by food name and by component	Updated regularly (each few years)—last update 21 January	European Food Safety Authority (2013) 'Food composition database for nutrient intake: selected vitamins and minerals in selected European countries'. Zenodo. doi: 10.5281/ZENODO.438313.
FAO	InFoods	http://www.fao.org/infoods/infoods/en/ accessed on 17 August 2021	InFoods is a network bringing together food composition compilers, data generators (e.g., chemists), and data users (e.g., nutritionists, food scientists), and decision makers	food composition database compilers retrieve analytical data on food composition for commonly consumed foods and complemented with other published sources	Datasets are downloadable in xls and pdf formats, as well as searchable with software tools for e.g., dietary assessment, labeling and food supply/availability data	Updated regularly	FAO. 2020. International Network of Food Data Systems (InFoods)
CIQUAL-ANSES	French Food Composition Database	https://ciqual.anses.fr/ accessed on 17 August 2021	Average nutritional composition of food consumed in France. Average value of each component, a minimum and a maximum, together with a confidence code (A = very reliable, D = less reliable). Information on a specific component (ex. list of food rich in calcium or poor in sodium)	Compilation of different sources: yearly sampling of around 60 to 80 foods in collaboration with subcontractor laboratory; data from OQALI; research programmes on food composition with external partners. Scientific literature and laboratory reports; foreign food composition tables	Easily searched by food item, food group, or by components	Released every 2 to 4 years	French Agency for Food, Environmental and Occupational Health and Safety. ANSES-CIQUAL French food composition table version 2020.

4. Main Whole Food Composition Databases

As referred above, the main food composition databases have been enriched with more and more information about food constituents and linkages to different databases. For example, FDBs discriminating nutrients and components of a given food, from fresh product to packed branded foodstuffs (e.g., EuroFIR), are linked to a second type of FDB, which is based on inspecting a wide range of foods for a given nutrient or a certain molecular family of compounds (as is the case of Phenol Explorer). A third type of specific FDB is the object of growing interest—that is, the case of HMDB (see below) exploring the interaction of food components, at the level of gut microbiota, and of metabolites, toxins, and specific compounds (biomarkers) at the cellular, organelle, or pathway level [31–34].

4.1. Food Data Central

The United States Department of Agriculture (USDA) manages and maintains Food-Data Central [15], a platform providing access to distinct types of data on nutrients and other food components, including Foundation Foods, National Nutrient Database for Standard Reference (SR Legacy), Food and Nutrient Database for Dietary Studies (FNDDS 2017–2018), and Experimental Foods. The DB platform is noteworthy for providing different types of searches (by component or by food), which may encompass a combination of databases. Metadata are provided, including the number of samples, sampling location, date of collection, analytical approaches used, and if appropriate, agricultural information (e.g., genotype and production practices—intensive, organic, etc.). In respect to Experimental Foods, it is noteworthy that they are meant for research purposes and described foods may not be available in the market. The corresponding database includes data from multiple sources to allow users to examine a range of factors that may affect the nutritional profiles of foods and resulting dietary intakes, as well as the sustainability of agricultural and dietary food systems. This FDB is available at https://agcros-usdaars.opendata.arcgis.com accessed on 17 August 2021, and the user is able to explore data (referring to US) by topic or by location, for example [15].

4.2. CIQUAL—French Food Composition Table

CIQUAL is an open access French FDB [16], covering a wide range of the most consumed foodstuffs in France. This reference database on the nutritional composition of foods is maintained by the Agency for Food, Environmental and Occupational Health and Safety (known by the acronym ANSES). This FDB was updated in 2020 and provides the levels of macro (lipids, fatty acids, carbohydrates) and micronutrients (vitamins, minerals, etc.) of more than 3185 foods and 67 components. The main axes targeted by CIQUAL are the input and management of a reference database relating to the composition of foods, the contribution to the assessment of nutritional risks, and the communication and dissemination of validated data to the greatest number of users (encompassing researchers, nutritionists, food manufacturers, and consumers). In the context of the present work, this database is herein described in more detail, to illustrate the general structure of whole food FDBs, sharing main features and functionalities, essential for interconnections between databases, as explained above.

According to ANSES, finding nutritional information can be carried out by looking for the food in question or by food category. Food categories are classified into eleven food groups:

- Starters and dishes, which in turn divide into six sub-groups: mixed salads (21), soups (46), dishes (159), pizzas, crepe and pies (47), sandwiches (40), savoury pastries, and other starters (24);
- Fruits, vegetables, legumes, and nuts: divided into vegetables (303), potatoes and other tubers (51), legumes (38), fruits (170), and nuts and seeds (52);
- Cereal products: pasta, rice, and grains (71), breads and similar (56), and savoury biscuits (18);

- Meat, egg, and fish: of which the largest sub-groups include cooked meat (133), raw meat (162), delicatessen meat and similar (173), other meat products (16), fish, cooked (63), fish, raw (106), seafood, cooked (24), seafood, raw (25), fish products (56), eggs (24), and meat substitutes (6);
- Milk and milk products are divided into four sub groups;
- Beverages, including water, alcoholic, and non-alcoholic drinks;
- Sugar and confectionery, including products such as jam, sweet biscuits, cakes, and pastry, etc.;
- Ice cream and sorbet, presented as ice cream (11), sorbet (5), and frozen desserts (12);
- Fats and oils (75), such as butters, vegetables oils, margarines, fish oils, and other fats;
- Miscellaneous group exhibit sauces (75), condiments (17), cooking aids (12), salts (6), spices (25), herbs (28), seaweed (17), foods for particular nutritional uses (5), and miscellaneous ingredients for vegetarians (26);
- Finally, the group of baby foods represented by four sub-groups: baby milk and beverages (17), baby dishes (13), baby deserts (5), and baby biscuits and cereals (4).

The nutritional information of each food product is given by a table either in detailed composition or in basic composition. In the case of detailed composition, the estimated energy provided from fibres is also included (based in Jones' factor). All the nutrients likely to be present in the food are provided by the table and are expressed in g/100 g or g/100 mL of the edible part. Lipids are detailed by the fatty acid profile (saturated and polyunsaturated). Fibres, water, starch, vitamins, and oligo-elements are all exposed, but not for all foods systematically, and the level of detail may vary. The data source of each compound is also mentioned by CIQUAL, and it may come from different sources, given the interlinkage between FDBs. Thus, data are a compilation between a sampling plan and analyses launched each year by ANSES on 60 to 80 foods in collaboration with subcontracted laboratories, plus data from OQALI (a French project, which aims at monitoring changes in processed foods supply available on the French market), research programs carried out jointly with external partners, information from scientific literature and laboratory, and finally, data from foreign food composition tables [16].

4.3. EuroFIR, European Food Information Resource

EuroFIR, European Food Information Resource [14], is an independent food composition resource in Europe bringing together food composition datasets from 26 European Countries, Canada, the US, New Zealand, and Japan. It is currently a non-profit international organization that resulted from a network project, Network of Excellence (NoE) comprising of 48 partners from academia, research organizations, and small- and medium-sized enterprises (SMEs). EuroFIR is a food composition table or database providing detailed information on the nutritional composition of foods, typically energy, macronutrients (e.g., protein, carbohydrate, fat) and their components (e.g., sugars, starch, fatty acids), minerals (e.g., calcium, iron, sodium), and vitamins [14]. One of its tools, Food Explorer, is an interface that allows to simultaneously search information about food composition data from most of the available databases from the EU, Canada, USA, New Zealand, and Japan. Food Explorer allows searches by food names or by nutritional groups with the unique ability to allow comparisons of attributes' values of foods from different countries. Another relevant tool in this FDB is Bioactive Substances in Food Information Systems (eBASIS), which is a compilation of food composition and their biological effects. Such data are extracted from peer-reviewed literature as raw data and critically evaluated, thus relying on the curation work of experts.

4.4. FoodDB

The Canadian database, FoodDB Version 1.0, 2021, is licensed under a Creative Commons Attribution-Non-Commercial 4.0 International License, and it is supported by the Canadian Institutes of Health Research, Canada Foundation for Innovation, and by The Metabolomics Innovation Centre (TMIC) [35].

This FDB supplies extensive data on food constituents, chemistry, and biology, providing information on both macronutrients and micronutrients, including many of the constituents that give foods their flavour, colour, taste, texture, and aroma with detailed compositional, biochemical, and physiological information (obtained from the literature). Searches can be made by food source, name, function, or concentrations, and the FDB content can be accessed from the Food Browse (listing foods by their chemical composition) or from the Compound Browse (listing chemicals by their food sources), according to the user's preferences. A section called 'reports' is noteworthy, since it concerns monographies of a list of foods featuring composition and nutritional and health benefits, based on scientific literature review [35].

4.5. Frida Food Data

The database Frida Food Data (frida.fooddata.dk), also known as DTU foods [17], is managed by the National Food Institute with the Technical University of Denmark (DTU) allowing public access to information about foods available in Denmark. The FDB also relies on the cooperation of stakeholders as food industries and retailers, as well as scholars and the Danish Veterinary and Food Administration. Metadata (as the number of samples and their source) are included in registries encompassing more than 1000 food items.

The information above is summarised in Table 1 presenting some features of the most utilized food composition databases, for whole foods, easily and freely accessible online, in English.

The FDBs listed in Table 1 follow international standards and are interconnected thus providing access to reliable, comprehensive information on foods serving most common purposes.

In view of the current transformation of food systems in meeting the 2030 agenda, average global data on food composition may not be enough, as consumers are being encouraged to prefer healthier foods respectful of their food cultures and the environment [18]. Such changes will sooner or later reflect the level of the usage of FDBs, and consequently, the inspection of food habits linked to traditional balanced diets may direct the spotlights towards certain FDBs of national ambit. The panorama is currently not so encouraging because of the great variation observed from country to country, as illustrated in the section below.

4.6. National Whole Food Composition Databases

National FDBs, where they exist, vary widely in the extent of provided information, standardisation at various levels (see Figure 1; Figure 2), and the ease of access (including the language). Thus, starting by the British food composition table, obviously in English, in the United Kingdom, Public Health England (PHE) is responsible for maintaining food composition data relating to nutrients (macronutrients, e.g., fats, protein, carbohydrates as well as their micronutrient content, which includes vitamins and minerals) mostly from analysing foods commonly consumed in the country. The results are published as McCance and Widdowson's 'The Composition of Foods'—the UK food composition tables. The Composition of Foods Integrated Dataset (CoFIDS) is a nutrient dataset for 2898 foods and 303 others in the 'old foods' file, comprising 185 individual nutrients. CoFIDS is searchable online and can be downloaded free of charge in MS Excel or Ascii format, and it was first published in 2008 (https://fdnc.quadram.ac.uk/ accessed on 17 August 2021), available online at the date of this publication.

PortFIR is the Portuguese national food composition database for the most consumed foods in Portugal. The data cover about 42 nutrients ex. energy, macronutrients, fatty acids, vitamins, and minerals (http://portfir.insa.pt/ accessed on 17 August 2021), available online at the date of this publication. The information is classified into groups and subgroups according to the FoodEx2 classification and description system (http://www.efsa.europa.eu/en/datex/datexfoodclass accessed on 17 August 2021), available online at the date of this publication from the EFSA. The PortFIR FDB is free online, displayed in English, and allows searches as well as downloading in Excel format [36].

Similarly, the Turkish food composition database, an open access digital platform, 'Türkomp' (http://www.turkomp.gov.tr/main accessed on 17 August 2021), available online at the date of this publication provides a considerable dataset and information related to the nutrients, composition, and energy values of processed or unprocessed agricultural products that are produced and consumed in Turkey. Türkomp exhibits 63,000 data entries on the nutritional and energy value of 100 food components belonging to 580 foods from 14 food groups [37].

As referred above, it is rare to find suitable food composition tables of reliable and updated contents from developing countries, and to illustrate such situations, a few examples are herein presented.

Thus, in Morocco, a country integrating the UNESCO's list of countries that safeguard the Mediterranean diet as intangible heritage of humankind [38], the development of a national composition table dates back to 1977 by the Ministry of Agriculture of Morocco and was revised in 1984 by El Khayate [39]. Since then, no updates have been made. Recently, a multidisciplinary team of Moroccan and international experts worked on updating the food composition table, in order to supplemented it with high quality composition data. The consolidated version includes information on 38 nutrients, from 587 food products commonly consumed in Morocco. This update represents a 79% addition of foods, and according to the authors, 7% of nutritional values come from Moroccan data sources and 93% from international data sources, mainly from Tunisia, West Africa, France, the United Kingdom, and the United States [40]. The updated version provides information on foods and dishes commonly consumed in Morocco and can be used as a tool to promote nutritional research and to design public health strategies.

Another common situation with national databases of developing countries can be illustrated by the Tunisian food composition table, which displays the 240 foods and dishes usually consumed by Tunisians. The table corresponds to 95% of the food needs of the entire Tunisian population. It includes, for each food, the energy value as well as the content in 34 nutrients, expressed per 100 g of the raw edible part. This table is presented in the form of a book produced by a group of nutritionists from the National Institute of Nutrition and Food Technology (INNTA) who were supported by French and Belgian experts within the framework of the European project 'Impact of transitions epidemiological studies on health in North African countries' [41]. Another common situation corresponds to the composition table of foods from the Republic of Bahrain, which is a printed book not so regularly updated and hardly available. This database brings together 150 raw and ready-to-eat foods and composite dishes according to standardized methods. This list includes cereals and grain products, bread and bread products, fruits, vegetables, legumes, nuts and seeds, meat, poultry and eggs, fish, milk and dairy products, fats and oils, herbs and spices, beverages, local and western fast foods, etc. The table provides data for proximate composition, three minerals (calcium, phosphorus, and iron), and five vitamins (retinol, thiamine, riboflavin, niacin and vitamin C) expressed per 100 g of edible portion [42].

Similarly, the Chinese food composition database is given by a printed book, not necessarily in English [43].

As the reader can easily deduce, the randomness of updates, the limited access, and the absence of English versions can be strong limitations to the use of national FDBs in disclosing specific food habits and/or the composition of particular food items.

In addition to free access institutional databases, a growing number of commercial customized applications have been appearing in the market. Such apps or so-called food databases mainly encompass different types of software to assist food formulation and labelling, dietary features, and recipe analysis, as well as fitness apps. The access is reserved and includes consultancy support services.

An example of a privately owned FDB, with an associated API, is offered by Edamam, a company that provides access to a food and grocery database with close to 900,000 basic foods, restaurant items, and consumer packaged foods available on the website, at the date of this publication, https://developer.edamam.com/food-database-api (accessed on 17 August 2021). The Food API provides a filter to sort data by diet and health, determining dietary, allergy, and nutrition labelling, based on the food's ingredients. Over 70+ claims are automatically generated such as peanut free, shellfish free, gluten free, vegan, and vegetarian.

Edamam also provide data for basic foods (as flour and eggs) for calories, fats, carbohydrates, protein, cholesterol, sodium, etc., for a total of 28 nutrients.

4.7. *Specific Purpose's Food Databases*

4.7.1. FDBs Directly Related with Human Metabolism

The food we ingest is expected to interact at the level of the gut microbiota, and thus, considering the scenario of metabolic pathways and the benefits of bioactive compounds in humans, Durazzo et al. [44] noted the database Human Metabolome Database (HMDB) version 4.0, also originating from Canada, and supported by the same organizations as FoodDB vs.1. This database, HMDB, contains detailed information about small molecule metabolites found in the human body aiming to be an input for studies in metabolomics, clinical chemistry, biomarker discovery, etc. This database encompasses data of different kinds: chemical, clinical, and molecular biology/biochemistry data, notably more than 100,000 metabolite entries (water-soluble and non-polar metabolites) either abundant (>1 μM) or rare (<1 nM), which are linked to almost 6000 protein sequences. Even if this database does not directly reflect food composition, it is of undoubted interest in nutritional studies to assess how a food or a diet might influence metabolism, either in a positive or unhealthy way. The HMDB supports text, sequence, chemical structure, and relational query searches, and it is linked to other databases whether on drugs, toxins, pollutants, or on nutrients and food additives [31–34]. At https://hmdb.ca accessed on 17 August 2021, available online at the date of this publication it is possible to browse metabolites, pathways, etc., as well as performing advanced searches based on molecular mass, chemical structure, or text queries.

Another freely accessible data resource of the same kind is MGnify, an EMBL-EBI online resource containing Human Gastrointestinal Protein catalogue and a dataset on the Human Gastrointestinal Genome, allowing researchers to compare their findings on microbial genomics and proteomics with existing datasets. MGnify has been growing, and promoters would like to close knowledge gaps, such as the variation in bacterial diversity across different human populations [45].

The Sydney University Glycaemic Index Research Service (SUGiRS) produced a free database that gives the glycaemic index of any food inserted on their search engine available on their website https://www.glycemicindex.com accessed on 17 August 2021, at the date of this publication and the Gluten-Free Food Database (Austria) provides quantitative information of macro- and micronutrients of the gluten-free products. This database can be accessed via the science collaboration platform, Open Science Framework, upon registration, and it also accepts contributions to the dataset [46].

4.7.2. FDBs Concerning Food Processing

In order to process safe food, several hours of research are needed when searching for the precise thermal processing parameters; D-value and z-value parameters that describe the characteristics of thermal death of food target microorganisms, for the ingredients or final food products, are not always easily found. The Lemgo D- and z-value Database for food, a project of the Institute for Food Technology NRW (ILT.NRW) at the OWL University of Applied Sciences and Arts, supplies information on these parameters, to design pasteurization or sterilization processes with a main focus on beverage spoiling microorganisms. Additional information is given on parameters known to have an effect on the D- and z-values like pH, Brix and a_w value. The data are sorted by the species of microorganism and their medium, and on the experiments from which these data originated or a cluster of relevant data [47].

Another very important database for food engineers is the Database of Physical Properties of Food is available online, at the date of the current publication http://www.nelfood.com, (accessed on 17 August 2021); nelfood.com grew out of the Physical Properties of Food Data Base project that started to collect and publish on the internet reliable and useful data on Physical Properties of Foods. This project was managed by Dr Paul Nesvadba with internet work done by NEL, and it was partly funded by the EU and partly sponsored by companies such as Nestle, RHM, and Unilever. It is only available to subscribed members that may search 11,094 bibliographic references, 1519 materials, and 1694 experiment datasets. These datasets range over 24 food categories encompassing 249 food subcategories and 260 physical properties. NELFOOD Database covers five main groups of physical properties: (1) Mechanical and Rheological Properties of Foods; (2) Sorption and Mass Diffusion Properties of Foods; (3) Electrical and Dielectric Properties of Foods, and (4) Optical Properties of Foods [48].

4.7.3. FDBs Concerning Environmental Impact of Foods

Generally speaking, current food systems are operating out of planetary boundaries, with agriculture being a top driver for biodiversity loss, using water above the natural capacity of replenishment, causing soil degradation, pollution, and more [49,50]. The urge of the food systems' transformation is such that, among many initiatives, the UN organised a food system summit in 2021 (https://www.un.org/en/food-systems-summit accessed on 17 August 2021), available online at the data of this publication and the European Union issued a climate law that binds the EU Institutions and the Member States to take the necessary measures to reach carbon neutrality by 2050. On the other hand, a growing awareness from consumers about the impact of their individual food choices in their health and the environment has been registered [51]. Shifts in food habits may fuel the desired changes, but the commitment of food producers is key. Business pledges need to be underpinned in well-established targets and robust metrics fed with comprehensive information on the food–environment nexus. Despite the still existing gaps, efforts in compiling information are many, and advancements of FDBs in integrating data on the environmental footprint of foods are to be expected.

In respect to the 2030 agenda, the Sustainable Development Report, by Sachs et al., [52] provides interactive dashboards with visual representation of performances by SDGs to identify priorities for action. One of such priorities is tackling food loss and waste for which the FAO maintains a database in connection to tools to track progress, available online at the date of this publication, http://www.fao.org/platform-food-loss-waste/flw-data/en/ (accessed on 17 August 2021). The food loss and waste database contains data and information from various sources, measuring food loss and waste across food products, stages of the value chain, and geographical areas, also presenting underlying causes, according to the literature [53].

Only a few FDBs present datasets on the environmental footprint of foods or are useful for its assessment. One of them is 'Experimental Foods' from USDA (see Section 4.1) that contains information on environmental inputs and outputs on the supply chains, etc.; however, it is not necessarily publicly available [15]. A dataset on food environmental impacts through producers and consumers was published by Poore and Nemecek in 2018 [54]. The ADEME (the French Agency for Ecological Transition) recently launched Agribalyse, a food database providing an environmental score (Ecoscore) for 2500 food products based on their life cycle analysis (LCA). However, this database has already been criticized, notably by institutions promoting organic agriculture, for favouring intensive farming systems and not taking into account the consequences on biodiversity, animal well-being, or the impact of pesticides [55]. More generally, LCA, on which Agribalyse is primarily based, has already been questioned for being unsuitable for comparing farming systems. Thus, an improvement of such a tool would be necessary to inform public policies [56].

5. Main Limitations of Food Databases: Missing Dimensions for Human Health?

First, beyond only nutrients, foods are interlinked with cultural identity while playing a key role in many local economies, as highlighted by Dembska et al. [18] in their double pyramids models connecting food culture, health, and climate. These authors and others [18,57–61] call attention to the need of leveraging the various dimensions of foods, which are closely related, under the so-called one-health approach [18,57].

If FDBs are specifically useful for balancing a diet for nutrient composition and fully addressing nutritional needs in human studies, they, however, reflect a reductionist view of foods, viewed as only the sum of nutrients [61], not considering the food matrix effect, and hence, the degree of processing [22]. Therefore, to be a relevant tool regarding human health in the long term, their data should not be used alone, but other parameters should be also considered, such as food form and degree of processing, together with other important food properties.

For example, the newly developed Siga score [59] is hierarchically combined with the first degree of processing, then the food matrix effect, added salt, fat, and/or sugar, and the number of markers of ultra-processing (including some cosmetic additives and non-additive markers) [60]. To be elaborated, this score typically needs not only the food composition data, but also the list of ingredients and the presence or not of added sugar, salt, and/or fat. Such a hierarchical and holistic score should be more considered, because, in the end, it is related to global (environmental and human) health [18,57,60,61]; whereas, food composition only is insufficient to address diets from the global or one-health perspective as needed (e.g., compliance with European Climate Law).

5.1. The Matrix Effect Is Not Considered

First, the whole food potential is not only reflected by its nutrient composition. Whole foods are first complex matrices, which govern the health effects of nutrients [22]. Besides, food form matters for human health, be it solid, semi-solid, or liquid. It should be emphasized that interactions between nutrients within the food matrix participate in a food's health potential, including notable food chewing and satiety [62], nutrient kinetics of release, and final bioavailability. For example, the calcium of dairy products is only 20–40% bioavailable; therefore, 120 mg of calcium in a yogurt corresponds to around 36 mg being bioavailable, with the remaining fraction reaching the colon [63]. The same is true for the lipid content of a whole almond, which is not fully available [64]. Otherwise, within an extruded-cooked breakfast cereal, wheat flour, and/or maize semolina behave close to simple sugars in human organism with a glycaemic index above 80 [65], and so on for most of nutrients, depending on the food form and on the impact of processing on the food matrix. Such fundamental physiological properties go beyond the simple nutritional composition, which leads to the hypothesis that chronic diseases have more to do with highly degraded and artificialized food matrices than with the food composition itself [22].

5.2. Some Important Bioactive Compounds and Food Properties Are Still Missing

Another limitation is often observed worldwide and consists of missing values for some important key nutrients, e.g., lipotropic compounds (such as choline, betaine, and myo-inositol) and phytic acid, but also for other characteristics of nutrients or foods, such as soluble and insoluble fibre (with different physiological effects), resistant starch, and glycaemic index [65]. It is true that some FDBs report choline content such as the USDA Database for the Choline Content of Common Foods, Release 2 (2008) [66], or the phytic acid content such as the FAO/INFOODS/IZiNCG, Global Food Composition Database for Phytate (2018) [67], or the glycaemic index [68], but this should be completed and extrapolated to other FDBs more broadly in the future, e.g., the French CIQUAL database [16].

5.3. The Important Dimension of the Degree of Food Processing

Therefore, FDBs must not be considered as a sufficient tool for reaching human health on a long term. Notably, one can fully address one's nutritional needs and become chronically ill, as is frequently observed in Western countries. This is notably due to the matrix quality of consumed calories, not only the quantity, and this quality may depend on the degree of food processing.

It is noteworthy that food engineers and food technologists have been, in the last three decades, dedicating a great part of their research to studies on reducing the processing load that can be achieved on the application of milder preservation technologies (still called emerging technologies). Such milder preservation technologies can be used alone or combined with less severe thermal treatments, such as high hydrostatic pressure [69], pulses of electric field [70], UV-c radiation [71], thermosonication [72], and others. The optimization of these processes aims at maximizing the retention of food nutrients such as vitamins [73], proteins, and sensory parameters such as texture, colour, and taste, while keeping the product safe [74,75].

On the other hand, in the circular economy model, food industries are expected to play a key role in tackling food loss and waste, which poses the double burden of depleting natural resources and wasting extra energy from production to disposal. Innovations that consist of using by-products of an industry as raw materials of another, as well as recovering nutrients that would otherwise be wasted are emerging tendencies within a biorefinery approach. An illustrative example is reported by Lucarni et al. 2020 [76], exposing a new class of ingredients that may not yet be adequately covered by FDB.

Considering industrially processed foods that are becoming dominant in our diets, in the future, FDBs should also distinguish between 'natural' and 'added' nutrients whatever they are and indicate the list of additives, as with the Open Food Facts database available online for industrial foods [77] or private pay-per-use food databases that also gives the list of ingredients, e.g., Alkemics and Num-Alim. More specifically, the Open Food Facts database is a collaborative database of food products and is licenced under the Open Database Licence (ODBL). For such foods, the list of ingredients tells us more about their whole nutritional quality (including environmental aspects) than the only composition.

Indeed, it should be underlined that no food is nutritionally balanced (except maternal milk for the growth of the infant), hence the recommendation to 'eat varied' at the level of the diet.

6. Emerging Applications and Trends of Food Databases

In view of the ongoing changes in food systems, needs for curated and organized information on the composition of food secondary raw-materials, novel foods, and/or sources for nutrients (as insects and microalgae) are expected to be met by FDBs. These challenges may exacerbate existing issues with food data composition. Thus, in addition to the intrinsic features of foods, parameters related to the extraction and analytical procedures should be considered, according to Durazzo et al., [44], as different extraction procedures and analytical techniques and methodologies may lead to different datasets. Moreover, still according to these authors, only a few compounds within a class are investigated, and there are knowledge gaps on appropriate analytical methods for food analysis. The acknowledged complexity of foods (in their multiple dimensions) calls for information on multiple relationships, as the nexus between public health and the environment, or consumer preference and health [51,52,78,79]. Ocké et al. [9], besides identifying some gaps herein mentioned, also refer to the need for FDBs' adaptation to the rapidly changing food landscape and the need for their improvement and harmonization to enable comparisons of research outputs at international level. More generally, in the near future, there is, therefore, an important need for more comprehensive and holistic FDB, not only addressing nutritional composition, but also other food properties. In this way, FDBs will, thus, constitute more robust tools for tackling global health, but this means a huge scientific work to gather all data, notably when thousands of new industrially processed foods are marketed each year worldwide.

7. Conclusions

Food composition data are fundamental information resources to many fields of work, as in formulating and labelling foods, as well as in public health and nutrition. Thus, food industrials, legislators, and consumers all need and/or use reliable data on food composition, provided from FDBs (Figure 3). In other words, nutritional and physico-chemical features of foods are valuable tools for medical doctors and dieticians in prescribing nutritionally balanced and/or low-GI diets, as well as for researchers and industry workers, notably in developing the most nutrient-dense foods.

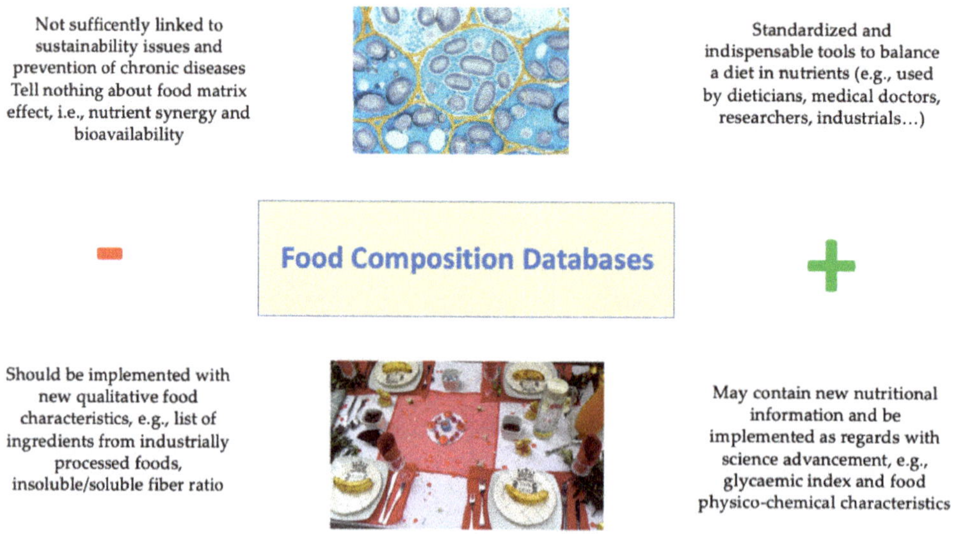

Figure 3. Main features, common uses, identified gaps, and expected trends of food composition databases (original figure by Anthony Fardet; photos: INRAE for bean food matrix under optical microscopy and Amélia Delgado for the meal table).

However, FDBs do have limitations, encompassing variability in the composition of foods between countries, from season to season; food composition depends on the cultivar or variety; manufactured foods of the same recipe may vary from brand to brand and between lots; missing values for some important food characteristics (e.g., list of ingredients for industrially processed foods), etc. In addition, FDBs can only provide an incomplete coverage of foods and/or their nutrients leading to gaps in values, as missing information on some minority compounds (from aromas to chemical contaminants). Despite efforts on updates, data ageing is inevitable due to limited resources.

Food databases have been following the advancements of science, as highlighted above (see Sections 3 and 4.7), and today's challenges include adding comprehensive information about the environmental impact of foods, health/sustainability linkages, as well as qualitative features, because food goes far beyond its composition (Figure 3).

Concerning the relevance of FDBs for human health, they only indirectly address a reductionist view of it and should not be used for other purposes than building a balanced diet to fully address nutritional needs and avoid nutritional deficiencies. However, other criteria should also be considered. Most importantly, food composition does not say anything about the nutrient kinetics of release and final bioavailability within the human organism and on health effects in the longer term. Otherwise, due to the increasing marketing of industrially processed foodstuffs worldwide, comprehensive FDBs should probably integrate more of these foods in a near future, together with their corresponding content in additives, aromas, and added fat, sugar, protein, fibre, and salt, to distinguish between the 'natural' and 'artificial' origins. In addition, other food health potential metrics or indicators such as the soluble/insoluble fibre ratio and/or glycaemic index would deserve to be added in FDBs whenever possible. This could be important issues for the future of this nutritional tool, and this will strengthen their link with human health.

In the end, if nutrient composition is a relevant tool for addressing nutrient needs, it is not sufficiently linked to global health and food system sustainability, and apart for organic plant/animal and some traditional foods that may contain higher nutritional densities (e.g., omega 3 fatty acids and antioxidants), the stronger connexion is between plant versus animal-based foods and with degree of food processing, i.e., at the level of complex foods, a higher scale of observation than nutrients, i.e., more in connection with reality.

Author Contributions: Conceptualization: A.D., M.I., M.C.V.; investigation, A.D., M.I., M.C.V., I.S.d.C., A.F., writing—original draft preparation, A.D., M.I., A.F.; writing—review and editing, A.D., M.C.V., I.S.d.C., A.F.; visualization, I.S.d.C. All authors have read and agreed to the published version of the manuscript.

Funding: This research received no external funding.

Institutional Review Board Statement: Not applicable.

Informed Consent Statement: Not applicable.

Conflicts of Interest: A.D., M.I., M.C.V. and I.S.d.C. declare no conflict of interest. Since 2017, Anthony Fardet has been a member of the Siga scientific committee.

References

1. Issaoui, M.; Delgado, A.M.; Caruso, G.; Micali, M.; Barbera, M.; Atrous, H.; Ouslati, A.; Chammem, N. Phenols, Flavors, and the Mediterranean Diet. *J. AOAC Int.* **2020**, *103*, 915–924. [CrossRef]
2. Gan, Z.; Wei, W.; Li, Y.; Wu, J.; Zhao, Y.; Zhang, L.; Wang, T.; Zhong, X. Curcumin and Resveratrol Regulate Intestinal Bacteria and Alleviate Intestinal Inflammation in Weaned Piglets. *Molecules* **2019**, *24*, 1220. [CrossRef] [PubMed]
3. Tsukamoto, Y.; Ikeda, S.; Uwai, K.; Taguchi, R.; Chayama, K.; Sakaguchi, T.; Narita, R.; Yao, W.L.; Takeuchi, F.; Otakaki, Y.; et al. Rosmarinic acid is a novel inhibitor for Hepatitis B virus replication targeting viral epsilon RNA-polymerase interaction. *PLoS ONE* **2018**, *13*, e0197664. [CrossRef] [PubMed]
4. Vieira, F.N.; Lourenço, S.; Fidalgo, L.G.; Santos, S.; Silvestre, A.; Jerónimo, E.; Saraiva, J.A. Long-Term Effect on Bioactive Components and Antioxidant Activity of Thermal and High-Pressure Pasteurization of Orange Juice. *Molecules* **2018**, *23*, 2706. [CrossRef] [PubMed]
5. Amini, A.; Liu, M.; Ahmad, Z. Understanding the link between antimicrobial properties of dietary olive phenolics and bacterial ATP synthase. *Int. J. Biol. Macromol.* **2017**, *101*, 153–164. [CrossRef]
6. Delgado, A.; Arroyo López, F.N.; Brito, D.; Peres, C.; Fevereiro, P.; Garrido-Fernández, A. Optimum bacteriocin production by Lactobacillus plantarum 17.2b requires absence of NaCl and apparently follows a mixed metabolite kinetics. *J. Biotechnol.* **2007**, *130*, 193–201. [CrossRef]
7. From Science to Practice: Research and Knowledge to Achieve the SDGs. Policy Brief 38 | July 2021, UNRIDS, Geneve, Switzerland. Available online: www.unrisd.org/rpb38 (accessed on 24 March 2021).
8. Dupraz, J.; Burnand, B. Role of Health Professionals Regarding the Impact of Climate Change on Health—An Exploratory Review. *Int. J. Environ. Res. Public Health* **2021**, *18*, 3222. [CrossRef]
9. Ocké, M.C.; Westenbrink, S.; van Rossum, C.T.M.; Temme, E.H.M.; van der Vossen-Wijmenga, W.; Verkaik-Kloosterman, J. The essential role of food composition databases for public health nutrition–Experiences from the Netherlands. *J. Food Compost. Anal.* **2021**, *101*, 103967. [CrossRef]
10. Greenfield, H.; Southgate, D.A.T. *Food Composition Data: Production, Management and Use*, 2nd ed.; FAO: Rome, Italy, 2003.
11. UN, Sustainable Development Goals, News, Sustainable Development Agenda, 24 September 2015. Available online: https://www.un.org/sustainabledevelopment/blog/2015/09/summit-charts-new-era-of-sustainable-development-world-leaders-to-gavel-universal-agenda-to-transform-our-world-for-people-and-planet/ (accessed on 11 July 2021).
12. Food and Agriculture Organization of the United Nations, FAO. International Network of Food Data Systems (INFOODS). Available online: http://www.fao.org/infoods/infoods/en/ (accessed on 24 March 2021).
13. FAO/WHO. Codex Alimentarius, International Food Standards. Codex Online Databases. Available online: http://www.fao.org/fao-who-codexalimentarius/codex-texts/dbs/en/ (accessed on 24 March 2021).
14. European Food Information Resource, EuroFIR. Available online: https://www.eurofir.org/about_eurofir/ (accessed on 20 March 2021).

15. U.S. Department of Agriculture, USDA. Agricultural Research Service. FoodData Central. 2019. Available online: https://fdc.nal.usda.gov (accessed on 10 March 2021).
16. ANSES-CIQUAL Food Composition Table. 2020. Available online: https://www.anses.fr/en/content/anses-ciqual-food-composition-table (accessed on 12 March 2021).
17. *DTU. Food Data (frida.fooddata.dk), Version 4*; National Food Institute, Technical University of Denmark: Lyngby, Denmark, 2019. Available online: https://frida.fooddata.dk/?lang=en (accessed on 12 March 2021).
18. Dembska, K.; Recanati, F.; Casari, S.; Antonelli, M.; Vaccaro, O.; Calabrese, I.; Giosuè, A.; Riccardi, G. *A One Health Approach to Food, the Double Pyramid Connecting Food Culture, Health and Climate*; Barilla Foundation & Research Unit on Nutrition, Diabetes and Metabolism, University of Naples Federico II: Parma, Italy, 2021.
19. Church, S.M. The history of food composition databases. *Nutr. Bull.* **2006**, *31*, 15–20. [CrossRef]
20. Food and Agriculture Organization of the United Nations. *Food Balance Sheets, a Handbook*; FAO: Rome, Italy, 2001.
21. Jacobs, D.R.; Tapsell, L.C.; Temple, N.J. Food Synergy: The Key to Balancing the Nutrition Research Effort. *Public Health Rev.* **2011**, *33*, 507–529. [CrossRef]
22. Fardet, A.; Rock, E. Exclusive reductionism, chronic diseases and nutritional confusion: Degree of processing as a lever for improving public health. *Crit. Rev. Food Sci. Nutr.* **2020**, 1–16. [CrossRef]
23. Biesalski, H.K.; Dragsted, L.O.; Elmadfa, I.; Grossklaus, R.; Müller, M.; Schrenk, D.; Walter, P.; Weber, P. Bioactive compounds: Definition and assessment of activity. *Nutrition* **2009**, *25*, 1202–1205. [CrossRef]
24. Heinrich, M. "Nutraceutical" In Encyclopedia Britannica. Available online: https://www.britannica.com/science/nutraceutical (accessed on 23 March 2021).
25. Regulation (EC) No 1924/2006 of the European Parliament and of the Council of 20 December 2006, on nutrition and health claims made on foods. *Off. J. Eur. Union* **2006**, *L404*, 9–25.
26. Langual. The International Framework for Food Description, Danish Food Informatics. 2017. Available online: https://www.langual.org/default.asp (accessed on 16 March 2021).
27. Møller, A.; Ireland, J. *LanguaL™ 2017-Multilingual Thesaurus (English–Czech–Danish–French–German–Italian–Portuguese–Spanish)*; Technical Report Danish Food Informatics; Danish Food Informatics: Roskilde, Danmark, 2018. [CrossRef]
28. FoodOn, 2021. Available online: https://foodon.org/about/ (accessed on 24 March 2021).
29. European Food Safety Authority. *The Food Classification and Description System FoodEx2 (Revision 2)*; EN-804; EFSA Supporting Publication: Parma, Italy, 2015; 90p.
30. Barabási, A.L.; Menichetti, G.; Loscalzo, J. The unmapped chemical complexity of our diet. *Nat. Food* **2020**, *1*, 33–37. [CrossRef]
31. Wishart, D.S.; Tzur, D.; Knox, C.; Eisner, R.; Guo, A.C.; Young, N.; Cheng, D.; Jewell, K.; Arndt, D.; Sawhney, S.; et al. HMDB: The Human Metabolome Database. *Nucleic Acids Res.* **2007**, *35*, D521–D526. [CrossRef] [PubMed]
32. Wishart, D.S.; Knox, C.; Guo, A.C.; Eisner, R.; Young, N.; Gautam, B.; Hau, D.D.; Psychogios, N.; Dong, E.; Bouatra, S.; et al. HMDB: A knowledgebase for the human metabolome. *Nucleic Acids Res.* **2008**, *37*, D603–D610. [CrossRef]
33. Wishart, D.S.; Jewison, T.; Guo, A.C.; Wilson, M.; Knox, C.; Liu, Y.; Djoumbou, Y.; Mandal, R.; Aziat, F.; Dong, E.; et al. HMDB 3.0—The Human Metabolome Database in 2013. *Nucleic Acids Res.* **2013**, *41*, D801–D807. [CrossRef] [PubMed]
34. Wishart, D.S.; Feunang, Y.D.; Marcu, A.; Guo, A.C.; Liang, K.; Vázquez-Fresno, R.; Sajed, T.; Johnson, D.; Allison, P.; Karu, N.; et al. HMDB 4.0: The human metabolome database for 2018. *Nucleic Acids Res.* **2018**, *46*, D608–D617. [CrossRef] [PubMed]
35. FoodDB Version 1.0. 2021. Available online: www.foodb.ca (accessed on 10 March 2021).
36. PortFir Version 4.1. 2019. Available online: http://portfir.insa.pt/ (accessed on 10 March 2021).
37. Turkish Food Composition Database. Available online: http://www.turkomp.gov.tr/main (accessed on 11 March 2021).
38. United Nations Educational Scientific and Cultural Organization, UNESCO. 2013. Intangible Cultural Heritage. Decision of the Intergovernmental Committee: 8.COM 8.10. Available online: https://ich.unesco.org/en/decisions/8.COM/8.10 (accessed on 18 March 2021).
39. El Khayate, R. Ebauche d'Une Table de Composition Alimentaire Marocaine. Ph.D. Thesis, University Hassan II, Institute of Agronomy and Veterinary Medicine, Casablanca, Morocco, 1984.
40. Khalis, M.; Garcia-Larsen, V.; Charaka, H.; Sidi-Deoula, M.M.; El Kinany, K.; Benslimane, A.; Charbotel, B.; Soliman, A.S.; Huybrechts, I.; Soliman, G.A.; et al. Update of the Moroccan Food Composition Tables: Towards a more reliable tool for nutrition research. *J. Food Compost. Anal.* **2020**, *87*, 103397. [CrossRef]
41. El Ati, J.; Béji, C.; Farhat, A.; Haddad, S.; Chérif, S.; Trabelsi, T.; Danguir, J.; Gaigi, S.; Le Bihan, G.; Landais, E.; et al. *Table de Composition des Aliments Tunisiens*; Institut National de Nutrition et de Technologie Alimentaire (INNTA): Tunis, Tunisia, 2007.
42. Musaiger, A.O. *Food Composition Tables for Kingdom of Bahrain*; Arab Center for Nutrition: Manama, Bahrain, 2011.
43. Bian, Y.Y.X.Z. *China Food Composition Table (Chinese Edition)*; Beijing Medical University Press: Beijing, China, 2004; Volume 2, ISBN 978-7810716789.
44. Durazzo, A.; Nazhand, A.; Lucarini, M.; Delgado, A.M.; De Wit, M.; Nyam, K.L.; Santini, A.; Ramadan, M.F. Occurrence of Tocols in Foods: An Updated Shot of Current Databases. *J. Food Qual.* **2021**, *2021*, 1–7. [CrossRef]
45. Almeida, A.; Stephen, N.; Boland, M.; Strozzi, F.; Beracochea, M.; Shi, Z.J.; Pollard, K.S.; Sakharova, E.; Parks, D.H.; Hugenholtz, P.; et al. A unified catalog of 204,938 reference genomes from the human gut microbiome. *Nat. Biotechnol.* **2020**, *39*, 105–114. [CrossRef]

46. Missbach, B.; Schwingshackl, L.; Billmann, A.; Mystek, A.; Hickelsberger, M.; Bauer, G.; König, J. Gluten-free food database: The nutritional quality and cost of packaged gluten-free foods. *Peer J.* **2015**, *3*, e1337. [CrossRef] [PubMed]
47. Lemgo D- and z-value Database for Food. OWL University of Applied Sciences and Arts. Available online: https://www.th-owl.de/fb4/ldzbase/index.pl?link=aboutus (accessed on 10 May 2021).
48. Database of Physical Properties of Food. Nelfood.com. Available online: http://www.nelfood.com (accessed on 10 May 2021).
49. European Commission. Sustainable Consumption and Production. Sustainable Food. Available online: https://ec.europa.eu/environment/eussd/food.htm (accessed on 12 May 2021).
50. TEEB, The Economics of Ecosystems and Biodiversity. *TEEB for Agriculture & Food: Scientific and Economic Foundations*; UN Environment: Geneva, Switzerland, 2018.
51. Guiné, R.P.F.; Bartkiene, E.; Florença, S.G.; Djekić, I.; Bizjak, M.Č.; Tarcea, M.; Leal, M.; Ferreira, V.; Rumbak, I.; Orfanos, P.; et al. Environmental Issues as Drivers for Food Choice: Study from a Multinational Framework. *Sustainability* **2021**, *13*, 2869. [CrossRef]
52. Sachs, J.; Schmidt-Traub, G.; Kroll, C.; Lafortune, G.; Fuller, G.; Woelm, F. *The Sustainable Development Goals and COVID-19. Sustainable Development Report 2020*; Cambridge University Press: Cambridge, UK, 2020.
53. Food and Agriculture Organization of the United Nations, FAO, 2021. Food Loss and Waste Database. Available online: http://www.fao.org/platform-food-loss-waste/flw-data/en/ (accessed on 12 May 2021).
54. Poore, J.; Nemecek, T. Supplementary materials for Reducing food's environmental impacts through producers and consumers. *Science* **2018**, *360*, 987–992. [CrossRef]
55. Agribalyse. Discover the Environmental Impact for Food Consumed Products. Available online: https://agribalyse.ademe.fr (accessed on 12 May 2021).
56. van der Werf, H.M.G.; Knudsen, M.T.; Cederberg, C. Towards better representation of organic agriculture in life cycle assessment. *Nat. Sustain.* **2020**, *3*, 419–425. [CrossRef]
57. FAO/WHO. *Sustainable Healthy Diets–Guiding Principles*; FAO and WHO: Rome, Italy, 2019.
58. Fardet, A.; Rock, E. Toward a new philosophy of preventive nutrition: From a reductionist to a holistic paradigm to improve nutritional recommendations. *Adv. Nutr.* **2014**, *5*, 430–446. [CrossRef]
59. Davidou, S.; Christodoulou, A.; Fardet, A.; Frank, K. The holistico-reductionist Siga classification according to degree of food processing: An evaluation of ultra-processed foods in French supermarkets. *Food Funct.* **2020**, *11*, 2026–2039. [CrossRef] [PubMed]
60. Davidou, S.; Christodoulou, A.; Frank, K.; Fardet, A. A study of ultra-processing marker profiles in 22,028 packaged ultra-processed foods using the Siga classification. *J. Food Compos. Anal.* **2021**, *99*, 103848. [CrossRef]
61. Fardet, A.; Rock, E. Ultra-processed foods and food system sustainability: What are the links? *Sustainability* **2020**, *12*, 6280. [CrossRef]
62. Chambers, L. Food texture and the satiety cascade. *Nutr. Bull.* **2016**, *41*, 277–282. [CrossRef]
63. Fardet, A.; Dupont, D.; Rioux, L.-E.; Turgeon, S.L. Influence of food structure on dairy protein, lipid and calcium bioavailability: A narrative review of evidence. *Crit. Rev. Food Sci. Nutr.* **2019**, *50*, 1987–2010. [CrossRef]
64. Grundy, M.M.; Lapsley, K.; Ellis, P. A review of the impact of processing on nutrient bioaccessibility and digestion of almonds. *Int. J. Food Sci. Technol.* **2016**, *51*, 1937–1946. [CrossRef] [PubMed]
65. Foster-Powell, K.; Miller, J.B. International tables of glycemic index. *Am. J. Clin. Nutr.* **1995**, *62*, 871S–890S. [CrossRef] [PubMed]
66. USDA Database for the Choline Content of Common Foods, Release 2 2008. Available online: https://data.nal.usda.gov/dataset/usda-database-choline-content-common-foods-release-2-2008 (accessed on 14 May 2021).
67. *FAO/INFOODS/IZiNCG Global Food Composition Database for Phytate Version 1.0. (PhyFoodComp1.0)—2018 User Guide*; FAO/IZiNCG: Rome, Italy, 2018.
68. Foster-Powell, K.; Holt, S.H.; Brand-Miller, J.C. International table of glycemic index and glycemic load values. *Am. J. Clin. Nutr.* **2002**, *76*, 5–56. [CrossRef] [PubMed]
69. Barba, F.J.; Tonello-Samson, C.; Puértolas, E.; Lavilla, M. (Eds.) *Present and Future of High Pressure Processing*; Elsevier: Amsterdam, The Netherlands, 2020. [CrossRef]
70. Nowosad, K.; Sujka, M.; Pankiewicz, U.; Kowalski, R. The application of PEF technology in food processing and human nutrition. *J. Food Sci. Technol.* **2021**, *58*, 397–411. [CrossRef] [PubMed]
71. Neves, F.I.G.; Silva, C.L.M.; Vieira, M.C. Combined pretreatments effects on zucchini (*Cucurbita pepo* L.) squash microbial load reduction. *Int. J. Food Microbiol.* **2019**, *305*, 108257. [CrossRef]
72. Cruz, R.M.S.; Vieira, M.C.; Silva, C.L.M. Effect of heat and thermosonication treatments on watercress (Nasturtium officinale) vitamin C degradation kinetics. *Innov. Food Sci. Emerg. Technol.* **2008**, *9*, 483–488. [CrossRef]
73. Vieira, M.C.; Teixeira, A.A.; Silva, C.L.M. Mathematical modelling of the thermal degradation kinetics of vitamin C in cupuaçu (Theobroma grandiflorum) nectar. *J. Food Eng.* **2000**, *43*, 1–7. [CrossRef]
74. Cruz, R.M.S.; Vieira, M.C.; Fonseca, S.C.; Silva, C.L.M. Impact of Thermal Blanching and Thermosonication Treatments on Watercress (Nasturtium officinale) Quality: Thermosonication Process Optimisation and Microstructure Evaluation. *Food Bioprocess Technol.* **2011**, *4*, 1197–1204. [CrossRef]
75. Raponi, F.; Moscetti, R.; Monarca, D.; Colantoni, A.; Massantini, R. Monitoring and Optimization of the Process of Drying Fruits and Vegetables Using Computer Vision: A Review. *Sustainability* **2017**, *9*, 2009. [CrossRef]

76. Lucarini, M.; Zuorro, A.; Di Lena, G.; Lavecchia, R.; Durazzo, A.; Benedetti, B.; Lombardi-Boccia, G. Sustainable Management of Secondary Raw Materials from the Marine Food-Chain: A Case-Study Perspective. *Sustainability* **2020**, *12*, 8997. [CrossRef]
77. Open Food Facts. Available online: https://world.openfoodfacts.org/ (accessed on 14 May 2021).
78. Średnicka-Tober, D.; Kazimierczak, R.; Rembialkowska, E. Organic food and human health—A review. *J. Res Appl. Agric. Eng.* **2015**, *60*, 102–107.
79. Fardet, A.; Rock, E. How to protect both health and food system sustainability? A holistic 'global health'-based approach via the 3V rule proposal. *Public Health Nutr.* **2020**, *23*, 3028–3044. [CrossRef] [PubMed]

nutrients

Article

An Integrated Control Plan in Primary Schools: Results of a Field Investigation on Nutritional and Hygienic Features in the Apulia Region (Southern Italy)

Vincenzo Marcotrigiano [1,*], Giacomo Domenico Stingi [1], Simona Fregnan [2], Pantaleo Magarelli [1], Pietro Pasquale [3], Samuele Russo [4], Giovanni Battista Orsi [5], Maria Teresa Montagna [6], Christian Napoli [4] and Christian Napoli [7]

1 Food Hygiene and Nutrition Service, Local Health Unit BT, Department of Prevention, 76125 Trani, Italy; giacomodomenico.stingi@aslbat.it (G.D.S.); pantaleo.magarelli@aslbat.it (P.M.)
2 Clinic for Diabetology and Endocrinology, Unique District, Local Health Unit BA, 70121 Bari, Italy; simona.fregnan@asl.bari.it
3 Apulia Region, Health and Wellness Promotion Section, Food Hygiene and Preventive Nutrition Organizational Structure, 70126 Bari, Italy; p.pasquale@regione.puglia.it
4 Department of Computer, Control, and Management Engineering, Sapienza University of Rome, 00185 Rome, Italy; samuele.russo@uniroma1.it (S.R.); cnapoli@diag.uniroma1.it (C.N.)
5 Department of Public Health and Infectious Diseases, Sapienza University of Rome, 00185 Rome, Italy; giovanni.orsi@uniroma1.it
6 Department of Biomedical Science and Human Oncology, University of Bari Aldo Moro, 70124 Bari, Italy; mariateresa.montagna@uniba.it
7 Department of Medical Surgical Sciences and Translational Medicine, Sapienza University of Rome, 00189 Rome, Italy; christian.napoli@uniroma1.it
* Correspondence: vincenzo.marcotrigiano@aslbat.it

Citation: Marcotrigiano, V.; Stingi, G.D.; Fregnan, S.; Magarelli, P.; Pasquale, P.; Russo, S.; Orsi, G.B.; Montagna, M.T.; Napoli, C.; Napoli, C. An Integrated Control Plan in Primary Schools: Results of a Field Investigation on Nutritional and Hygienic Features in the Apulia Region (Southern Italy). *Nutrients* **2021**, *13*, 3006. https://doi.org/10.3390/nu13093006

Academic Editors: Massimo Lucarini and Alessandra Durazzo

Received: 29 July 2021
Accepted: 25 August 2021
Published: 28 August 2021

Abstract: Data concerning overweight and obesity in children and adolescent populations are alarming and represent one of the most serious public health problems of our time. Moreover, it is demonstrated that the school environment may play an important role in health promotion with regard to nutritional aspects. This article reports the results of a study conducted in the Apulia region (Southern Italy), aimed at providing an integrated surveillance of the behaviors related to nutrition habits in students and the hygienic and nutritional conditions of the school's canteens attended by enrolled students. To this purpose, a sample of 501 students attending primary school (third class—children approximately eight years old) replied to a validated questionnaire, and official controls (OC), of both food and nutritional safety, were performed in 22 primary schools. A team of healthcare professionals carried out the study, and the implementation of all the prescribed improvement actions were subsequently verified through follow-up OC. The results of our study show a critical situation in the student sample, with 41.3% of children having a weight excess (overweight or obesity). With regard to the children's behaviors, only 59.8% of children ate at least one fruit or had a fruit juice for breakfast, and 10.8% did not have breakfast at all. Overall, 40.1% of the total children played outdoors the afternoon before the survey and 45% reported going to school on foot or by bicycle. During the afternoon, 83.5% of the sample watched television or used video games/tablets/mobile phones, while 42.3% played sports. The schools had an internal canteen with on-site preparation of meals in 36.4%, the remaining 63.6% received meals from external food establishments. With regard to OC, for the hygienic–sanitary section, eleven prescriptions were issued, in the great part related to the structure and organization of the canteen. For the nutritional section, nine corrective actions were prescribed, mainly related to official documents and management. The follow-up OC showed that all prescriptions were subsequently addressed. Eating at school was less frequent among obese and overweight students compared with those with normal weight. Although this evidence needs to be further confirmed, it highlights the potential role that the school canteens may play in health promotion and prevention of nutritional disorders. On the other hand, in order to fulfill its health promotion task, the school canteens have to comply with official regulations and guidelines; therefore, OC during the management of the food service at school are needed.

Keywords: nutritional knowledge; eight-year-old students; health promotion; official control; school canteen

1. Introduction

The spread of non-communicable diseases (NCDs), such as obesity, has become a paramount concern in the world health panorama. In order to limit the spread of such diseases, the World Health Organization (WHO) has identified different strategic areas of intervention for health promotion [1]; however, such actions first require a precise assessment of the state of health in the target population. In Italy, where obesity and overweight have been identified among the priority areas for intervention, a national surveillance project for nutritional status assessment and related factors in children at the third year of primary school started in 2008 [2]. The last survey was conducted in 2019 and showed that 20.4% of eight-year-old children were overweight, while 9.4% were affected by pathological obesity [3]. In particular, males were more affected, especially in the southern regions of Italy, as well as in families with a disadvantaged economic status [3].

The selected age group (eight-year-old) is considered a main target for prevention; in fact, the choices, attitudes, and behaviors adopted at this age have life-long repercussions and can be further amplified by other factors, especially in the adolescent period [4]. An international study conducted in 2018, investigated the Italian population aged 11, 13, and 15 years. This latter study highlighted that 16.6% of the subjects of this group were overweight, while 3.2% were obese. Overweight and obesity seem to affect males more often than females, and there is a confirmed higher prevalence in the southern regions than in the central–northern ones [4]. These data are also in line with the nutritional status among university students: a study in southern Italy reported a 17.8% and 3.4% prevalence of overweight and obese students, respectively [5].

Comparison with international data shows that weight excess is more common in Italian youth (in all age groups) with respect to the European average [6]. In addition, the COVID-19 pandemic also worsened this situation, aggravating lifestyle habits, limiting physical activities, and creating psychological distress [7,8].

It is also demonstrated that nutritional status in school-aged children is related to unhealthy behaviors (such as skipping breakfast, low intake of fruits and vegetables, and unhealthy snacking) and insufficient physical activity, which may be reflected in cognitive performance too [9,10]. In Italy, with regard to these aspects, the last national survey on nutritional aspects showed that 44.3% of children skip or have an inadequate breakfast, 24.3% do not eat fruit or vegetables daily, unhealthy snacks are consumed by 48.3%, and 20.3% do not practice any physical activity [3]. At the same time, the increasing prevalence of childhood overweight and obesity is often related to school environment features [9].

Children spend six or more hours per day at school for more than six months per year; therefore, an education intervention focused on improving nutrition knowledge, attitudes, and practices among primary school children must also involve the food quality available at school canteens. Health promotion interventions for school canteens, aimed at offering healthier food, seem promising and effective in improving the school food environment [11].

In light of these data, the Apulia region has identified schools to be one of the most important settings for health promotion, also from a nutritional point of view. Therefore, starting from the forecasts of the Regional Prevention Plan (PRP) 2010–2013 with the "School in Health" program, a "Regional Strategic Plan for Health Promotion in Schools: Catalogue" is also periodically issued [12,13]. Subsequently, the 2014–2018 PRP further focused its attention on the need to also provide guidelines on nutritional safety [14]. Therefore, the Regional Decree no. 1435 of 2 August 2018 has drawn up the "Guidelines for School and Corporate Catering" to ensure the adoption of correct eating habits for health promotion and disease prevention, fighting against an incorrect diet [15].

Starting from the new regional guidelines, this article shows the results of an integrated project aiming to (i) evaluate anthropometric data and behaviors in terms of nutrition habits in a sample of students attending primary school (third class—children approximately eight years old); (ii) carry out, in the same schools, the official controls (OC) in terms of both hygiene conditions and nutritional normative aspects foreseen by the DGR no. 1435 of 2 August 2018; and (iii) verify, through follow-up OC, the implementation of all the prescribed improvement actions.

2. Materials and Methods

The study was conducted in the Province BAT during the 2018–2019 school year (September 2018–June 2019) by the Food Hygiene and Nutrition Service (SIAN) of the Prevention Department of the Local Health Unit (LHU). The investigation was performed in accordance with the World Medical Association Declaration of Helsinki and did not include any experiments involving human or biological human samples, nor research on identifiable human data. Regardless, the study protocol was approved, also with regard to ethical issues, by the Regional Sanitary Authority (approval no. 238_2018).

2.1. Behaviors in Terms of Nutrition and Anthropometric Data

A reduced and adapted version of the questionnaire used during the 2016 national surveillance study was used in order to evaluate behavior with respect to nutrition habits, as well as to evaluate anthropometric data, in a sample of students attending the third year of primary school (children 7 to 9 years of age) [16,17]. The body mass index (BMI) was calculated and subsequently expressed as a percentile obtained with respect to age-and sex-related growing reference data provided by WHO for the European region [18]. In order to classify children in terms of weight status categories, the reference percentile ranges provided by the US Centers for Disease Control and Prevention (CDC) were used [19]: underweight (less than the 5th percentile); normal or healthy weight (from the 5th percentile to less than the 85th percentile); overweight (from the 85th to less than the 95th percentile); obese (equal to or greater than the 95th percentile). Additional factors such as general information on physical activity and use of mobile devices were also investigated. The proposed version of the questionnaire was validated during a pilot study in a sample of 100 eight-year-old students (data not reported or included in the study). In the same pilot sample, the reliability index was assessed using Cronbach's alpha (internal consistency coefficient): the alpha value achieved was 0.79, showing a satisfactory level of reliability [20]. Moreover, the modified version of the questionnaire was also validated during the pilot study in terms of intelligibility: the students were asked to assign a score to each item of the questionnaire on a 7-point-scale (from 1: not meaningful to 7: very meaningful). Moreover, in order to guarantee variability in the answers, the original questionnaire was modified in the pilot version: 12 further items (FI) reporting errors (grammatical and/or semantic) were added to the items (OI) belonging to the original questionnaire. OI reported a mean score for each question >6 (almost the maximum); FI showed a mean score ≤ 1. These data confirmed that the content of the questionnaire was clear to the readers.

The population investigated was represented by eight-year-old children attending the third grade of primary school, selected through a cluster survey design, using the class as a sampling unit, a method also recommended by the WHO and widely used in international surveys [21,22]. First of all, the study was described to the enrolled children with regard to the objectives of the study and the instructions for survey completion. Additionally, parents were fully informed and provided their consent for study enrollment.

The questionnaire was composed of two macro-sections:

(a) Anthropometric form, in which a child's weight and height were reported; measurement was performed at school by LHU healthcare professionals, previously calibrated. Weight was measured to the nearest 0.1 kg by an electronic balance. Children's height was measured to the nearest 0.1 cm by a precision stadiometer;

(b) Children's questionnaire. In the classroom, the children themselves completed the questionnaire, which was divided into three subsections asking:

First section referring to the day of compilation:

1. If they had breakfast;
2. If they ate at least one fruit or drank fruit juice for breakfast;
3. If they watched TV before going to school;
4. If they went to school on foot or by bicycle;
5. If they had a snack at school;
6. If they have eaten at least one fruit or fruit juice as a snack;
7. If they eat lunch in the school canteen.

Second section referred to the previous afternoon:

8. If they played video games, computers, tablets, or mobile phones;
9. If they watched TV programs;
10. If they played outdoors;
11. If they played sports.

Third section referring to the previous evening:

12. If they played video games, computers, tablets, or mobile phones after dinner;
13. If they watched TV after dinner;
14. If they brush their teeth after dinner.

Since the whole reference population included 3620 students, a sample of at least 348 individuals would have been required to investigate the selected variables, assuming a response proportion of 50%, a 95% confidence level, and a 5% margin of error.

The nutritional status in terms of obese/overweight and normal/underweight was compared between the two sex groups using the chi-squared test. A $p \leq 0.05$ was considered statistically significant.

In order to evaluate the association between variables, a standard binary correlation matrix was used, showing correlation coefficients between variables (a score of 1 is the maximum association coefficient). A table was built in which each cell of the adopted matrix (crossing an abscissa and ordinate value) describes the level of correlation between the variables in abscissa and the variable in ordinate expressed as the Pearson's correlation index, which is the ratio between the covariance and the standard deviations of each pair of variables.

2.2. Official Controls (OC) in Terms of Both Hygiene Conditions and Nutritional Guidelines

Parallel to the investigations directly conducted on the children, OC were carried out in school canteens. These OC were aimed at verifying both the hygienic–sanitary conditions and nutritional rules compliance. The controls were performed by a direct inspection of the canteens as well as by inspection of documentation completeness and maintenance [23,24].

The OC were performed by a team of official inspectors of the LHU: one medical doctor expert in human nutrition, one medical doctor expert in public health, one dietician, and one environmental health officer. The inspection team had to file a standardized report with one checklist, divided in two sections, provided by the Integrated Regional Control Plan to assess the elements of structural, procedural, and managerial compliance in the field of food safety in school catering (cooking centers, canteens with on-site preparation, and school refectories) [25]. The inspection was aimed at evaluating compliance with the requirements foreseen by law. In detail:

Hygienic–sanitary section:

1. Presence of accurate documentation held by the Food Business Operator (FBO) with respect to the activities carried out;
2. Presence of an adequate Hazard Analysis and Critical Control Points plan (HACCP);
3. Respect of status and hygienic conditions of systems, equipment, tools, premises, and structures;

4. Presence of raw materials, ingredients, and any other product intended for consumption;
5. Presence of semi-finished products, finished products and materials, and objects intended to come into contact with food;
6. Presence of disinfection procedures, ordinary and extraordinary cleaning, and maintenance;
7. Presence of production technological processes and transformation of food products;
8. Labeling, food products presentation, and preservation means, with particular attention to substances that cause food allergies and intolerances;
9. Previous non-compliance and corrective actions adopted;
10. Compliance with the regional guidelines for school and company catering.

As required by the guidelines for food safety OC, findings were recorded according to a four-point-scale that took into account full compliance (YES), partial compliance (yes), inadequacy/minor non-compliance (no) and major non-compliance (NO) [26].

Nutritional section:

1. Presence of tender specifications, canteen committee, and a plan for users who have allergies, intolerances, and/or adopt ethical/religious diets or diets adopted for non-health reasons;
2. Presence of a food safety training for kitchen and administration staff;
3. Presence of a nutritional table, menu validation, and correspondence between meals scheduled on the inspection day and the foods actually prepared; presence of allowed frozen or deep-frozen foods intended for preparations;
4. Presence of organic food and ingredients coming from a short supply chain;
5. Presence of IV or V range and/or canned foods;
6. Use of extra virgin olive oil and iodized salt;
7. Single- or multi-portion meal packaging;
8. Presence of a standardized plan for carrying out any food transport from external food establishments and transport time.

Regarding this section, the answers provided were dichotomous (yes/no).

2.3. Implementation of All the Prescribed Improvement Actions

Following the first cycle of OC and the actions and measures prescribed by the inspection team in case of non-compliance to official rules, follow-up OC were carried out to verify the implementation of corrective measures addressing the prescriptions given.

3. Results

3.1. Behaviors in Terms of Nutrition and Anthropometric Data

Overall, 26 classes from 22 different schools distributed across the provincial territory were enrolled. Of the 563 students attending the 26 classes enrolled, 501 completed the questionnaire (89%): 47.5% were male and 52.5% female, with an average age of 8.3 years (range 7–9 years). Weight status categories by sex are reported in Table 1.

Table 1. Weight status categories by sex.

Range of BMI * Percentile	Total (*n*/%)	Female (*n*/%)	Male (*n*/%)
>95th percentile (obese)	76 (15.2%)	37 (14.1%)	39 (16.4%)
85th to 95th percentile (overweight)	131 (26.1%)	70 (26.6%)	61 (25.6%)
5th to 85th percentile (normal weight)	290 (57.9%)	153 (58.2%)	137 (57.6%)
<5th percentile (underweight)	4 (0.8%)	3 (1.1%)	1 (0.4%)

* BMI: body mass index.

Overall, 41.3% of children had a weight excess, which included both overweight and obesity. With regard to sex, males showed a lower percentage of overweight and normal weight but a higher percentage of obesity. Nevertheless, the statistical analysis did not show a significant association between sex and nutritional status.

In our sample, only 59.8% of children ate a qualitatively adequate breakfast that included at least one fruit or a fruit juice, and 10.8% did not have breakfast at all. Only 23.7% of the children consumed an adequate mid-morning snack, which included a fruit or fruit juice. Most children took an inadequate snack, and 2.8% did not consume one; 40.1% of the total children played outdoors the afternoon before the survey. Overall, 45% of children, on the morning of the survey, reported that they went to school on foot or by bicycle; conversely, 55% used a public or private means of transport. In our sample, 59.6% of children watched TV in the morning before going to school. Overall, 83.5% and 83.7% of children watched television or used video games/tablets/mobile phones the afternoon and the evening of the previous day, respectively. In addition, 71.5% of the children reported that they had brushed their teeth the evening before the survey. Answers to the fourteen questions are presented in Figure 1. Table 2 reports the results of the answers to the items of the questionnaire, stratified by nutritional status of enrolled children.

Figure 1. Answers distribution in the sample of students. The answer "NO" is represented in green, while the answer "YES" is represented in yellow.

Table 2. Answers provided by students, stratified by nutritional status.

Question	Nutritional Status/ Response to the Questions	Total		Obese (BMI > 95th Percentile) (% of 501 People)		Overweight (BMI between 95th and 85th Percentile) (% of 501 People)		Normal Weight (BMI between 85th and 5th Percentile) (% of 501 People)		Underweight (BMI < 5th Percentile) (% of 501 People)	
		Yes	No	Yes	No	Yes	No	Yes	No	Yes	No
	First section referring to the day of compilation										
(1)	they had breakfast	447 (89.2%)	54 (10.8%)	68 (13.6%)	8 (1.6%)	117 (23.3%)	14 (2.8%)	258 (51.5%)	32 (6.4%)	4 (0.8%)	0 (0.0%)
(2)	they ate at least one fruit or juice for breakfast	300 (59.8%)	201 (40.2%)	36 (7.2%)	40 (8.0%)	66 (13.1%)	65 (13.0%)	196 (39.1%)	94 (18.8%)	2 (0.4%)	2 (0.4%)

Table 2. *Cont.*

Question	Nutritional Status/ Response to the Questions	Total		Obese (BMI > 95th Percentile) (% of 501 People)		Overweight (BMI between 95th and 85th Percentile) (% of 501 People)		Normal Weight (BMI between 85th and 5th Percentile) (% of 501 People)		Underweight (BMI < 5th Percentile) (% of 501 People)	
		Yes	No	Yes	No	Yes	No	Yes	No	Yes	No
(3)	before going to school they watched TV	299 (59.6%)	202 (40.4%)	45 (9.0%)	31 (6.2%)	78 (15.6%)	53 (10.6%)	174 (34.7%)	116 (23.1%)	2 (0.4%)	2 (0.4%)
(4)	they went to school on foot or by bicycle	225 (45%)	276 (55%)	12 (2.4%)	64 (12.8%)	36 (7.2%)	95 (19.0%)	175 (34.9%)	115 (22.9%)	2 (0.4%)	2 (0.4%)
(5)	they had a snack at school	487 (97.2%)	14 (2.8%)	73 (14.6%)	3 (0.6%)	127 (25.3%)	4 (0.8%)	283 (56.5%)	7 (1.4%)	4 (0.8%)	0 (0.0%)
(6)	they ate fruit or juice as a snack	119 (23.7%)	382 (76.3%)	3 (0.6%)	73 (14.6%)	8 (1.6%)	123 (24.5%)	107 (21.4%)	183 (36.5%)	1 (0.2%)	31, (0.6%)
(7)	they eat lunch in the school canteen	225 (45%)	276 (55%)	18 (3.6%)	58 (11.6%)	40 (8.0%)	91 (18.1%)	165 (32.9%)	125 (25.0%)	2 (0.4%)	2 (0.4%)
	Second section referring to the previous afternoon:										
(8)	they played video games, computers, tablets, or mobile phones	201 (40.1%)	300 (59.9%)	31 (6.2%)	45 (9.0%)	53 (10.6%)	78 (15.6%)	115 (22.9%)	175 (34.9%)	2 (0.4%)	2 (0.4%)
(9)	they watched a program on TV	217 (43.4%)	284 (56.6%)	29 (5.8%)	47 (9.4%)	64 (12.8%)	67 (13.4%)	122 (24.3%)	168 (33.5%)	2 (0.4%)	2 (0.4%)
(10)	they played outdoors	201 (40.1%)	300 (59.9%)	12 (2.4%)	64 (12.8%)	44 (8.8%)	87 (17.4%)	143 (28.5%)	147 (29.3%)	2 (0.4%)	2 (0.4%)
(11)	they played sports	212 (42.3%)	289 (57.7%)	7 (1.4%)	69 (13.8%)	31 (6.2%)	100 (20.0%)	172 (34.3%)	118 (23.5%)	2 (0.4%)	2 (0.4%)
	Third section referring to the previous evening:										
(12)	after dinner they played video games, computers, tablets, or mobile phones	222 (44.4%)	279 (55.6%)	36 (7.2%)	40 (8.0%)	64 (12.8%)	67 (13.4%)	120 (23.9%)	170 (33.9%)	2 (0.4%)	2 (0.4%)
(13)	after dinner they watched TV	197 (39.3%)	304 (60.7%)	27 (5.4%)	49 (9.8%)	50 (10.0%)	81 (16.2%)	118 (23.5%)	172 (34.3%)	2 (0.4%)	2 (0.4%)
(14)	after dinner they brushed their teeth	358 (71.5%)	143 (28.5%)	50 (10.0%)	26 (5.2%)	85 (16.9%)	46 (9.2%)	220 (43.9%)	70 (14.0%)	3 (0.6%)	1 (0.2%)

Figure 2 shows the correlation between the ratio of positive answers to the different questions in order to summarize data, and it gives a hint at possible clusters of co-occurring answers to groups of questions. Large values (close to 1) in this matrix indicate possible collinearity between the variables involved. Yellow cells show the maximum level of association between questions.

Our analysis suggests a mild correlation between the different children's habits and lifestyles: the association between affirmative answers such as eating lunch at the school canteen correlates with the positive answers of playing sport, but it also correlates with playing video games, computers, tablets, or mobile phones and with watching TV programs. It also emerged that the morning breakfast with fruits or juices was done while watching TV. On the other hand, it is interesting to notice that "playing video games, computers, tablets, or mobile phones" were habits not associated with "going to school on foot or by bicycle", analogously "play sport or play sport outdoor" was not associated with "going to school on foot or by bicycle". It came to light that healthy family habits and proper lifestyles, according to our analytic model, may play an essential role in children's health status.

In Figure 3 is reported the correlation matrix between answers compatible with a healthier lifestyle and nutritional status. A yellowish color shows a possibly unhealthier lifestyle. As it is notable, the large part of the yellower cells is distributed among obese and overweight subclasses: in particular, unhealthy habits were registered with regard to question nos. 4, 6, 7, 10, 11, and 14 in the obese/overweight classes. On the contrary, the greener cells, indicating healthy habits, are more represented in the subclass of those with normal weight.

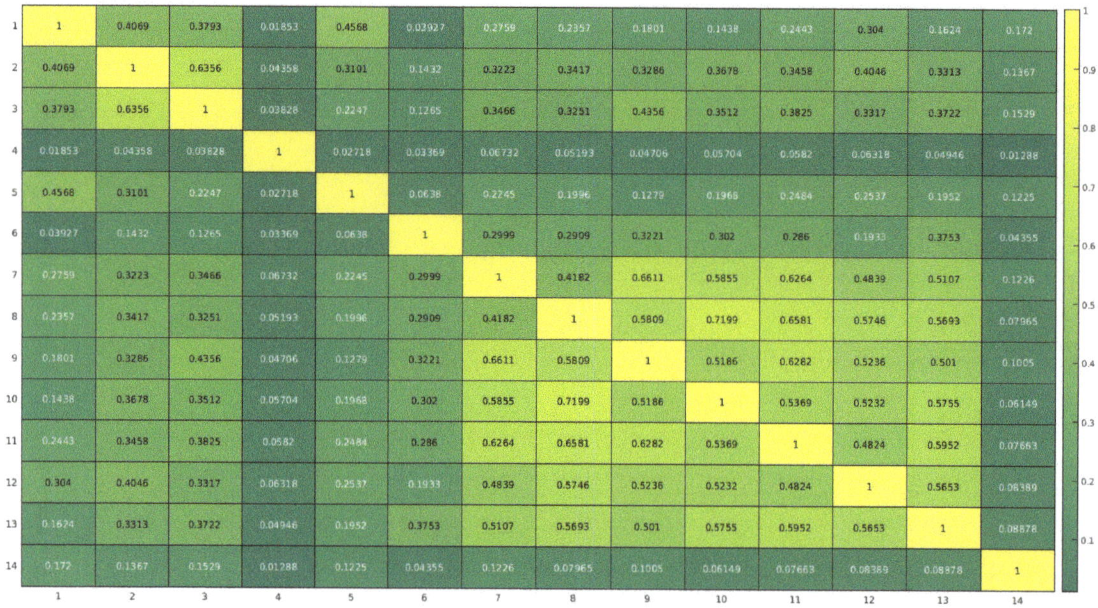

Figure 2. Correlation matrix showing the association between the answers to the questions numbered progressively in Table 2.

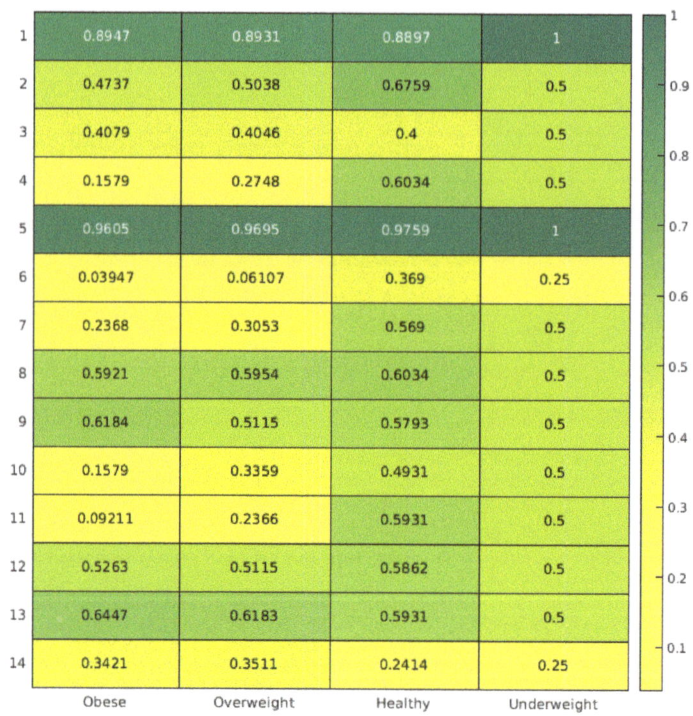

Figure 3. Correlation matrix showing the association between answers to the questions and nutritional status.

3.2. Official Controls (OC) in Terms of Both Hygiene Conditions and Nutritional Guidelines

Hygienic–sanitary and nutritional OC were carried out in the same 22 school complexes and 26 classes where students were enrolled for goal 1. Of the 22 schools, in 8 (36.4%) school canteens, meals were prepared on site, while 14 (63.6%) were managed by a specialized private company, and meals came from an external cooking center. A general compliance with the minimum requirements in terms of food safety was verified; however, some non-conformities were found in eight facilities. Some inadequacies were found only regarding the wear-out of some work-tops, the presence of mosquito nets that were not properly maintained, and the absence of lockers divided into dirty and clean compartments. Moreover, workspace areas were quite small with respect to the requirements (given for by the regional guidelines for school catering). In such cases, the Competent Authority issued mandatory provisions to the FBOs (Table 3—Hygienic–sanitary section).

Table 3. OC—Hygienic–sanitary section (classified in 4 levels).

		YES	Yes	No	NO
(a)	congruity of documentation	22 (100%)	-	-	-
(b)	HACCP	21 (95%)	-	1 (5%)	-
(c)	hygienic conditions	20 (91%)	-	2 (9%)	-
(d)	raw materials and ingredients	22 (100%)	-	-	-
(e)	finished products and food contact materials	22 (100%)	-	-	-
(f)	disinfection, cleaning, and maintenance	17 (77%)	-	5 (23%)	-
(g)	production processes	22 (100%)	-	-	-
(h)	labeling	22 (100%)	-	-	-
(i)	resolution of previous non-conformities	22 (100%)	-	-	-
(j)	compliance regional guidelines for school catering	19 (86%)	-	3 (14%)	-

A general compliance with the minimum requirements in terms of nutritional features was also confirmed, but in 27% of the enrolled facilities, an in-depth documentary investigation concerning the nutritional requirements of the menus proposed for special and non-special diets was needed. Furthermore, in 9% of cases, inconsistencies were found between the meals scheduled on the day of the OC and the foods actually prepared, due to the difficulty in finding on market and from retail sources the ingredients needed by the menu. The sample investigated also showed that 86% of FBOs have a specific certification of attendance to training courses on special diets, with particular reference to the methods of preparing and administering meals for people affected by celiac disease (Table 4—Nutritional section).

Table 4. OC—Nutritional section (classified in 2 levels).

		Compliant	Not Compliant
(a1)	tender specifications	22 (100%)	-
(a2)	canteen commission	22 (100%)	-
(a3)	plan for users with food allergies/intolerances/ethical–religious diets	22 (100%)	-
(b)	food safety training	19 (86%)	3 (14%)
(c1)	presence of nutritional table and menu validation	16 (73%)	4 (27%)
(c2)	correspondence of the meals scheduled and foods prepared	20 (91%)	2 (9%)
(c3)	presence of allowed frozen or deep-frozen ingredients	22 (100%)	-
(d)	presence of organic food/short supply chain	22 (100%)	-
(e)	presence of IV or V range and/or canned foods	22 (100%)	-
(f)	use of extra virgin olive oil and iodized salt	22 (100%)	-
(g)	single- or multi-portion packaging of the meal	22 (100%)	-
(h)	plan for external transport (if applicable)	14 (100%)	-

In total, for the hygienic–sanitary section, the number of prescriptions issued were as follows: documentary 1 (5%), structural 5 (23%), management 2 (9%), and organizational 3 (14%).

For the nutritional section, the number of actions that were taken to ensure full compliance with the sector regulations were, in particular: documentary 4 (27%), managerial 3 (14%), and organizational 2 (9%).

With regard to both nutritional and hygienic evaluations, no differences were reported between schools having internal or external services.

3.3. Implementation of All the Prescribed Improvement Actions

To allow the verification of compliance with the given prescriptions, eight follow-up OC were carried out during the four months following the first round of OC, aiming at verifying the resolution of 20 inadequacies and minor non-conformities previously detected in 8 of 22 schools enrolled. Despite some difficulties for FBOs related to finding the economic resources to be allocated to extraordinary maintenance interventions—which made it necessary to grant some time extensions to complete the execution of the works that involved structures or to encourage the purchase of materials—full compliance with the requirements and the checklists were concluded favorably during the scholastic year under analysis.

4. Discussion

The results of our study highlight a critical situation in our sample of eight-year-old students with regard to the percentages of overweight (26.1%) and obese (15.2%), which were greater than those detected at the national and regional level [3]. As a matter of fact, according to Italian surveillance, the national and regional levels of pathological obesity are 9.4% and 15.1%, respectively. The overweight percentages do not offer better data: 20.4% and 21.6% at national and regional level, respectively [3]. With regard to sex, although the differences among nutritional status are non-significant, in our study males showed a lower percentage of overweight vs females at the local level as well as at regional level, and males showed a higher percentage of obesity at the local level as well as at national level [3].

It has been confirmed that data concerning the population of children and adolescents with respect to obesity and overweight are alarming and represent one of the most serious public health problems of our time [27]. This situation is correlated with bad habits, such as the consumption of processed foods rich in simple sugars and fats, and with high calorie diets associated with a sedentary lifestyle and with the growth of mechanized transport, urbanization, and information technology [5,7,28]. With regard to incorrect habits, our sample, compared with national population, showed a high level of skipping breakfast (10.8% vs. 8.7%), low intake of fruits and vegetables (40.2% vs. 24.3%), and unhealthy snacking (76.3% vs. 55.2%) [3]. The latter habits have been associated not only with bad nutritional status but also with low cognitive performance [9]. With regard to physical activities, 55% of the sample went to school by walking or cycling (compared with 26.6% at the national level), and 40.1% in the afternoon watch TV or play videogames/tablet/cellphones (compared with 44.5% at the national level). Although the local situations seem better than those at national level, unhealthy habits were still frequent in our sample and need attention.

In our sample, healthy lifestyles and correct food habits were not always correlated, and also, while the consumption of fruit for breakfast was considerable, this habit was not related to playing sports or other activities outdoors. This is a very important aspect since a low level of physical activity in young students leads to a reduction in physical activity/sport practice experienced adults, highlighting the necessity of promoting sports in this school-age period of life [8]. Therefore, the implementation of targeted interventions of education and health promotion in primary schools can undoubtedly favor the spread of healthy habits, which represents, especially in children, a useful investment in the prevention of the development of NCDs.

With regard to the correlation matrix in Figure 3, it confirms that unhealthy habits are more common among students who are obese or overweight. However, it is particularly interesting to underline that eating at school was more common among students with normal weight and, on the contrary, that obese and overweight subclasses were associated with students who did not eat at school. Although this evidence needs to be studied in depth, it highlights the potential role that the school environments may play in health promotion to prevent nutritional disorders [9]. Consuming a nutritionally correct meal, one that is adequate to the needs of children and adolescents in the school context, may represent a qualitative and quantitative guarantee with respect to the energy needs of this target population: often the school canteen is the only time when the meal consumed meets the macro- and micronutrient needs of children. More studies are needed in order to further analyze and to set the canteen menus to the energy expenditure of children.

At the same time, in order to fulfill its health promotion task, the school canteen has to completely respect the official rules; therefore, OC during the management of food service at school are needed. An adequate and more effective OC planning could contribute to achieving better results in terms of the capabilities of the inspections performed and preventive interventions adopted, especially in school environments.

Previous data report that 58.9% of the schools have their own internal school canteen, and in 52% of cases, the menu drafting is carried out by LHU dieticians; in the remaining 48%, the menu is edited by external professionals [16]. In our experience, in only 36.4% of the cases are meals prepared within the school, and this aspect has pros and cons: the presence of an internal canteen favors meals that are produced on site and immediately served, which guarantees the organoleptic qualities, consistency, and minimal alteration of foods [15]. On the other hand, the external cooking centers, managed by large companies, guarantee standardized procedural aspects, but it is necessary to consider that the transport phase in food delivery bags has a few critical points (e.g., with respect to hot or cold chains) [15]. From our OC on food hygiene and nutritional safety, although there is substantial compliance with the regulatory requirements, some prescriptions aimed at conforming structural aspects were issued and, in some cases, it was necessary to investigate specific nutritional aspects at later stages. No difference was reported between schools having an internal or external food service. The constant and targeted control system for this type of activity is able to detect substantial and formal deficiencies and promote timely corrective actions, even potentially related to reducing the risks of foodborne diseases. Moreover, in our study, all the registered non-conformities were solved during the same scholastic year. This demonstrates that it is possible to obtain full compliance to the rules of law only by constant monitoring. The current local organization of OC can allow a single access made by a team of different healthcare professionals (medical doctor, dietician, environmental health officers, food technologist, etc.), each with a different training background, useful for creating favorable synergies with FBOs and for improving verification in the field, assessing both nutritional and hygienic–sanitary aspects jointly. To be completely efficient, these OC should also foresee laboratory test of environmental matrices and food, such as is done in other human environments [29].

To our knowledge, this is the first study reporting integrated data on children's nutritional habits and OC in school canteens, which jointly investigated many aspects by different healthcare professional in a single inspection, an approach that is favored by regional reference laws, which are innovative in this regard [15,23].

The authors are aware of some limitations of the study. First, lifestyles were not investigated in depth, in order to avoid an excessive length of the questionnaire, the compilation of which could have favored rejection. This could have hidden important information, such as the children's energy expenditure, as well as other sociodemographic variables that were not collected. Furthermore, this study was targeted to a sample of students attending schools that are not representative of the whole population of students in Italy. Therefore, our study can be considered as preliminary research. Due to the limitation in representativeness, further studies are needed to deepen the investigation in this subpopulation.

Furthermore, the study was conducted before the COVID-19 pandemic. Different studies have demonstrated that sedentary behaviors increased and that all physical activities decreased significantly during the lockdown [8]. Therefore, incorrect habits highlighted during the present study can be further worsened by the preventive measures adopted in response to the pandemic emergency. Since promoting physical activity during non-pandemic periods may also have positive effects in case of a lockdown [8], greater emphasis will have to be given to school interventions to promote healthy lifestyles, including those associated with OC conducted in school contexts. Finally, it should be noted that, in contrast with infectious diseases—where surveillance systems are already implemented for both health risk assessment and early detection also in critical situations [30,31]—continuous surveillance systems for risk factors of NCDs, such as overweight and physical inactivity, are very difficult to implement and maintain.

5. Conclusions

This study reports some critical issues regarding nutritional status and habits in eight-year-old students and some features of school canteen services. It is noteworthy to underline how eating at school was less frequent among obese and overweight students compared with those with normal weight. Although this evidence needs to be further confirmed, it highlights the contribution that the school canteens may provide for health promotion and prevention of nutritional disorders. On the other side, in order to fulfill its health promotion task, school canteens have to comply with official regulations and guidelines at every step of food chain; therefore, OC on school food management services are needed.

Data obtained from the present study may be useful in developing and implementing effective policies able to integrate nutrition education and OC for a healthier school environment.

Author Contributions: Conceptualization, V.M., G.D.S., and C.N. (Christian Napoli 7); methodology, G.D.S., P.M., S.F. and P.P.; software, validation, and formal analysis, V.M., S.R., C.N. (Christian Napoli 4) and C.N. (Christian Napoli 7); investigation, S.F., G.D.S. and P.M.; data curation, V.M., G.D.S., S.R., C.N. (Christian Napoli 4) and C.N. (Christian Napoli 7); writing—original draft preparation, V.M. and C.N. (Christian Napoli 7); writing—review and editing, G.B.O., M.T.M., P.P., V.M., G.D.S., C.N. (Christian Napoli 4) and C.N. (Christian Napoli 7); supervision, G.D.S., P.M., P.P. and M.T.M.; project administration, G.D.S., P.M. and S.F. All authors have read and agreed to the published version of the manuscript.

Funding: This research received no external funding.

Institutional Review Board Statement: Not applicable.

Informed Consent Statement: Not applicable.

Data Availability Statement: The data presented in this study are available on request from the corresponding author. The data are not publicly available due to privacy and ethical reasons.

Conflicts of Interest: The authors declare no conflict of interest.

References

1. Carinci, F. Implementation of the Tallinn Charter in the WHO European region: Where is the evidence? A systematic comparison of online information before and after the end of the year of signature. *Health Policy* **2012**, *104*, 215–221. [CrossRef] [PubMed]
2. Okkio alla Salute—Indagine 2008. Available online: https://www.epicentro.iss.it/okkioallasalute/indagine-2008 (accessed on 18 August 2021).
3. Okkio alla Salute—Indagine 2019. Available online: https://www.epicentro.iss.it/okkioallasalute/indagine-2019 (accessed on 18 August 2021).
4. La Sorveglianza HBSC 2018—Health Behaviour in School-Aged Children: Risultati dello Studio Italiano tra i Ragazzi di 11, 13 e 15 Anni. Available online: https://www.epicentro.iss.it/hbsc/indagine-2014 (accessed on 31 January 2021).
5. Gallè, F.; Sabella, E.A.; Da Molin, G.; Liguori, G.; Montagna, M.T.; Orsi, G.B.; Valerio, G.; Napoli, C. A cross-sectional study investigating lifestyle and weight perception of undergraduate students in southern Italy. *BMC Public Health* **2019**, *19*, 1316. [CrossRef] [PubMed]
6. Spotlight on Adolescent Health and Well-Being. Volume 2. Key Data. Available online: https://www.euro.who.int/en/health-topics/Life-stages/child-and-adolescent-health/health-behaviour-in-school-aged-children-hbsc/hbsc-2020 (accessed on 31 January 2021).

7. Gallè, F.; Sabella, E.A.; Ferracuti, S.; De Giglio, O.; Caggiano, G.; Protano, C.; Valeriani, F.; Parisi, E.A.; Valerio, G.; Liguori, G.; et al. Sedentary Behaviors and Physical Activity of Italian Undergraduate Students during Lockdown at the Time of COVID-19 Pandemic. *Int. J. Environ. Res. Public Health* **2020**, *17*, 6171. [CrossRef] [PubMed]
8. Roma, P.; Monaro, M.; Colasanti, M.; Ricci, E.; Biondi, S.; Di Domenico, A.; Verrocchio, M.C.; Napoli, C.; Ferracuti, S.; Mazza, C. A 2-Month Follow-Up Study of Psychological Distress among Italian People during the COVID-19 Lockdown. *Int. J. Environ. Res. Public Health* **2020**, *17*, 8180. [CrossRef] [PubMed]
9. Teo, C.H.; Chin, Y.S.; Lim, P.Y.; Masrom, S.A.H.; Shariff, Z.M. Impacts of a School-Based Intervention That Incorporates Nutrition Education and a Supportive Healthy School Canteen Environment among Primary School Children in Malaysia. *Nutrients* **2021**, *13*, 1712. [CrossRef]
10. Hayashi, C.; Krasevec, J.; Kumapley, R.; Mehra, V.; de Onis, M.; Borghi, E. Levels and Trends in Child Malnutrition. UNICEF/WHO/World Bank Group Joint Child Malnutrition Estimates: Key Findings of the 2017 Edition. 2017. Available online: https://www.who.int/nutgrowthdb/2018-jme-brochure.pdf (accessed on 18 August 2021).
11. Mensink, F.; Schwinghammer, S.A.; Smeets, A. The Healthy School Canteen programme: A promising intervention to make the school food environment healthier. *J. Environ. Public Health* **2012**, *2012*, 415746. [CrossRef] [PubMed]
12. Deliberazione Giunta Regionale Puglia 28 Dicembre 2010, n. 2994. Atto di Programmazione. Intesa Stato Regione del 29 Aprile 2010. Piano Regionale della Prevenzione 2010–2012. Approvazione. Available online: http://burp.regione.puglia.it/documents/10192/4968866/N13_25_01_11.pdf/b9c3b4eb-d6dd-4232-ab17-cab45158afae (accessed on 22 August 2021).
13. Deliberazione Giunta Regionale Puglia 1 luglio 2013, n. 1196. Accordo Stato Regione del 07/02/2013. Rep. Atti 53/CSR. Approvazione "Piano Regionale della Prevenzione 2013". Available online: http://burp.regione.puglia.it/documents/10192/4885056/DELIBERAZIONE+DELLA+GIUNTA+REGIONALE+1+luglio+2013%2C%20n.+1196+%28id+4885094%29/c4991f82-6882-462d-9cee-e75db9ebd6ff (accessed on 22 August 2021).
14. Deliberazione Giunta Regionale Puglia 22 marzo 2016, n. 302. Intesa Stato-Regioni 13/11/2014, rep. Atti 156/CSR. Piano Regionale della Prevenzione 2014–2018 adottato con DGR n. 1209 del 27/05/2015. Integrazioni e Modifiche. Approvazione. Available online: http://burp.regione.puglia.it/documents/10192/6130777/DGR302.pdf/749d9828-26e1-4037-9841-eb77d444448d (accessed on 23 August 2021).
15. Deliberazione Giunta Regionale Puglia 2 Agosto 2018, n. 1435. Linee di Indirizzo Regionali per la Ristorazione Scolastica ed Aziendale. Aggiornamento alla DGR n. 276 del 19 marzo 2002. Available online: http://burp.regione.puglia.it/documents/10192/31452346/DEL_1435_2018.pdf/d7a1363e-3350-41c8-a5a1-d5e4783f1bd5 (accessed on 23 August 2021).
16. Okkio alla Salute—Indagine 2016. Available online: https://www.epicentro.iss.it/okkioallasalute/indagine-2016 (accessed on 18 August 2021).
17. Nardone, P.; Spinelli, A.; Buoncristiano, M.; Lauria, L.; Pierannunzio, D.; Galeone, D. Surveillance System OKkio alla Salute: 2016 Results. Istituto Superiore di Sanità. 2018. Available online: https://www.epicentro.iss.it/okkioallasalute/pdf/ONLINE_OKKIO_ALLA_SALUTE.pdf (accessed on 20 August 2021).
18. World Health Organization; Regional Office for Europe. Body Mass Index—BMI-for-Age (5–19 Years). Available online: https://www.who.int/tools/growth-reference-data-for-5to19-years/indicators/bmi-for-age (accessed on 16 August 2021).
19. Centers for Disease Control and Prevention. Healthy Weight, Nutrition and Physical Activity. How is BMI Calculated for Children and Teens? Available online: https://www.cdc.gov/healthyweight/assessing/bmi/childrens_bmi/about_childrens_bmi.html#HowIsBMICalculated (accessed on 16 August 2021).
20. Taber, K.S. The Use of Cronbach's Alpha When Developing and Reporting Research Instruments in Science Education. *Res. Sci. Ed.* **2018**, *48*, 1273–1296. [CrossRef]
21. World Health Organization. *Vaccination Coverage Cluster Surveys: Reference Manual*; World Health Organization: Geneva, Switzerland, 2018; Available online: https://apps.who.int/iris/bitstream/handle/10665/272820/WHO-IVB-18.09-eng.pdf?sequence=1&isAllowed=y (accessed on 16 August 2021).
22. European Commission. Survey Sampling Reference Guidelines—Introduction to Sample Design and Estimation Techniques. 2008. Available online: https://ec.europa.eu/eurostat/documents/3859598/5901961/KS-RA-08-003-EN.PDF.pdf/833f7740-0589-47e1-99a5-c14878a2c1a8?t=1414781399000 (accessed on 16 August 2021).
23. Deliberazione della Giunta Regionale Puglia 12 dicembre 2017, n. 2194. Piano Regionale Integrato in Materia di SICUREZZA Alimentare, Mangimi, Sanità e Benessere Animale anni 2017–2018. Available online: http://burp.regione.puglia.it/documents/10192/20077226/DEL_2194_2017.pdf/1cf775a6-e050-4f42-bbe3-c9d77f25f881 (accessed on 23 August 2021).
24. Deliberazione della Giunta Regionale Puglia 7 marzo 2019, n. 387. Recepimento Intesa Rep atti n. 155/CSR del 6 Settembre 2018. Proroga per l'anno 2019, Modifica ed Integrazione del Piano Regionale dei Controlli Ufficiali in Materia di Sicurezza Alimentare, Mangimi, Sanità e Benessere Animale di cui alla DGR 2104/17. Available online: https://trasparenza.regione.puglia.it/provvedimenti/provvedimenti-della-giunta-regionale/130648 (accessed on 23 August 2021).
25. Deliberazione della Giunta Regionale Puglia 19 marzo 2020, n. 370. Piano di Controllo Nazionale Pluriennale 2020–2022. Recepimento Intesa Stato Regioni rep. Atti 16/CSR del 20/02/2020 ed individuazione del Punto di Contatto Regionale. Available online: http://burp.regione.puglia.it/documents/10192/51893115/DEL_370_2020.pdf/04be83c4-42eb-42ae-a782-38857b6d053a (accessed on 23 August 2021).

26. Presidenza del Consiglio dei Ministri. Conferenza Permanente per i Rapporti tra lo Stato, le Regioni e le Province Autonome di Trento e Bolzano. Intesa, ai sensi dell'articolo 8, Comma 6, della Legge 5 Giugno 2003, n.131, tra il Governo, le Regioni e le Province Autonome di Trento e Bolzano sul Documento Concernente "Linee Guida Per Il Controllo Ufficiale ai Sensi dei Regolamenti (CE) n. 882/2004 e 854/2004". Rep. Atti n. 212/CSR del 10/11/2016. Available online: http://www.regioni.it/download/news/488553/ (accessed on 23 August 2021).
27. HBSC Stili di Vita e Salute dei Giovani Italiani tra 11 e 15 Anni Regione Puglia. Available online: https://www.epicentro.iss.it/hbsc/pdf/report/puglia-2018.pdf (accessed on 31 January 2021).
28. Hruby, A.; Hu, F.B. The Epidemiology of Obesity: A Big Picture. *Pharmacoeconomics* **2015**, *33*, 673–689. [CrossRef] [PubMed]
29. Pasquarella, C.; Veronesi, L.; Castiglia, P.; Liguori, G.; Montagna, M.T.; Napoli, C.; Rizzetto, R.; Torre, I.; Masia, M.D.; Di Onofrio, V.; et al. SItI working group Hygiene in Dentistry. Italian multicentre study on microbial environmental contamination in dental clinics: A pilot study. *Sci. Total Environ.* **2010**, *408*, 4045–4051. [CrossRef] [PubMed]
30. Napoli, C.; Salcuni, P.; Pompa, M.G.; Declich, S.; Rizzo, C. Estimated imported infections of Chikungunya and Dengue in Italy, 2008 to 2011. *J. Travel Med.* **2012**, *19*, 294–297. [CrossRef] [PubMed]
31. Napoli, C.; Riccardo, F.; Declich, S.; Dente, M.G.; Pompa, M.G.; Rizzo, C.; Rota, M.C.; Bella, A. National Working Group. An early warning system based on syndromic surveillance to detect potential health emergencies among migrants: Results of a two-year experience in Italy. *Int. J. Environ. Res. Public Health* **2014**, *11*, 8529–8541. [CrossRef] [PubMed]

nutrients

Article

Identifying Nutrient Patterns in South African Foods to Support National Nutrition Guidelines and Policies

Yusentha Balakrishna [1,2,*], Samuel Manda [2,3,4], Henry Mwambi [2] and Averalda van Graan [5,6]

1 Biostatistics Research Unit, South African Medical Research Council, Durban 4001, South Africa
2 School of Mathematics, Statistics and Computer Science, University of KwaZulu-Natal, Pietermaritzburg 3201, South Africa; samuel.manda@mrc.ac.za (S.M.); mwambih@ukzn.ac.za (H.M.)
3 Biostatistics Research Unit, South African Medical Research Council, Pretoria 0001, South Africa
4 Department of Statistics, University of Pretoria, Pretoria 0001, South Africa
5 Biostatistics Research Unit, SAFOODS Division, South African Medical Research Council, Cape Town 8001, South Africa; averalda.vangraan2@mrc.ac.za
6 Division of Human Nutrition, Department of Global Health, Stellenbosch University, Cape Town 8001, South Africa
* Correspondence: yusentha.balakrishna@mrc.ac.za; Tel.: +27-31-203-4855

Abstract: Food composition databases (FCDBs) provide the nutritional content of foods and are essential for developing nutrition guidance and effective intervention programs to improve nutrition of a population. In public and nutritional health research studies, FCDBs are used in the estimation of nutrient intake profiles at the population levels. However, such studies investigating nutrient co-occurrence and profile patterns within the African context are very rare. This study aimed to identify nutrient co-occurrence patterns within the South African FCDB (SAFCDB). A principal component analysis (PCA) was applied to 28 nutrients and 971 foods in the South African FCDB to determine compositionally similar food items. A second principal component analysis was applied to the food items for validation. Eight nutrient patterns (NPs) explaining 73.4% of the nutrient variation among foods were identified: (1) high magnesium and manganese; (2) high copper and vitamin B_{12}; (3) high animal protein, niacin, and vitamin B_6; (4) high fatty acids and vitamin E; (5) high calcium, phosphorous and sodium; (6) low moisture and high available carbohydrate; (7) high cholesterol and vitamin D; and (8) low zinc and high vitamin C. Similar food patterns (FPs) were identified from a PCA on food items, yielding subgroups such as dark-green, leafy vegetables and, orange-coloured fruit and vegetables. One food pattern was associated with high sodium levels and contained bread, processed meat and seafood, canned vegetables, and sauces. The data-driven nutrient and food patterns found in this study were consistent with and support the South African food-based dietary guidelines and the national salt regulations.

Keywords: food composition database; nutrient pattern; nutrient composition; principal component analysis; food-based dietary guideline; salt intake; South Africa

Citation: Balakrishna, Y.; Manda, S.; Mwambi, H.; van Graan, A. Identifying Nutrient Patterns in South African Foods to Support National Nutrition Guidelines and Policies. *Nutrients* **2021**, *13*, 3194. https://doi.org/10.3390/nu13093194

Academic Editors: Massimo Lucarini and Alessandra Durazzo

Received: 19 August 2021
Accepted: 11 September 2021
Published: 14 September 2021

1. Introduction

Public health nutrition focuses on promotion and improvement of optimal health of a population through nutrition-related health dietary guidelines and policies. In the sub-Saharan African region, public health challenges such as the increasing burden of malnutrition, diabetes mellitus and cardiovascular diseases, can potentially be addressed with adequate nutrition interventions [1,2]. However, to implement effective nutritional interventions in the region, the nutritional situation of the targeted population needs to be known. This requires reliable food consumption data.

Food composition databases (FCDBs) are essential to public health nutrition and associations between diet and health have been shown at the levels of dietary patterns, food groups, foods, and nutrients [3]. They are used together with dietary intake studies to develop food frequency questionnaires and assess relationships between diet and disease. FCDBs also provide insight into food groups and foods containing low or high nutrient levels. Once these relationships have been determined, food-based dietary guidelines (FBDGs) and nutrition policies can be implemented. FBDGs translate recommended dietary allowances to food-related guidelines for improved public health nutrition and guidance [4]. South Africa first developed the FBDGs in 2003 and revised the guidelines in 2012. The South African FBDGs have since been adopted by the National Department of Health as the 'official' dietary recommendations for the country in people aged 5 years or older [4]. The eleven guidelines aim to promote a change in the dietary habits of South Africans to address nutrition-related public health diseases such as malnutrition and obesity. The guidelines encourage dietary diversity and highlight foods that should be limited such as fats, sugar, and salt. Other public health nutrition measures to improve health such as food fortification [5], salt regulations [6] and taxes on sugar-sweetened beverages [7], have also been implemented in South Africa.

Fruits, vegetables, legumes, dairy, and meat are just a few of the common food groups found in FCDBs and accepted by nutritionists. Food items within these food groups generally provide similar amounts of macronutrients. However, while nutritional composition may be similar within these groupings, subgroups may be identifiable and compositional similarity may also be found across these groupings. The growing number of food items in FCDBs presents consumers with dietary choices that need to be based on nutrition, availability, cost, and preference. Classifying food items into nutritionally homogenous groups allows consumers to select alternative food items whilst maintaining a similar nutritional intake. Identifying compositionally similar food items guides dietary recommendations, assists in consumer education, and informs product reformulation. With the ever-expanding food market and inclusion of country-specific foods, it can also aid the categorization of a new food item by grouping it with similar foods that are already known [3]. The identification of unhealthy food items that may not be immediately apparent, also becomes possible.

Several studies have investigated the clustering of food items [8–12] and nutrient co-occurrence patterns [13,14] using statistical methods, but only one was found to use data from Africa [15]. More specifically, the study of nutrient patterns in South Africa has been limited to consumption data [16–19]. Thus, there is a need to develop capacity in methods applicable to the African scenario to help inform consumers and public health policy makers in food nutrient patterns and composition.

Using statistical methods, this study aims to identify compositionally similar food items and nutrient co-occurrence patterns within the South African Food Composition Database (SAFCDB) [20]. The results of this study will provide data-driven evidence that may support the current dietary guidelines and nutritional policies or offer an alternative view.

2. Materials and Methods

2.1. Data

The 2017 SAFCDB [20] (available at http://safoods.mrc.ac.za/products.html, accessed on 9 September 2021) contained nutrient information on 1667 food items and 169 food components. This consisted of both uncooked and cooked food items, as well as composite dishes. Fortified food items were described as such. Table 1 provides a detailed description of the food items by food group. For ease of reference, we will use the term 'nutrients' to encompass the nutrients, minerals and vitamins used in the analysis. All nutrient values were expressed per 100 g. The most common nutrients with a minimal quantity of missing values were selected for analysis (n = 28; Table 2). In our selection of the nutrients, we also ensured that nutrients were non-collinear. For example, because total carbohydrate is the sum of available carbohydrate and dietary fibre, we opted to include available carbohydrate and dietary fibre instead of total carbohydrate. Nine macronutrients, nine minerals, and ten vitamins were analysed. Due to the standard principal component analysis (PCA) technique requiring complete data for all variables, all food items that had complete nutrient information for the selected 28 nutrients were included in the principal component analysis (n = 971).

Table 1. Number of foods per food group.

Food Group	n	%
Cereals and Cereal Products	273	16.38
Vegetables	312	18.72
Fruit	143	8.58
Legumes and Legume Products	37	2.22
Nuts and Seeds	27	1.62
Milk and Milk Products	76	4.56
Eggs	30	1.80
Meat and Meat Products	172	10.32
Fish and Seafood	61	3.66
Fats and Oils	50	3.00
Sugar, Syrups and Sweets	48	2.88
Soups, Sauces, Seasonings and Flavourings	76	4.56
Beverages	52	3.12
Infant and Paediatric Feeds and Foods	250	15.00
Therapeutic/Special/Diet Products	32	1.92
Miscellaneous	28	1.68
Total	1667	100.00

Table 2. Nutrients included in the analysis with their unit of measurement and corresponding abbreviations used in figures.

Macronutrients	Minerals	Vitamins
Moisture (g), moist	Calcium (mg), ca	Vitamin A (RE) (μg), vita_re
Plant protein (g), pl_prot	Iron (mg), fe	Thiamin (mg), thiamin
Animal protein (g), an_prot	Magnesium (mg), mg	Riboflavin (mg), ribofl
Saturated fatty acids (g), satfat	Phosphorous (mg), p	Niacin (mg), niacin
Mono-unsaturated fatty acids (g), mufat	Potassium (mg), k	Vitamin B_6 (mg), vit_b6
Polyunsaturated fatty acids (g), pufat	Sodium (mg), na	Vitamin B_{12} (μg), vit_b12
Cholesterol (mg), choles	Zinc (mg), zn	Pantothenate (mg), pantothn
Carbohydrate, available (g), cho_avail	Copper (mg), cu	Vitamin C (mg), vit_c
Total dietary fibre (g), fib_tot	Manganese (μg), mn	Vitamin D (μg), vit_d
		Vitamin E (mg), vit_e

Abbreviations: g = grams; mg = milligrams; μg = micrograms; RE = retinol equivalents.

2.2. Methods

Statistical methods that consider the correlated nature and presence of multiple nutrients within a food item are needed to evaluate the nutrient patterns amongst food items. Principal component analysis is one of the oldest and simplest dimension-reduction techniques available [21] and is applicable to correlated variables. When applied to food composition data, PCA allows the analysis of multiple nutrients simultaneously. PCA aims to describe the maximum amount of variation in the dataset using the least number of principal components (PCs). The PCs are uncorrelated linear combinations of the original variables that capture most of the variation within the first few components. PCA aids data reduction by explaining the covariation amongst the variables using a few linear combinations. PCA also aids data interpretation by finding features that explain the covariation. The contribution of each variable to a component is called the loading and high loadings indicate important variables. Rotation methods can be applied to enhance interpretability by producing loadings that are as close to zero or one as possible. For each PC, observations have a score that combines each of the variables. The score indicates how much each observation is related to a PC [22]. Factor analysis is also a common multivariate dimension reduction technique but has slight differences to PCA. While PCA describes the relationships among the observed variables in a simpler way, factor analysis finds latent factors that influence the observed variables. Hence, the application of factor analysis is more suited to the analysis of consumption data as it will be able to generate latent factors, that is, dietary patterns, which predict food choices [23]. Figure 1 presents the methodology and rationale.

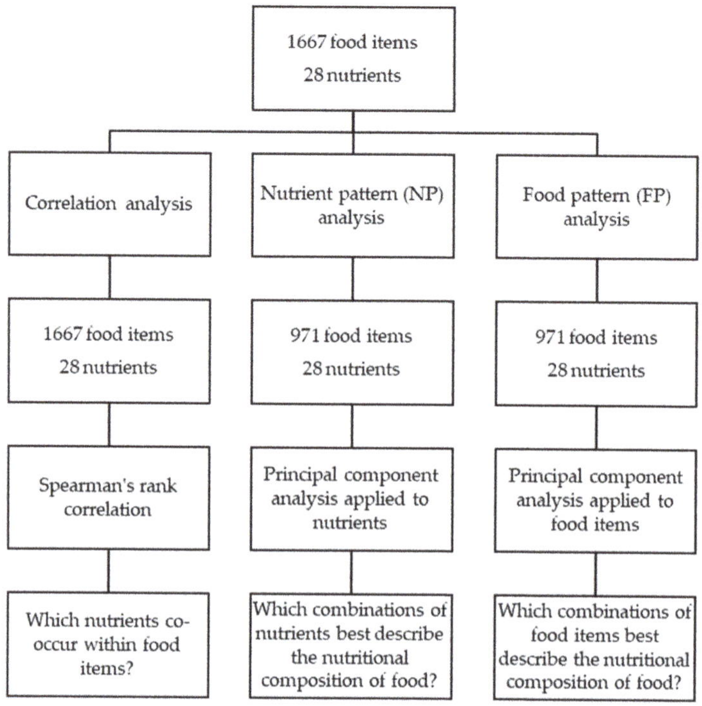

Figure 1. Flow chart of methodology and rationale.

2.2.1. Correlation Analysis

For each nutrient, some foods contained exceptionally high values. For example, oysters were especially high in zinc and amaranth leaves were especially high in magnesium. Due to these outliers, we calculated pairwise-complete Spearman correlations for the complete dataset ($n = 1667$), to determine nutrient co-occurrence patterns.

2.2.2. Nutrient Pattern Analysis

For the sub-sample ($n = 971$), we explored the data using PCA with orthogonal varimax rotation and Kaiser normalization to enhance interpretability. PCA was applied to the correlation matrix due to the scale differences between nutrients. Components were retained considering the scree plot, eigenvalues greater than 1 (the average of the eigenvalues when using the correlation matrix) and interpretability. High-loading nutrients were defined as having an absolute loading of at least 0.4 and were used to interpret the component. To enhance and support the interpretation, nutrients with absolute loadings between 0.3 and 0.4 were also considered. Food items were allocated to groups corresponding to their highest PC score. The chi-square test was used to test for an association between the FCDB and PCA groupings. The Kruskal–Wallis test was used to test for differences in nutrient values between the PC groupings. The PCs identified by the nutrient analysis were termed 'nutrient patterns (NPs)'.

2.2.3. Food Pattern Analysis

We also applied PCA to the food items to confirm the components found during the nutrient pattern analysis. Components with eigenvalues greater than 1 and that accounted for at least 1% of the variation were retained. The highest absolute loading within each component ranged between 0.07 and 0.16. Hence, absolute loadings greater than 0.05 were used to interpret the component. The PCs identified by the food item analysis were termed 'food patterns (FPs)'.

Trace values (values below the limit of detection) accounted for 1.2% of the data and were imputed with half of the limit of detection for each nutrient [24]. Results were considered significant for $p < 0.05$ and Bonferroni-adjusted significance levels were used to account for multiple testing. All analyses were done in Stata version 16 (StataCorp, College Station, TX, USA) and R (available at https://www.R-project.org/, accessed on 9 September 2021).

3. Results

3.1. Correlation Analysis

The correlations between the nutrients are presented in Figure 2. Overall, correlations were mostly positive indicating frequent nutrient co-occurrences. Negative correlations occur when the increase in one nutrient results in the decrease of another nutrient. Most negative correlations were found between moisture and all other nutrients, except vitamin C ($r = 0.25$, $p < 0.001$). Animal protein, fatty acids, and cholesterol positively correlated with phosphorous, sodium, zinc, riboflavin, vitamin B_{12}, pantothenic acid and vitamin D ($p < 0.001$). In contrast, animal protein, saturated fatty acids, mono-unsaturated fatty acids, and cholesterol, negatively correlated with total fibre and vitamin C ($p \leq 0.001$). Vitamin E had the highest positive correlations with fatty acids ($r = 0.46$, $r = 0.52$, $r = 0.67$, $p < 0.001$) and vitamin D ($r = 0.64$, $p < 0.001$). Plant protein had the highest positive correlations with total fibre ($r = 0.82$, $p < 0.001$) and manganese ($r = 0.67$, $p < 0.001$). Plant protein and total fibre both negatively correlated with animal protein, cholesterol, and vitamin B_{12} ($p < 0.001$). Vitamin B_{12} and vitamin D exhibited similar patterns, both negatively correlating with plant protein, total fibre, and manganese ($p < 0.001$). Strong, positive correlations among the minerals and vitamins were also found. Iron, magnesium, and copper were connected by positive correlations ($p < 0.001$) as well as thiamin, riboflavin, niacin, and vitamin B_6 ($p < 0.001$). Positive correlations were also evident between animal-derived micronutrients such as phosphorus, zinc, and pantothenic acid ($p < 0.001$).

3.2. Nutrient Pattern Analysis

Eight nutrient patterns had an eigenvalue greater than 1 (Figure S1) that explained 73.4% of the total nutritional variation in the data. The rotated loadings are presented in Table A1 in Appendix A. A characterisation of the patterns, using the nutrients that loaded highly (absolute loadings >0.4) on each, is shown in Table 3, along with the supporting nutrients that had absolute loadings between 0.3 and 0.4. At least two nutrients per pattern had high loadings. PC scores were calculated for each food item and the highest score determined pattern membership. NP 1 was characteristic of food items high in plant protein, total fibre, magnesium, potassium, and manganese. Iron also featured on NP 1 but had a loading of 0.27—less than our threshold of 0.3. Wheat products, dark leafy greens, legumes, nuts, and seeds scored highest on this pattern. NP 2 was found to be high in vitamin A, copper, riboflavin, and vitamin B_{12} and linked with foods such as kidney, liver, mussels, and oysters. Meat, meat products, crab, and oily fish scored high on NP 3 as they shared high levels of animal protein, niacin, and vitamin B_6. Fortified bread was also included due to its increased vitamin B_6 content. Saturated fatty acids, mono-unsaturated fatty acids, polyunsaturated fatty acids, and vitamin E characterised NP 4 and were found to be highest in fats and oils, avocados, some nuts (almonds, pecans, walnuts, macadamias, and coconuts), and sauces. Foods made or fried with oil or margarine also scored highly on this pattern, as well as chicken skin and processed meats. NP 5 identified foods high in calcium, phosphorous and sodium such as milk, milk products, canned vegetables, biltong, and shrimp. Foods made with milk and cheese were also found to be associated with NP 5. While most nutrients had positive loadings, moisture and zinc had a negative loading on NPs 6 and 8, respectively. NP 6 had positive loadings of available carbohydrate and thiamin, correlating with baked items, dried fruit, jams, as well as sugar and sweets, while NP 8 had positive loadings of vitamin A and vitamin C. Fruits and vegetables related mostly to NP 8, as well as soft maize meal. High cholesterol and vitamin D content characterised NP 7. Foods associated with this pattern were eggs, composite dishes using eggs, fish, offal, and tripe. Fortified milk powder and breastmilk substitutes also scored high on this nutrient pattern. Pantothenic acid featured on NP 2 and NP 7 but, like iron, had loadings below the absolute value threshold of 0.3.

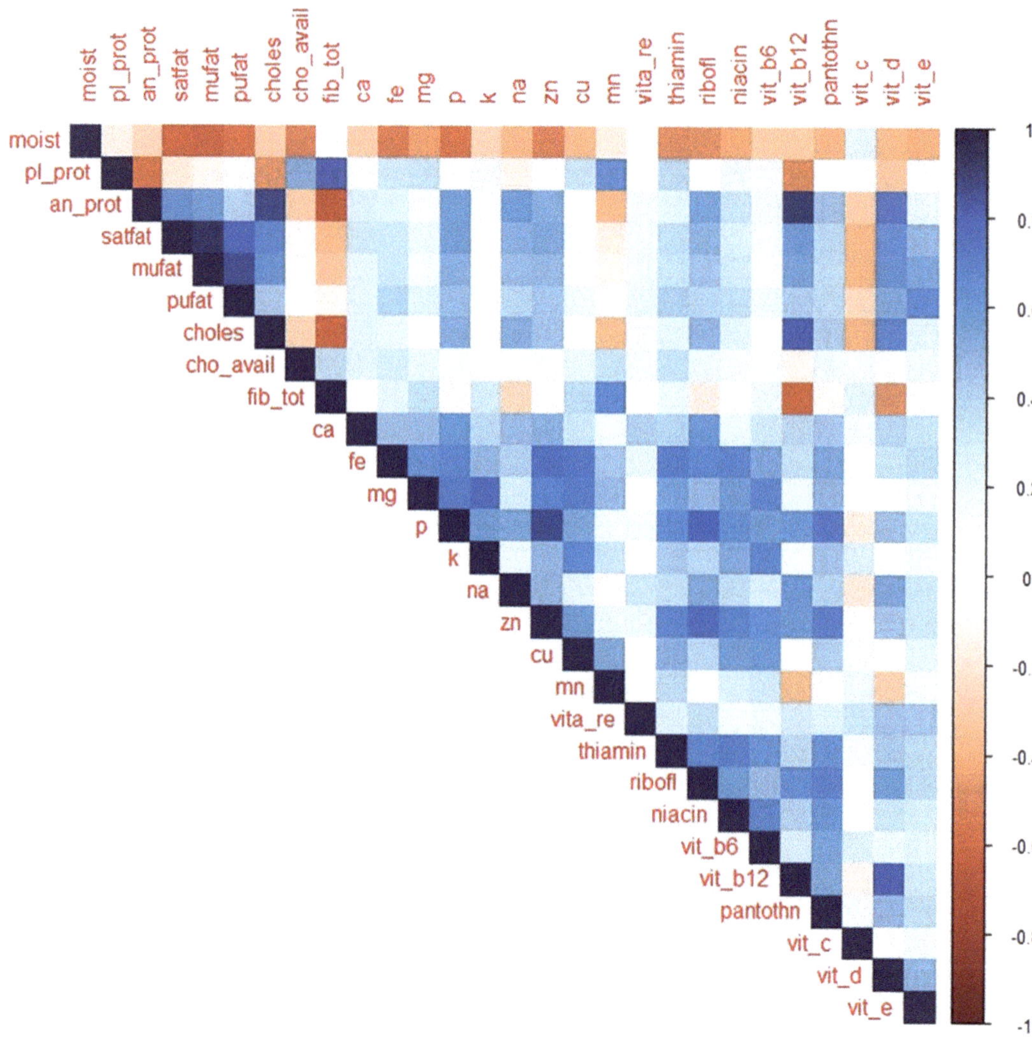

Figure 2. Spearman's rank correlations between the nutrients.

Table 3. Characterisation of nutrient patterns (NP).

NP	Nutrients with Absolute Loadings >0.3 and >0.4	Examples of Food Items That Scored Highly on Pattern
1	high in plant protein, total fibre, *magnesium*, potassium, and *manganese*	wheat products, oats, brown rice, dark leafy greens, peas, dehydrated green beans, dehydrated cabbage, dehydrated cauliflower, legumes and legume products, nuts, seeds
2	high in vitamin A, *copper*, riboflavin, and *vitamin* B_{12}	kidney, liver, giblets, mussels, oyster, mushroom
3	high in *animal protein*, *niacin*, and *vitamin* B_6	meat and meat products, crab, oily fish, fortified bread/rolls
4	high in *fatty acids* and *vitamin E*	foods made or fried with oil or margarine, chicken skin, processed meats, fats and oils, avocado, nuts, sauces
5	high in *calcium*, *phosphorous*, and *sodium*	milk and milk products (including foods made with milk and cheese), canned vegetables, biltong, shrimp/prawn
6	low in *moisture*, high in *available carbohydrate*, and thiamin	bread, breakfast cereals, cakes, cookies, puddings, pasta, pastries, maize and maize meal (stiff and crumbly), white rice, pies, dried fruit, jam/marmalade, honey, sugar, sweets
7	high in *cholesterol* and *vitamin D*	eggs and foods using eggs (e.g., custard, choux pastry), fortified milk powder, breastmilk substitutes, offal, tripe, battered/crumbed fish, fishcake made with egg, salmon, sardine, salad dressing
8	low in *zinc*, high in vitamin A, and *vitamin C*	fruit, vegetables, fruit juices, soft maize meal

Nutrients with absolute loadings >0.4 are indicated in italic.

Table 4 compares the food categories in the SAFCDB to the groupings found by the PCA. PCA groupings consisted of food items across the SAFCDB groupings, and the grouping structures were significantly associated ($p < 0.001$). Vegetables and legumes contributed 42% and 23% of NP 1, respectively. Meat and seafood together accounted for 91.8% of NP 3. All food items in the category 'Eggs' were grouped under NP 7, together with composite dishes from 'Cereals and cereal products' that were made with eggs. Most of the food items within 'Legume and legume products', 'Milk and milk products', 'Fats and oils' and 'Sugar, syrups and sweets' remained together under the PCA groupings.

Table 4. Food group percentage for each nutrient pattern (NP); *n* (%).

Food Group	Nutrient Pattern [a]								Total
	NP 1	NP 2	NP 3	NP 4	NP 5	NP 6	NP 7	NP 8	
Cereals and Cereal Products	17 (17)		4 (3.64)	15 (17.05)	17 (21.52)	116 (70.73)	19 (21.35)	7 (2.16)	195 (20.08)
Vegetables	42 (42)	4 (23.53)	1 (0.91)	12 (13.64)	4 (5.06)	2 (1.22)		180 (55.56)	245 (25.23)
Fruit	3 (3)			1 (1.14)	1 (1.27)	20 (12.2)		107 (33.02)	132 (13.59)
Legumes and Legume Products	23 (23)			1 (1.14)		2 (1.22)			26 (2.68)
Nuts and Seeds	11 (11)			8 (9.09)		1 (0.61)			20 (2.06)
Milk and Milk Products					32 (40.51)	1 (0.61)	8 (8.99)		41 (4.22)
Eggs							27 (30.34)		27 (2.78)
Meat and Meat Products		9 (52.94)	89 (80.91)	15 (17.05)	2 (2.53)	2 (1.22)	3 (3.37)		120 (12.36)
Fish and Seafood		3 (17.65)	12 (10.91)	2 (2.27)	3 (3.8)		16 (17.98)		36 (3.71)
Fats and Oils				20 (22.73)	1 (1.27)	1 (0.61)	4 (4.49)		26 (2.68)
Sugar, Syrups and Sweets	1 (1)			3 (3.41)	1 (1.27)	11 (6.71)		1 (0.31)	17 (1.75)
Soups, Sauces, Seasonings and Flavouring	3 (3)	1 (5.88)		9 (10.23)	7 (8.86)	1 (0.61)	3 (3.37)	6 (1.85)	30 (3.09)
Beverages					8 (10.13)	3 (1.83)	1 (1.12)	15 (4.63)	27 (2.78)
Infant and Paediatric Feeds and Foods						3 (1.83)	7 (7.87)		10 (1.03)
Therapeutic/Special/Diet Products			4 (3.64)		2 (2.53)		1 (1.12)		7 (0.72)
Miscellaneous				2 (2.27)	1 (1.27)	1 (0.61)		8 (2.47)	12 (1.24)
Total	100 (100)	17 (100)	110 (100)	88 (100)	79 (100)	164 (100)	89 (100)	324 (100)	971 (100)

[a] Blank cells represent 0 (0).

Table A2 in Appendix A reports the median (IQR) nutrient values for each principal component grouping. Median nutrient values for each NP agreed with the characterisation of the patterns and are graphically represented in Figure 3. Table A2 in Appendix A and Figure 3 represent the expected nutritional composition of an average food item from each pattern. A randomly selected food item from NP 3 will, on average, contain the highest amount of animal protein and niacin than a food item from any of the other NPs. Sodium content can be expected to be lowest in foods from NP 8, and highest in foods from NP 5 and NP 7. NP 2 and NP 8 had the greatest cumulative quantity of vitamins, with vitamin C contributing the largest proportion of the composition. Niacin was also a large contributor of vitamin content in NP 2. The first three patterns had the highest quantity of minerals made up largely from phosphorous and potassium.

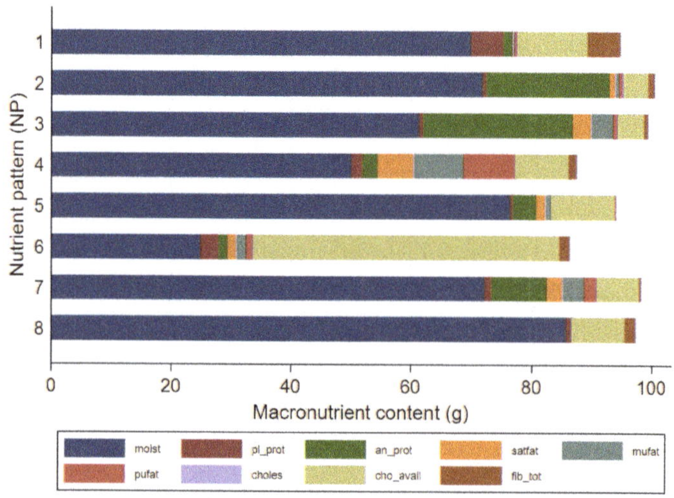

(**a**)

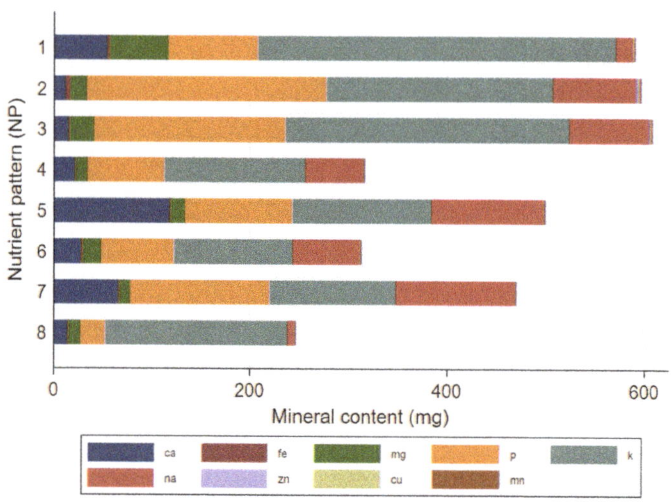

(**b**)

Figure 3. *Cont.*

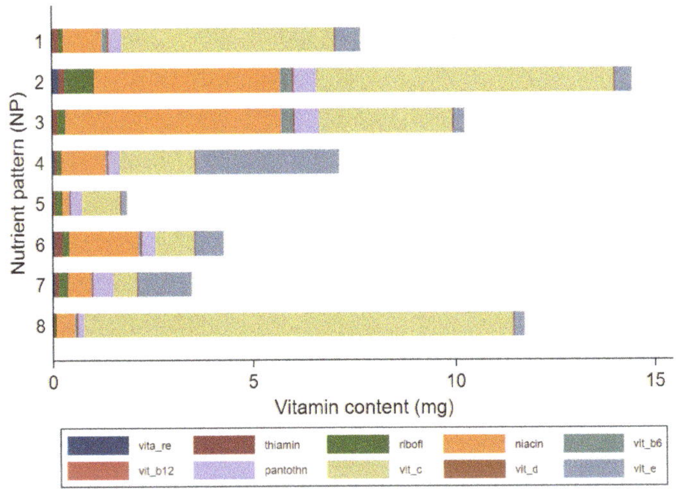

(c)

Figure 3. Median nutrient values per nutrient pattern (NP) for (**a**) macronutrients; (**b**) minerals and (**c**) vitamins.

3.3. Food Pattern Analysis

We also applied PCA to the 971 food items to confirm the nutrient patterns we found. Seven food patterns had both an eigenvalue greater than 1 and accounted for at least 1% of the variation in the dataset (Figure S2). The seven FPs explained 97.4% of the total variation in the data (Table 5). The maximum absolute loading for components ranged from 0.07 to 0.16, hence, we interpreted the components using absolute loadings that were greater than 0.05. The patterns found when analysing the food items reflected the patterns found when analysing the nutrients thus, confirming the presence of nutrient patterns and validating our results. Both analyses identified a pattern grouping together wheat products, leafy vegetables, and legumes (NP 1 and FP 2) as well as patterns for milk and milk products (NP 5 and FP 3), and eggs and food items using eggs (NP 7 and FP 4). However, applying PCA to the food items enabled better discrimination of fruit and vegetables by vitamin C (FP 1) and beta-carotene (FP 6) content. Orange-coloured fruit and vegetables were identified with FP 6 in the food item analysis. In addition, greater discrimination was apparent among dark leafy greens, which were split between FP 2 and FP 6 compared to being grouped together under the nutrient analysis. Composite dishes using distinctive ingredients were also able to be identified and grouped with the raw versions of the ingredient. For example, carrot cake grouped with carrots and pastries made using eggs grouped with eggs. Rusks made with wholewheat flour (FP 2) and rusks made with white flour (FP 5) were also able to be identified and separated. Processed meat such as luncheon meat and sausages were separated from meat and meat products, and instead grouped with processed cheese (FP 5) and processed fish. Sodium scored high on this food pattern. The last pattern in the food item analysis (FP 7) separated soft maize meal from the stiff and crumbly versions based on its higher moisture content, similar to the results of the nutrient analysis. Soft maize meal was grouped together with other moisture rich food items such as beverages, cabbage and brinjal.

Table 5. Characterisation of food patterns (FP).

FP	Explained Variance (%)	Cumulative Explained variance (%)	Food Item with Absolute Loadings >0.05	Nutrients Which Scored High on Pattern
1	56.5	56.5	white and sweet potatoes, squash, celery, cucumber, cauliflower, brinjal, mushroom, green pepper, citrus fruits, stone fruits, and grapes in raw, canned, dried, and juice versions	potassium, magnesium, vitamin C
2	14.5	71.0	oats, rice, wheat, maize, barley, rye, wholewheat products, spinach and amaranth leaves, peas, green beans, berries (including pineapples), legumes and legume products, nuts and seeds, oyster, chocolate	manganese
3	10.4	81.4	milk and milk products (including foods made with milk and milk beverages), breastmilk substitutes, canned sardine, canned salmon	calcium
4	6.7	88.1	eggs and foods made with eggs (e.g., custard, choux pastry, sauces), meat and meat products, battered/crumbed fish	animal protein, cholesterol, phosphorous
5	3.9	92.0	bread, breakfast cereals, pastries, breadcrumbs, rusks made with white flour, canned vegetables, processed cheese, feta cheese, processed meat, meat pies, processed fish and seafood, sauces, icing for cakes	sodium
6	3.0	95.0	fortified maize meal, carrot (including carrot cake), pumpkin, butternut, hubbard squash, orange flesh sweet potato, leafy greens (e.g., lambquarters, sow thistle, cat's whiskers leaves), apricot, mango, naartjie, beef kidney and liver, chicken liver and giblets, offal	vitamin A
7	2.4	97.4	soft maize meal, marrow, cabbage, brinjal, apple, pear, lemon, lime, fruit canned in syrup, trotters, tripe, miscellaneous (water, tea, coffee, alcohol)	moisture, plant protein, fatty acids, available carbohydrate, total fibre, iron, zinc, copper, thiamin, riboflavin, niacin, vitamin B_6, vitamin B_{12}, pantothenic acid, vitamin D, vitamin E

The patterns found supported the South African FBDGs [4], as shown in Table 6. Guideline 1 aims to facilitate balanced nutrient intake by encouraging the consumption of a variety of foods. As the nutrient patterns obtained differ in nutritional composition, consuming foods from different patterns supports this guideline. Starchy foods, as described in Guideline 2, such as bread, rice, cereals, and pasta were associated with NP 6. NP 6 also contained products high in sugar content such as cakes, cookies, and sweets and reflects Guideline 10. Similarly, other nutrient patterns were able to be matched to the South African FBDGs.

Guideline 11 was best captured by FP 5 and reflected categories targeted by the national sodium regulation [6], as highlighted in Table 7. Foods affected by the regulation were all found within FP 5.

Table 6. Comparison between the South African food-based dietary guidelines and principal component analyses.

South African Food-Based Dietary Guidelines [4]	Corresponding Pattern
1. Enjoy a variety of foods.	The nutrient patterns obtained differ in nutritional composition.
2. Be active!	Not applicable
3. Make starchy foods part of most meals.	NP 6
4. Eat plenty of vegetables and fruit every day.	NP 8
5. Eat dry beans, split peas, lentils, and soya regularly.	NP 1
6. Have milk, maas, or yoghurt every day.	NP 5
7. Fish, chicken, lean meat, or eggs can be eaten daily.	NP 3, NP 7
8. Drink lots of clean, safe water.	Not applicable
9. Use fats sparingly. Choose vegetable oils, rather than hard fats.	NP 4
10. Use sugar and foods and drinks high in sugar sparingly.	NP 6
11. Use salt and food high in salt sparingly.	FP 5

Table 7. Comparison between the food categories targeted by the national sodium regulation and foods associated with FP 5.

Food Category as per the National Sodium Regulation [6]	Corresponding Foods Associated with FP 5
1. Bread	All bread types (pumpernickel, raisin, rye, sweetcorn, brown and white bread/rolls, breadcrumbs) except wholewheat bread/rolls
2. All breakfast cereals and porridges, whether ready-to-eat, instant or cook up, hot or cold	All breakfast cereals (puffed rice, puffed corn) except homemade muesli
3. All fat spreads and butter spreads	Mixed butter and hard margarine, brick/hard margarine, polyunsaturated margarine
4. Ready-to-eat savoury snacks, excluding salt-and-vinegar flavoured savoury snacks 5. Flavoured potato crisps, excluding salt-and-vinegar flavoured potato crisps 6. Flavoured, ready-to-eat, savoury snacks and potato crisps, salted and salt-and-vinegar only	Potato crisps
7. Processed meat, cured 8. Processed meat, uncured	Bacon, biltong, corned beef, ham, luncheon meat, meatloaf, pastrami, pork/beef sandwich spread
9. Raw-processed meat sausages (all types) and similar products	Frankfurter, pepperoni, salami, sausages
10. Dry savoury soup powders 11. Dry gravy powders and savoury sauce powders 12. Dry savoury powders with dry instant noodles 13. Stock cubes, stock powders, stock granules, stock emulsions, stock pastes or stock jellies	Soups, sauces

4. Discussion

Public health practitioners and policy makers rely on FCDBs to assess nutrient availability and provide information to link dietary data with nutrient intake for nutritional epidemiology. They also utilize FCDBs for developing nutrition interventions and for informing consumer education. Policies impact food product composition to address dietary shortfalls, but the full potential of food composition is often not recognized [25]. In South Africa, studies have been limited to determining consumption habits among populations [16–19] but our study aims to examine the nutrient patterns present within the food items consumed by the population. More specifically, we aimed to examine the nutrient patterns present among food items listed in the SAFCDB [20] using correlation and PCA. FCDBs are often country-specific due to the influence of environmental, genetic, and processing factors on the nutrient content of food. National FCDBs also include country-specific foods and recipes, reflecting the unique consumption patterns of the country [26]. Therefore, analysing foods contained in the SAFCDB would provide information on the nutrient levels of foods consumed by the South African population.

Significant correlations between the nutrients were identified. Nutrients obtained primarily from plant-based foods, such as total fibre and available carbohydrates, exhibited a strong positive correlation with plant protein. Nutrients obtained primarily from animal products, such as cholesterol and vitamin B_{12}, were strongly associated with animal protein. These plant-derived nutrients negatively correlated with animal-derived nutrients, confirming what is known about nutrient co-occurrence. Our results are also consistent with the correlations found elsewhere among raw foods [13] and raw plant foods [9], suggesting that similar nutrient patterns are evident among cooked and composite dishes as well, which were included in our analysis. The underlying correlation structure contributes to features that distinguish between nutrient-based food groupings. This must be accounted for in any statistical analyses undertaken using multivariate methods. In addition, the high correlation implies better prediction models which are useful in estimating values of missing nutrients, a problem common to FCDBs. While the 2017 SAFCDB contained nutrition data for 1667 food items, only 971 food items could be analysed due to missing data. In addition, missingness also excluded biotin and folate from the analysis, which are both vital B-vitamins that are sourced from food [27]. Methods to impute missing values in food composition data have been investigated [28–30] and further research in this area could facilitate the completeness of FCDBs.

Our study affirmed that some food items are more compositionally alike than others, by identifying eight nutrient patterns that were consistent with existing knowledge. All analysed nutrients, except iron and pantothenic acid, featured on a pattern. Although iron and pantothenic acid did not meet our threshold for a high loading, both stood out on nutrient patterns that contained their expected sources. Vitamin A featured on two nutrient patterns, due to its availability in foods of both plant and animal origin. A study [14] conducted in Finnish foods, identified four nutrient content patterns using factor analysis and was able to group wheat products with legumes, and mushrooms with offal foods—a common finding in our study as well. Although the study was able to include 106 nutrients, the patterns were comparable to the patterns found in our study, suggesting that only a few key nutrients are needed to successfully determine nutrient patterns.

We also validated our results by applying the dimension reduction technique to the food items themselves. Results of both analyses were similar, and a large amount of the nutritional variation was able to be explained by a few patterns. The patterns included food items from across different food groups, suggesting compositional similarity despite conceptual dissimilarity. Hence, applying clustering techniques within each conceptual FCDB group may reveal more intricate groupings. However, this approach may suffer from high dimensionality with small sample size issues. Two studies applied clustering techniques within FCDB food groups. The first study [15] found six subgroups within the 'Cereals' category of the West Africa Food Composition Table. These subgroups separated grains by type and preparation methods. For example, pearl millet separated from other grains, and maize was separated across three clusters depending on whether it was raw, boiled, or prepared as a porridge. Likewise, our analysis differentiated between white and brown rice, and soft maize meal and stiff or crumbly maize meal. The second study [9] applied clustering techniques within five food categories (fruits, vegetables, nuts and seeds, legumes, and cereal grains) of the U.S. Department of Agriculture (USDA) National Nutrient Database for Standard Reference (SR) Legacy (2018). The study found that similar foods were not necessarily from the same category. For example, wheat germ was found to cluster with legumes, a finding repeated in our analysis as well. Another similar finding was almonds and coconuts, macadamias, pecans, and walnuts separating from other nuts in the database. Chestnuts were also isolated from other nuts. Our results suggest that statistical methods can be used to create a natural food exchange list to accommodate different dietary preferences.

Dark leafy greens such as spinach and other leaves (amaranth, blackjack, cowpea, etc.) were differentiated from other vegetables in the database. The application of PCA to food items had greater discernability than PCA applied to nutrients. Under the food pattern analysis, dark leafy greens were further divided into spinach and amaranth leaves and other leafy greens. Similarly, orange-coloured fruit and vegetables grouped together, which was not seen under the nutrient pattern analysis. This type of clustering was also identified in Pennington et al. [10]. The daily consumption of dark-green leafy vegetables and orange-coloured fruit and vegetables is recommended as per the South African FBDGs [4] and the Dietary Guidelines for Americans [31] and is important for a healthy diet as they are rich sources of vitamins and minerals [4]. Classifications that are based on nutritional similarity are useful to nutritionists, researchers, and consumers for the development of dietary guidance materials, development of food frequency questionnaires and reporting of consumption studies, and adherence to dietary guidelines [32].

The PCA method was also able to separate canned vegetables and vegetables fried in oil from the other vegetables. This is helpful in determining food preparation characteristics from nutrient information. Both analyses were able to identify foods made with egg, such as choux pastry and custard, and group these items together with eggs. However, the nutrient analysis additionally included milk and savoury tarts, which are traditionally made with egg. Similarly, both analyses were able to identify foods made with milk and cheese, such as malted milk beverages, puddings, yoghurt, and cheese sauces. Employing a principal component analysis may additionally be helpful in identifying ingredients for composite dishes in a FCDB.

Our results provide data-driven evidence to support the existing knowledge of food and nutrient patterns, as well as South African food-based dietary guidelines and nutrition policies. Each of the nutrient patterns identified corresponded to a guideline and supports the consumption of a variety of foods and moderation of other foods. High sodium levels in food items have led to the current promulgated salt regulation and reduction of salt content of food items in the country [6]. Food items belonging in the high sodium food pattern closely mirrored the categories identified in the regulation. Under the food item analysis, canned vegetables grouped together with other processed food items on a high sodium pattern. Canned vegetables, processed meat, processed cheese, bread, and sauces are suggested to have similar levels of sodium, and this is consistent with research showing these categories to have the highest median sodium levels, based on packaged foods in South Africa [33]. This analysis supports the regulation and can be used in a similar fashion to identify foods with a high sugar content. FBDGs are developed in response to a public health problem [34] and requires identifying rich sources of nutrients that are of public health importance [35]. The patterns identified in our results each describe foods that are rich sources of specific nutrients. Foods providing these nutrients are recommended to be either limited or increased, as appropriate, and implementation of the FBDGs should then be accompanied by monitoring and evaluation of the effects. Food systems [36] are dynamic and are influenced by key drivers such as regulatory frameworks, consumer influence, technological innovations, concerns for food safety, and growing attention paid to diet and health [37,38]. Thus, continuous updates of a FCDB are essential to reflect the changes not only in the types of food provided but also the composition thereof [39]. The evaluation of the effects can be based on changes in food composition [34] and some studies have applied statistical methods to different versions of FCDBs to determine changes over time in composition of fruits and vegetables [40–42]. Therefore, repeating our analysis on past versions and future updates of food composition data could assess whether the implementation of FBDGs and regulations have impacted the reformulation of products. The SAFCDB is updated every three years as updates are resource-intensive and can be challenging to regularly implement, as updates are applied to all database-related tools and products such as publications, software programs, and applications.

The research of innovative statistical methods tailored towards food composition data has the potential to provide improved evidence for dietary guidelines and policy. In addition, it can also support the dietary patterns found in consumption studies. Makura-Kankwende et al. [17] showed that the animal driven dietary pattern, characterised by animal protein and saturated fat, was associated with an increased body mass index amongst black South African women. From our results, foods high in animal protein and saturated fat correspond to meat and meat products, processed meat, and fried foods. These foods are generally present in the Western diet, and the animal driven pattern found may be suggestive of a shift towards this diet [17]. Another study, Visser et al. [19] found that a dietary pattern featuring vitamin A and vitamin B_{12} was associated with lower odds of anaemia in 5–12-year-old South African children. This dietary pattern is reflected in our results which identified foods containing this combination of nutrients, mainly, organ meat such as kidney and liver [19].

Some limitations need to be considered. Missing nutrient values excluded essential nutrients such as folate and biotin from the analysis and contained our sample to 58% of foods available in the SAFCDB. With respect to the PCA method, subjective decisions on the data matrix, rotation method, number of retained components and loading threshold need to be made [16,18]. However, our results are consistent with existing knowledge and has strengths in presenting nutrient and food patterns among South African foods that support food-based dietary guidelines, nutrition policies and consumption studies. We are currently working on developing K-Means and Gaussian Mixtures (GMs) clustering models to identify food items that are more like each other. We are aware that several food items contain missing nutrient values in the database, so we will incorporate multiple imputation techniques to account for missing data. We believe the development and application of these models to food composition databases will contribute to an understanding of nutritional uptake in the population and monitoring adherence to national nutritional prevailing regulation and guidelines.

5. Conclusions

To our knowledge, this is the first study to investigate nutrient patterns in food items using South African food composition data. This analysis provides an overview of the inherent groups available within South African foods. The nutrient co-occurrence patterns identified using data-driven methods are consistent with current knowledge and comparable to similar studies from other countries. The results support current dietary guidelines and nutritional policies.

Supplementary Materials: The following are available online at https://www.mdpi.com/article/10.3390/nu13093194/s1, Figure S1: Screeplot for the principal component analysis of nutrients, Figure S2: Screeplot for the principal component analysis of food items.

Author Contributions: Conceptualization, Y.B. and S.M.; methodology, Y.B. and S.M; formal analysis, Y.B.; writing—original draft preparation, Y.B.; writing—review and editing, all authors; visualization, Y.B.; supervision, S.M., H.M. and A.v.G.; project administration, Y.B. All authors have read and agreed to the published version of the manuscript.

Funding: Y.B., S.M., and A.v.G. time on this research was funded by the South African Medical Research Council.

Institutional Review Board Statement: Not applicable.

Informed Consent Statement: Not applicable.

Data Availability Statement: No new data were created or analysed in this study. Data sharing is not applicable to this article.

Conflicts of Interest: The authors declare no conflict of interest.

Appendix A

Table A1. Explained variance and rotated principal component loadings for the first eight nutrient patterns (NPs) identified by the analysis of nutrients.

Component	NP 1	NP 2	NP 3	NP 4	NP 5	NP 6	NP 7	NP 8
Explained variance (%)	14.9	10.1	9.6	9.3	9	8	7.6	4.9
Cumulative explained variance (%)	14.9	25	34.6	43.9	52.9	60.9	68.5	73.4
Nutrients								
Moisture	0.0084	0.0133	−0.0682	−0.2164	−0.0549	**−0.5215**	0.0035	−0.0133
Plant protein	**0.3867**	−0.0242	−0.0419	0.045	−0.0535	0.127	−0.0618	−0.0681
Animal protein	−0.1189	−0.0821	**0.4862**	0.0282	0.0676	−0.1463	0.1563	−0.1198
Saturated fatty acids	−0.1144	−0.0298	0.1156	**0.3581**	0.0453	0.0524	0.0531	−0.0496
Monounsaturated fatty acids	0.0125	−0.0222	0.0587	**0.5269**	−0.0025	−0.0095	−0.0378	−0.0669
Polyunsaturated fatty acids	0.0376	0.0355	−0.0781	**0.5328**	−0.0255	0.0038	−0.0191	0.0308
Cholesterol	−0.0064	0.0604	−0.025	−0.008	−0.028	−0.0622	**0.5883**	−0.0559
Carbohydrate, available	−0.0884	0.03	−0.1202	−0.142	−0.0158	**0.7057**	−0.0234	0.0669
Total dietary fibre	**0.3627**	−0.0134	−0.0418	−0.0491	−0.0484	0.0324	−0.0696	0.2578
Calcium	0.0703	0.0182	−0.1097	−0.0302	**0.5985**	−0.0281	0.0271	0.0584
Iron	0.2691	0.1124	0.0588	−0.0642	0.0972	0.1351	0.1487	−0.0592
Magnesium	**0.4515**	−0.0281	0.0234	0.0396	0.0523	−0.1023	0.028	0.001
Phosphorous	0.0421	0.0322	−0.0445	0.0024	**0.5864**	0.0047	0.01	−0.0436
Potassium	**0.3209**	−0.0459	0.1839	−0.0612	0.0478	−0.0488	0.0185	0.264
Sodium	−0.1109	−0.0525	0.1434	0.0302	**0.5085**	0.0352	−0.0727	−0.0058
Zinc	0.1675	0.1157	0.1339	−0.0286	−0.0123	−0.0914	−0.081	**−0.4198**
Copper	0.0509	**0.573**	−0.0206	0.0442	−0.0229	−0.032	−0.1268	−0.0586
Manganese	**0.4394**	−0.0077	−0.0655	0.0133	−0.0217	−0.0541	0.0014	−0.1554
Vitamin A (RE)	−0.018	**0.3839**	−0.0597	0.0217	0.0298	−0.0291	−0.0148	**0.3456**
Thiamin	0.1796	−0.0436	0.1771	−0.0286	−0.0454	**0.3188**	0.0556	−0.1524
Riboflavin	−0.0038	**0.3799**	0.0698	−0.0388	0.0141	0.1108	0.2049	−0.024
Niacin	−0.0464	0.0425	**0.5414**	0.0395	−0.0341	0.0695	−0.1054	−0.0615
Vitamin B_6	0.1054	0.0272	**0.4134**	−0.0376	−0.0218	0.0863	−0.0923	0.1435
Vitamin B_{12}	−0.0782	**0.5583**	0.0201	−0.0098	−0.0014	−0.0439	0.0106	−0.0643
Pantothenic acid	0.016	0.075	0.2889	−0.0291	−0.0384	−0.009	0.2697	0.154
Vitamin C	−0.0237	−0.0304	0.1344	0.0111	−0.0189	−0.073	−0.0439	**0.6353**
Vitamin D	−0.0103	−0.092	−0.0612	0.0223	0.0033	0.0671	**0.6351**	0.0271
Vitamin E	0.0854	0.0302	−0.1135	**0.4652**	−0.0215	−0.0181	0.1437	0.1294

Absolute loadings greater than 0.3 are shown in bold.

Table A2. Median (IQR) nutrient values by nutrient pattern (NP).

Nutrient	NP 1	NP 2	NP 3	NP 4	NP 5	NP 6	NP 7	NP 8	*p*-Value
Moisture (g)	69.9 (10.3–82.1)	71.8 (66.8–85.1)	61.4 (55.1–69.4)	50.1 (23.2–69.2)	76.4 (65.3–83.6)	24.9 (12–43.8)	72.2 (59.7–78)	85.7 (80.8–90.3)	<0.001
Plant protein (g)	5.3 (3.3–13.7)	0.7 (0.2–2)	0.7 (0.6–0.8)	1.9 (0.9–3.6)	0.5 (0.3–0.8)	3 (2.1–4.9)	1.1 (0.5–1.7)	0.9 (0.6–1.4)	<0.001
Animal protein (g)	1.4 (0–2.1)	20.4 (14.6–24.5)	24.7 (19.2–28.5)	2.5 (0.8–9.9)	3.8 (2.9–9.4)	1.4 (0.6–2.4)	9.1 (3.7–13.5)	0 (0–0.1)	<0.001
Saturated fatty acids (g)	0.2 (0.1–0.9)	1.1 (0.3–1.5)	3.1 (1.5–6.1)	6.1 (2.2–12)	1.6 (0.7–4.2)	1.6 (0.2–3.2)	2.7 (1.8–3.4)	0 (0–0.1)	<0.001
Mono-unsaturated fatty acids (g)	0.2 (0.1–1.8)	0.6 (0.2–1.4)	3.7 (2.1–6)	8.1 (3.4–16.6)	0.8 (0.3–3)	1.6 (0.3–4.7)	3.6 (1.6–4.4)	0 (0–0.1)	<0.001
Polyunsaturated fatty acids (g)	0.5 (0.2–2.8)	0.6 (0.3–1.1)	0.8 (0.3–2.4)	8.6 (3.5–11.7)	0.1 (0.1–0.8)	1.3 (0.4–3.2)	2.1 (0.7–3.6)	0.1 (0–0.2)	<0.001
Cholesterol (mg)	44.5 (8.1–49.4)	343 (165–426.8)	81 (57–97)	47.6 (21–91)	9.8 (6.5–37)	29.3 (18.5–49.5)	86.3 (62.6–280)	5.7 (2.2–7.6)	<0.001
Carbohydrate, available (g)	11.5 (4.8–23)	3.9 (2.5–5.9)	4.2 (1.6–6.3)	8.7 (3.6–19.8)	10.6 (4.9–15.4)	50.7 (38–67.5)	7 (2.4–14.9)	8.8 (4.3–14.7)	<0.001
Total dietary fibre (g)	5.5 (3.6–9.3)	1.1 (0.4–2.1)	0.7 (0.3–1.2)	1.4 (0.8–3.1)	0.2 (0.1–0.7)	1.7 (0.7–3.7)	0.3 (0.1–0.4)	1.8 (1.4–2.6)	<0.001
Calcium (mg)	54.5 (26.4–119.2)	12 (6–19)	15.2 (11–24)	21.4 (9.7–73.3)	118.9 (102.7–212)	27.7 (12.4–68.3)	65.8 (39–93.9)	15 (8–27)	<0.001
Iron (mg)	2.4 (1.5–4.7)	4.9 (1.1–6.2)	1.3 (0.8–2.2)	1 (0.5–1.7)	0.4 (0.2–0.8)	1.8 (0.9–2.9)	1.2 (0.7–1.8)	0.4 (0.3–0.6)	<0.001
Magnesium (mg)	59.5 (30–134)	16 (12–19.4)	24 (20–27)	11.5 (9–21.5)	14 (11.4–18.7)	18.2 (9.7–37.4)	11 (9.3–18.7)	11.4 (9–15)	<0.001
Phosphorous (mg)	92 (62–311.5)	244.7 (73–327)	196 (158.8–223)	79.1 (43.1–122.5)	110 (85.3–192.7)	75 (52.4–120)	142.8 (94.7–186)	26 (16–35)	<0.001
Potassium (mg)	362.5 (206–652)	229 (198–265)	287 (247–337)	143.4 (98–232.5)	141.2 (117–185)	120.9 (74.2–250.5)	128.4 (98–176)	186 (131–256.7)	<0.001

Table A2. *Cont.*

Nutrient	NP 1	NP 2	NP 3	NP 4	NP 5	NP 6	NP 7	NP 8	*p*-Value
Sodium (mg)	17 (6–66.5)	85 (69–106.6)	81 (59–204.8)	59 (12.6–230.8)	113.2 (51.4–530)	68.5 (11–159.5)	119.4 (80.9–156.8)	7 (3–25.7)	<0.001
Zinc (mg)	1 (0.6–2.5)	3 (0.8–4.2)	2.5 (1.5–4.1)	0.4 (0.3–1.1)	0.5 (0.4–1)	0.5 (0.4–0.8)	0.6 (0.5–1.1)	0.2 (0.1–0.3)	<0.001
Copper (mg)	0.2 (0.2–0.5)	0.6 (0.4–4.5)	0.1 (0.1–0.2)	0.1 (0–0.1)	0 (0–0.1)	0.1 (0.1–0.2)	0.1 (0–0.1)	0.1 (0.1–0.1)	<0.001
Manganese (μg)	1189 (416.5–1894)	199 (133–345.5)	22 (14–70.1)	118.4 (29.5–254.4)	22.3 (4.9–66.5)	253.6 (156–360)	46.8 (38–110)	120.4 (70–200.6)	<0.001
Vitamin A (RE) (μg)	25 (4.5–163.5)	175.8 (41–1753)	14.6 (8–25.4)	64.1 (14.4–178.9)	43.1 (22.1–98.7)	55.9 (10.6–119.3)	68.6 (48–108.7)	22 (4.6–62.5)	<0.001
Thiamin (mg)	0.2 (0.1–0.4)	0.2 (0.1–0.2)	0.1 (0.1–0.2)	0.1 (0–0.2)	0 (0–0.1)	0.2 (0.1–0.3)	0.1 (0.1–0.1)	0 (0–0.1)	<0.001
Riboflavin (mg)	0.1 (0–0.2)	0.7 (0.2–2.8)	0.2 (0.1–0.3)	0.1 (0–0.2)	0.2 (0.1–0.2)	0.1 (0.1–0.2)	0.2 (0.2–0.3)	0 (0–0)	<0.001
Niacin (mg)	1 (0.6–2.4)	4.6 (1.2–9.6)	5.4 (4.2–7.9)	1.1 (0.4–2.8)	0.2 (0.1–0.5)	1.8 (0.7–2.6)	0.6 (0.1–1.5)	0.5 (0.3–0.7)	<0.001
Vitamin B_6 (mg)	0.2 (0.1–0.4)	0.3 (0.1–0.7)	0.3 (0.2–0.4)	0.1 (0–0.1)	0 (0–0.1)	0.1 (0–0.2)	0 (0–0.1)	0.1 (0–0.1)	<0.001
Vitamin B_{12} (μg)	0.1 (0–0.3)	18.9 (8.8–48.8)	1.5 (0.5–2.2)	0.2 (0.1–0.4)	0.4 (0.3–0.6)	0.2 (0.1–0.3)	0.9 (0.4–1.5)	0 (0–0)	<0.001
Pantothenic acid (mg)	0.3 (0.1–0.8)	0.6 (0.2–3.3)	0.6 (0.3–1.1)	0.3 (0.1–0.5)	0.3 (0.3–0.4)	0.4 (0.2–0.4)	0.5 (0.4–1.1)	0.2 (0.1–0.3)	<0.001
Vitamin C (mg)	5.3 (1.2–14)	7.4 (2.7–10.4)	3.3 (1–8.3)	1.9 (0.9–4.3)	1 (0.8–2.2)	1 (0.2–3.8)	0.6 (0.4–4.4)	10.7 (4.4–24)	<0.001
Vitamin D (μg)	0.3 (0.3–0.9)	1.1 (0.6–1.2)	0.6 (0.2–1)	0.8 (0.4–1.9)	0.1 (0–0.5)	1 (0.5–1.7)	2.9 (1.4–5.9)	0.3 (0.2–0.3)	<0.001
Vitamin E (mg)	0.6 (0.2–2.2)	0.4 (0.1–0.8)	0.3 (0.2–0.5)	3.6 (1–6.9)	0.1 (0.1–0.6)	0.7 (0.3–1.3)	1.3 (0.7–3.4)	0.3 (0.1–0.6)	<0.001

References

1. Ene-Obong, H.; Schönfeldt, H.C.; Campaore, E.; Kimani, A.; Mwaisaka, R.; Vincent, A.; El Ati, J.; Kouebou, P.; Presser, K.; Finglas, P.; et al. Importance and use of reliable food composition data generation by nutrition/dietetic professionals towards solving Africa's nutrition problem: Constraints and the role of FAO/INFOODS/AFROFOODS and other stakeholders in future initiatives. *Proc. Nutr. Soc.* **2019**, *78*, 496–505. [CrossRef]
2. Onyango, A.W.; Jean-Baptiste, J.; Samburu, B.; Mahlangu, T.L.M. Regional Overview on the Double Burden of Malnutrition and Examples of Program and Policy Responses: African Region. *Ann. Nutr. Metab.* **2019**, *75*, 127–130. [CrossRef]
3. Ocké, M.C.; Westenbrink, S.; van Rossum, C.T.M.; Temme, E.H.M.; van der Vossen-Wijmenga, W.; Verkaik-Kloosterman, J. The essential role of food composition databases for public health nutrition—Experiences from the Netherlands. *J. Food Compos. Anal.* **2021**, *101*, 103967. [CrossRef]
4. Vorster, H.; Badham, J.; Venter, C. An introduction to the revised food-based dietary guidelines for South Africa. *S. Afr. J. Clin. Nutr.* **2013**, *26*, S5–S12.
5. Department of Health, Republic of South Africa. *Regulations Relating to the Fortification of Certain Foodstuffs (No. R. 504 of 2003)*; Department of Health, Republic of South Africa: Pretoria, South Africa, 2003; pp. 3–26.
6. South Africa Government. *Government Gazette: No. R. 214 Foodstuffs, Cosmetics and Disinfectants Act, 1972 (Act 54 of 1972) Regulations Relating to the Reduction of Sodium in Certain Foodstuffs and Related Matters*; South Africa Government: Cape Town, South Africa, 2013.
7. Health Promotion Levy on Sugary Beverages. Available online: https://www.sars.gov.za/customs-and-excise/excise/health-promotion-levy-on-sugary-beverages/ (accessed on 18 August 2021).
8. Akbay, A.; Elhan, A.; Ozcan, C.; Demirtas, S. Hierarchical cluster analysis as an approach for systematic grouping of diet constituents on basis of fatty acid, energy and cholesterol content: Application on consumable lamb products. *Med. Hypotheses* **2000**, *55*, 147–154. [CrossRef]
9. Li, Y.; Bahadur, R.; Ahuja, J.; Pehrsson, P.; Harnly, J. Macro- and micronutrients in raw plant foods: The similarities of foods and implication for dietary diversification. *J. Food Compos. Anal.* **2021**, *102*, 103993. [CrossRef]
10. Pennington, J.A.T.; Fisher, R.A. Classification of fruits and vegetables. *J. Food Compos. Anal.* **2009**, *22*, S23–S31. [CrossRef]
11. do Prado, S.B.R.; Giuntini, E.B.; Grande, F.; de Menezes, E.W. Techniques to evaluate changes in the nutritional profile of food products. *J. Food Compos. Anal.* **2016**, *53*, 1–6. [CrossRef]
12. Windham, C.T.; Windham, M.P.; Wyse, B.W.; Hansen, R.G. Cluster-Analysis to Improve Food Classification within Commodity Groups. *J. Am. Diet Assoc.* **1985**, *85*, 1306–1314. [PubMed]
13. Kim, S.; Sung, J.; Foo, M.; Jin, Y.S.; Kim, P.J. Uncovering the nutritional landscape of food. *PLoS ONE* **2015**, *10*, e0118697. [CrossRef]
14. Similä, M.; Ovaskainen, M.-L.; Virtanen, M.J.; Valsta, L.M. Nutrient content patterns of Finnish foods in a food composition database. *J. Food Compos. Anal.* **2006**, *19*, 217–224. [CrossRef]
15. Atsa'am, D.D.; Oyelere, S.; Balogun, O.S.; Wario, R.; Blamah, N.V. K-means cluster analysis of the West African species of cereals based on nutritional value composition. *Afr. J. Food Agric. Nutr.* **2021**, *21*, 17195–17212. [CrossRef]
16. Chikowore, T.; Pisa, P.T.; van Zyl, T.; Feskens, E.J.; Wentzel-Viljoen, E.; Conradie, K.R. Nutrient Patterns Associated with Fasting Glucose and Glycated Haemoglobin Levels in a Black South African Population. *Nutrients* **2017**, *9*, 9. [CrossRef] [PubMed]
17. Makura-Kankwende, C.B.T.; Gradidge, P.J.; Crowther, N.J.; Norris, S.A.; Chikowore, T. Nutrient Patterns and Body Composition Parameters of Black South African Women. *Nutrients* **2020**, *13*, 6. [CrossRef] [PubMed]
18. Pisa, P.T.; Pedro, T.M.; Kahn, K.; Tollman, S.M.; Pettifor, J.M.; Norris, S.A. Nutrient patterns and their association with socio-demographic, lifestyle factors and obesity risk in rural South African adolescents. *Nutrients* **2015**, *7*, 3464–3482. [CrossRef]

19. Visser, M.; Van Zyl, T.; Hanekom, S.M.; Baumgartner, J.; van der Hoeven, M.; Taljaard-Krugell, C.; Smuts, C.M.; Faber, M. Nutrient patterns and their relation to anemia and iron status in 5- to 12-y-old children in South Africa. *Nutrition* **2019**, *62*, 194–200. [CrossRef] [PubMed]
20. SAFOODS. *SAMRC Food Composition Tables for South Africa*, 5th ed.; South African Medical Research Council, Ed.; South African Medical Research Council: Cape Town, South Africa, 2017.
21. Jolliffe, I.T.; Cadima, J. Principal component analysis: A review and recent developments. *Philos. Trans. R. Soc. A Math. Phys. Eng. Sci.* **2016**, *374*. [CrossRef]
22. Everitt, B.; Hothorn, T. Multivariate Data and Multivariate Analysis. In *An Introduction to Applied Multivariate Analysis with R*; Springer: New York, NY, USA, 2011; pp. 61–102.
23. de Oliveira Santos, R.; Gorgulho, B.M.; de Castro, M.A.; Fisberg, R.M.; Marchioni, D.M.; Baltar, V.T. Principal Component Analysis and Factor Analysis: Differences and similarities in Nutritional Epidemiology application. *Rev. Bras. Epidemiol.* **2019**, *22*, e190041. [CrossRef] [PubMed]
24. Greenfield, H.; Southgate, D.A.T. *Food Composition Data. Production Management and Use*, 2nd ed.; Food and Agriculture Organization of the United Nations: Rome, Italy, 2003.
25. Schönfeldt, H.C.; Hall, N.; Pretorius, B. The important role of food composition in policies and programmes for better public health: A South African case study. *Food Chem.* **2018**, *238*, 94–100. [CrossRef]
26. INFOODS: Food Composition Challenges. Available online: http://www.fao.org/infoods/infoods/food-composition-challenges/en/ (accessed on 18 August 2021).
27. Schellack, G.; Harirari, P.; Schellack, N. B-complex vitamin deficiency and supplementation. *SA Pharm. J.* **2015**, *82*, 28–33.
28. Ispirova, G.; Eftimov, T.; Korošec, P.; Seljak, B.K. MIGHT: Statistical methodology for missing-data imputation in food composition databases. *Appl. Sci.* **2019**, *9*, 4111. [CrossRef]
29. Ispirova, G.; Eftimov, T.; Seljak, B.K. Evaluating missing value imputation methods for food composition databases. *Food Chem. Toxicol.* **2020**, *141*, 111368. [CrossRef] [PubMed]
30. Westrich, B.J.; Altmann, M.A.; Potthoff, S.J. Minnesota's Nutrition Coordinating Center uses mathematical optimization to estimate food nutrient values. *Interfaces* **1998**, *28*, 86–99. [CrossRef]
31. U.S. Department of Agriculture; U.S. Department of Health and Human Services. *Dietary Guidelines for Americans, 2020–2025*, 9th ed.; U.S. Department of Agriculture and U.S. Department of Health and Human Services: Washington, DC, USA, 2020.
32. Pennington, J.A.T.; Fisher, R.A. Food component profiles for fruit and vegetable subgroups. *J. Food Compos. Anal.* **2010**, *23*, 411–418. [CrossRef]
33. Peters, S.A.E.; Dunford, E.; Ware, L.J.; Harris, T.; Walker, A.; Wicks, M.; van Zyl, T.; Swanepoel, B.; Charlton, K.E.; Woodward, M.; et al. The Sodium Content of Processed Foods in South Africa during the Introduction of Mandatory Sodium Limits. *Nutrients* **2017**, *9*, 404. [CrossRef] [PubMed]
34. World Health Organization. *Joint FAO/WHO Consultation on Preparation and Use of Food-Based Dietary Guidelines (1995: Nicosia, Cyprus). Preparation and Use of Food-Based Dietary Guidelines/Report of a Joint FAO/WHO Consultation*; World Health Organization: Geneva, Switzerland, 1998.
35. EFSA Panel on Dietetic Products, Nutrition, and Allergies (NDA). Scientific Opinion on establishing Food-Based Dietary Guidelines. *EFSA J.* **2010**, *8*, 1460. [CrossRef]
36. von Braun, J.; Afsana, K.; Fresco, L.; Hassan, M.; Torrero, M. Food Systems—Definition, Concept and Application for the UN Food Systems Summit. 2020. Available online: https://www.un.org/sites/un2.un.org/files/food_systems_concept_paper_scientific_group_-_draft_oct_26.pdf (accessed on 9 September 2021).
37. Béné, C.; Prager, S.D.; Achicanoy, H.A.E.; Toro, P.A.; Lamotte, L.; Cedrez, C.B.; Mapes, B.R. Understanding food systems drivers: A critical review of the literature. *Glob. Food Secur.* **2019**, *23*, 149–159. [CrossRef]
38. Kraemer, K.; Cordaro, J.B.; Fanzo, J.; Gibney, M.; Kennedy, E.; Labrique, A.; Steffen, J.; Eggersdorfer, M. Chapter 1.1 Ten Forces Shaping the Global Food System. In *Good Nutrition: Perspectives for the 21st Century*; Karger: Basel, Switzerland, 2016; pp. 19–30.
39. Delgado, A.; Issaoui, M.; Vieira, M.C.; de Carvalho, I.S.; Fardet, A. Food Composition Databases: Does It Matter to Human Health? *Nutrients* **2021**, *13*, 2816. [CrossRef]
40. Davis, D.R.; Epp, M.D.; Riordan, H.D. Changes in USDA Food Composition Data for 43 Garden Crops, 1950 to 1999. *J. Am. Coll. Nutr.* **2004**, *23*, 669–682. [CrossRef]
41. Mayer, A.M. Historical changes in the mineral content of fruits and vegetables. *Br. Food J.* **1997**, *99*, 207–211. [CrossRef]
42. White, P.J.; Broadley, M.R. Historical variation in the mineral composition of edible horticultural products. *J. Hortic. Sci. Biotechnol.* **2005**, *80*, 660–667. [CrossRef]

nutrients MDPI

Article

Sodium Content of Foods Sold in the Spanish Market. Results from the BADALI Project

Marta Beltrá [1], Fernando Borrás [2] and Ana B. Ropero [1,*]

[1] Institute of Bioengineering, Miguel Hernández University, 03202 Elche, Spain; beltra@umh.es
[2] Department of Statistics, Mathematics and Informatics, Miguel Hernández University, 03202 Elche, Spain; f.borras@umh.es
* Correspondence: ropero@umh.es

Abstract: High sodium/salt intake is a risk factor for Non-Communicable Diseases (NCDs). Excess sodium intake has been associated with high coronary heart disease, stroke and high blood pressure. The sodium daily intake is above the recommendations in the world as well as in Spain. Reducing salt content in processed foods and ready meals is one of the main strategies for reducing sodium intake. The aim of the present work is to characterise the presence of sodium in foods sold in the Spanish market. We also study a possible shift in sodium content in products over the last few years. For this purpose, 3897 products included in the BADALI food database were analysed, classified into 16 groups (G). We found that 93.3% of all foods displayed the sodium/salt content in the nutrition declaration. Meat—processed and derivatives (G8) had the highest mean and median values for sodium content, followed by snacks (G15) and sauces (G14). Only 12.7% of foods were sodium-free (≤5 mg/100 g or 100 mL), 32.4% had very low sodium (≤40 mg/100 g or 100 mL) and 48.2% were low in sodium (≤120 mg/100 g or 100 mL). On the contrary, 47.2% were high in sodium according to the Pan American Health Organisation Nutrient Profile Model (PAHO-NPM), while there were 31.9% according to the Chile-NPM. The agreement between the two NPMs was considered 'substantial' (κ = 0.67). When sodium content was compared over the years, no decrease was observed. This analysis was performed in the entire food population, by food group and in matched products. Therefore, more effort should be made by all parties involved in order to decrease the sodium/salt intake in the population.

Keywords: salt content; nutrient composition; nutritional claims; nutrient profile/profiling models; changes in sodium content; food database; public health

Citation: Beltrá, M.; Borrás, F.; Ropero, A.B. Sodium Content of Foods Sold in the Spanish Market. Results from the BADALI Project. *Nutrients* **2021**, *13*, 3410. https://doi.org/10.3390/nu13103410

Academic Editor: Pedro Moreira

Received: 13 August 2021
Accepted: 25 September 2021
Published: 27 September 2021

1. Introduction

Reduction in salt intake was considered by the World Health Organization (WHO) in 2018 as one of the best investments to reduce Non-Communicable Diseases (NCDs) [1]. NCDs are the leading cause of death in the world. It is estimated that they are responsible for 41 million deaths in the world each year, which represents 71% of all deaths [2]. Cardiovascular diseases (CVD) and cancer account for most of those deaths [2]. In Spain, the Global Burden Disease (GBD) 2019 and data from WHO showed that 9 out of the 10 main causes of death are NCDs [3–5].

Raised blood pressure is the leading metabolic risk factor in the world contributing to NCDs [2]. High sodium/salt intake has been associated with high blood pressure and is a risk factor for NCDs [6,7]. In addition, excess sodium intake has been related to coronary heart disease and stroke [6].

According to the World Cancer Research Foundation (WCRF), there is strong evidence that consuming foods preserved by salting is a cause of stomach cancer [8]. High dietary salt has also been shown to adversely affect the vasculature, heart, kidneys, skin, brain and bone [9].

As a consequence of all the evidence, WHO stablished the maximum recommended sodium intake for adults in 2 g/d in 2012 [10]. According to the GBD, the global mean intake of sodium was 3.95 g/d in 2010 [11]. Salt/sodium intake in the Spanish population is also higher than recommended. The last estimation performed by 24 h urinary sodium excretion was of 9.8 g salt/d, with 88.2% of the subjects with intakes above 5 g/d [12]. More recent data from the ANIBES study, following a three-day food records, also showed an excess of sodium intake in Spain [13].

Randomized trials demonstrate that salt reduction lowers blood pressure in normotensive, as well as in hypertensive individuals additively to antihypertensive treatments [14]. Studies have shown that decreasing salt intake is associated with reduced risk of CVD, all-cause mortality, kidney disease, stomach cancer and osteoporosis [14]. A recent study estimated the impact of the salt reduction program in England. Salt intake decreased from 2000 to 2018 [15]. Authors calculated that maintaining the salt intake at 2018 levels would reduce considerable the cases of premature ischemic heart disease and strokes [15]. This would generate more than half million of extra quality-adjusted life-years and £1640 million health care cost savings for the adult population in England [15].

Most sodium intake in Europe and Northern American countries comes from salt added in manufactured foods (around 75% of the total intake) [16]. Therefore, reducing salt content in processed foods and ready meals is one of the main strategies for decreasing sodium intake in the population [17]. Salt reducing programs have been ongoing for some years in countries such as UK [18], Canada [19], Argentina [20], Brazil [21,22], Italy [23] and South Africa [24]. Recently, WHO released global sodium benchmarks depending on the food category [25]. Maximum sodium values were set in those programs for food groups such as bread, processed meat and fish, canned vegetables and legumes, snacks, breakfast cereals, sauces, among others.

In 2018, the Spanish Agency for Food Safety and Nutrition (AESAN), along with food professionals, released the Plan for the Improvement of the Composition of Food, Beverages and Other Measures 2020 for the period 2017–20 [26]. Reducing added sugar, salt and trans fatty acids content in foods were the main targets [26]. Snacks, processed meat, sauces, vegetable purees, ready to eat and precooked foods were the groups included in the plan for a 5–16% reduction in sodium content [26]. Joining the plan was voluntary.

In the last few years, the sodium content of foods has been studied over time. Results are diverse and depend on the food category and the country of study [27–34]. In Spain, a government report in 2015 showed a decrease in sodium content in some food categories in 2012 compared to 2009 [35]. No scientific publication has been released so far with the results of the Plan 2020.

The aim of the present work is to characterize the presence of sodium in foods sold in the Spanish market in recent years and to analyse a possible reduction over time. This study will focus on food groups.

2. Materials and Methods

2.1. BADALI Database of Food Products Available in the Spanish Market

The data used in this work come from the BADALI database project [36,37]. Details about the food and brand selection process can be found in Ropero et al., 2020 [38]. In short, the information used in this study was obtained from the manufacturers' web pages, including the nutrient composition and ingredients. Serving size for precooked and ready-to-eat foods was also obtained from online supermarkets (June 2021).

Nutrient composition of foods was extracted by the researchers and inconsistent information was not used for further analysis. For the purpose of this study and in order to reduce heterogeneity, foods were classified following similarities in the main ingredients, use and/or sodium content (Table S1). Fresh foods were poorly represented in the database, the main exception being fish and seafood (included in G10). For the calculation of the percentage of sodium daily intake, 2 g sodium/d was applied [10].

Two versions of the database were utilised for the present study. The oldest version was used only for the baseline sodium content in the comparative study. It included foods collected from June 2014 to April 2019. The newest version of the database was used for all the analyses throughout this work. It is comprised of all the foods in the previous version, except for those collected before January 2017, which were removed. In addition, the information on some foods was updated, and new information was added (from October 2020 to May 2021).

2.2. Classification of Products According to Their Sodium Content

For the classification of foods as "low in sodium", "very low in sodium" or "sodium-free", the criteria established in the Regulation (EC) No 1924/2006 and the Codex Alimentarius for the respective nutrition claims were used (Table 1) [39,40].

Table 1. Criteria used to classify foods according to their sodium content.

Criteria	Claim	Threshold-Sodium
Regulation (EC) No 1924/2006 and Codex Alimentarius [39,40]	Sodium-free Very low in sodium Low in sodium	≤5 mg/100 g or 100 mL ≤40 mg/100 g or 100 mL ≤120 mg/100 g or 100 mL
PAHO-NPM [41]	Excessive in sodium	≥1 mg/kcal
Chile-NPM [42]	High in sodium	Solids: >400 mg/100 g Liquids: >100 mg/100 mL

Two criteria were applied to classify foods as high in sodium (Table 1). On one hand, the Pan American Health Organization Nutrient Profile Model (PAHO-NPM) [41] and, on the other hand, the Chilean warning label system established by the Minister for Health (Chile-NPM) [42]. These NPMs have been previously used to determine the "healthiness" of foods, based on their content of several nutrients [43–45]. In addition, their criteria for sodium/salt was also used to classify foods as high in sodium, independently of the presence of other nutrients [46].

According to PAHO, the food and beverage products that should be evaluated with their NPM are limited to processed and ultra-processed products, which typically contain elevated amounts of sodium, free sugars, saturated fat, total fat and trans-fatty acids added by the manufacturer. There is no reason to apply the PAHO-NPM to unprocessed or minimally processed foods [41]. As for the Chile-NPM, the labelling of products as high in sodium, energy, total sugar or saturated fats is not compulsory for foods without added sugar, honey, sodium or saturated fats [42]. In spite of these restrictions, we decided to apply both NPMs to all foods in the database as this is a research project and not a public health initiative.

2.3. Comparison of Sodium Content over the Years

The two versions of the database described in Section 2.1 were used. Since some items were present in both versions, duplicates were removed prior to the analysis. For the matching comparison, identical products were chosen in different years (2–6 years gap). Small differences were permitted, given that the product didn't undergo major changes.

2.4. Statistics

The Kruskal-Wallis test is useful as a general nonparametric test for comparing more than two independent samples. It can be used to test whether such samples come from the same distribution. This test is a powerful alternative to the one-way analysis of variance. Nonparametric ANOVA has no assumption of normality of random error but the independence of random error is required. If the Kruskal-Wallis statistic is significant, the nonparametric multiple comparison tests are useful methods for further analysis.

Pairwise agreement between both NPMs in the proportions of foods classified as "high in sodium" was assessed across all foods using the κ statistics, as follows: 0.01–0.20 'slight'; 0.21–0.40 'fair'; 0.41–0.60 'moderate'; 0.61–0.80 'substantial'; 0.81–0.99 'near perfect'. When agreement is high, the κ statistics either cannot be calculated or provides inconsistent values. Therefore, for some groups the agreement was assessed by using the disagreement probability (0 to 1). When this parameter was >0.1, it was considered 'substantial'; <0.1 'near perfect' and 0 'perfect'. The statistical analysis of the application data in this work was performed with Microsoft Excel and Google Colab with Jupyter Notebooks, libraries scikit-learn 0.22.2.post1, Pandas v0.25.3, and Matplotlib Python v3.2.0. The significance level was set as $p < 0.05$ in all statistical analyses.

3. Results

3.1. Data Description and General Overview

A total of 3897 products were collected from 2017 to 2021, belonging to 169 well identified brands and classified into groups as in Table S1. As shown in Table 2, the most abundant food groups were dairies and substitutes (G5), sweets (G16) and the one-type of ingredient group (G10). The least abundant was that of fats (G6).

Table 2. Foods included in the study and foods with sodium/salt content displayed.

Food Groups	No Foods	% of Total Foods	No Foods with Sodium/Salt Content	% Foods with Sodium/Salt Content [1]
Total	3897	100	3634	93.3 [2]
G1—Bread and bread-like cereal derivatives	148	3.8	146	98.6
G2—Canned vegetables	112	2.9	104	92.9
G3—Cereal sweet derivatives	363	9.3	356	98.1
G4—Cheese	182	4.7	167	91.8
G5—Dairies and substitutes	474	12.2	465	98.1
G6—Fats	64	1.6	58	90.6
G7—Fish/seafood—canned, processed and derivatives	273	7.0	255	93.4
G8—Meat—processed and derivatives	299	7.7	281	94
G9—Non-alcoholic drinks	249	6.4	246	98.8
G10—One type of ingredient	403	10.3	285	70.7
G11—Other processed and plant based derivatives	131	3.4	129	98.5
G12—Pasta	140	3.6	136	97.1
G13—Precooked and ready-to-eat food	224	5.7	224	100
G14—Sauces	88	2.3	75	85.2
G15—Snacks	279	7.2	276	98.9
G16—Sweets	468	12.0	431	92.1

[1] Calculated as: No foods with sodium/salt content within each group × 100/Total No foods within each group. [2] Calculated as: Total No foods with sodium/salt content × 100/Total No foods surveyed.

Of the total population, 93.3% displayed the sodium or salt content (Table 2), while 263 foods did not. Precooked and ready-to-eat food (G13) displayed the sodium/salt content in all the items, while the one-type of ingredient group (G10) only in 70.7% of the cases. In 18 foods (0.5%) an error was detected. Therefore, a total of 3616 foods were subsequently used for further analysis.

The sodium content fell below 2500 mg in all items, except for 10 foods (Figure S1). For these, the values were in the range of 3480–5200 mg sodium/100 g and eight of them were canned anchovies (Figure S1—insert). Median sodium content was highest for meat—processed and derivatives (G8), followed by snacks (G15) and sauces (G14) (Table 3). Five groups had median sodium values below 50 mg/100 g: dairies and substitutes (G5), non-alcoholic drinks (G9), one-type of ingredient (G10), pasta (G12) and sweets (G16) (Table 3). These groups displayed a narrow dispersion of values (Figure S1). In fact, foods in G10 (one type of ingredient) and G12 (pasta) had no added salt. Two powdered milk in G5 (dairies and substitutes) and four tomato juices in G9 (non-alcoholic drinks) (with added salt) were exceptions with unusually high sodium content (Figure S1).

Table 3. Sodium content by food group.

Food Groups	No Foods	Mean Sodium (mg/100 g)	SD (mg/100 g)	Sodium Percentiles (mg/100 g)				
				Min	25th	50th (Median)	75th	Max
Total	3616	327.8	431.9	0	32	172	520	5200
G1	144	442.3	236.3	4	339	440	560	1240
G2	104	239.5	119.5	0	169	256	345	480
G3	355	276.8	200.7	0	152	252	356	1120
G4	165	606.2	328.3	32	400	560	748	2100
G5	465	53.2	35.1	4	40	48	60	520
G6	58	186.1	244.6	0	32	52	295	1000
G7	255	692	786.6	40	400	560	600	5200
G8	280	999.2	414.6	400	720	840	1230	3480
G9	246	13.8	42.1	0	0	4	12	400
G10	276	22.7	33.1	0	4	12	30.5	212
G11	129	324.3	284.4	0	140	320	400	2000
G12	136	22.9	22.9	0	12	12	27	120
G13	223	408.1	219.3	20	248	392	520	1440
G14	75	691.8	320.8	44	480	600	816	1960
G15	274	691.1	360	0	440	680	920	2500
G16	431	45.9	68.5	0	0	28	64	576

SD: Standard deviation.

Some other interesting results could be observed in Table 3. The sodium content of some particular food types was also calculated and shown in Table S2. In spite of their sweet taste, cereal sweet derivatives (G3) had considerable amounts of sodium, corresponding to added salt. Breakfast cereals and bars had the lowest values within this group (Table S2). Cheese (G4) could be classified into two types: fresh/soft cheese had lower sodium content than mature (Table S2). Dairies had slightly higher sodium content than substitutes and emulsion-based sauces had lower values than the rest of the foods in this group (Table S2). The sodium content of cereal-based snacks was higher than for nuts and vegetables (Table S2).

For precooked and ready-to-eat food (G13), the sodium content per serving recommended by the manufacturer could be calculated for 141 products (Table S3). The median sodium content per serving was 507.5 mg. The percentage of the daily reference intake (RI) was determined for all of them and the median was 25.4% (Table S3).

3.2. Food Classification According to Their Sodium Content

As shown in Figure 1A, only 12.7% of foods complied with the conditions for the nutrient claim sodium-free and most of them belonged to G9 (non-alcoholic drinks), G10 (one type of ingredient) and G16 (sweets) (Table S4). Around one third of foods could be categorized as very low in sodium (32.4%), while 48.2% of foods were low in sodium (Figure 1A).

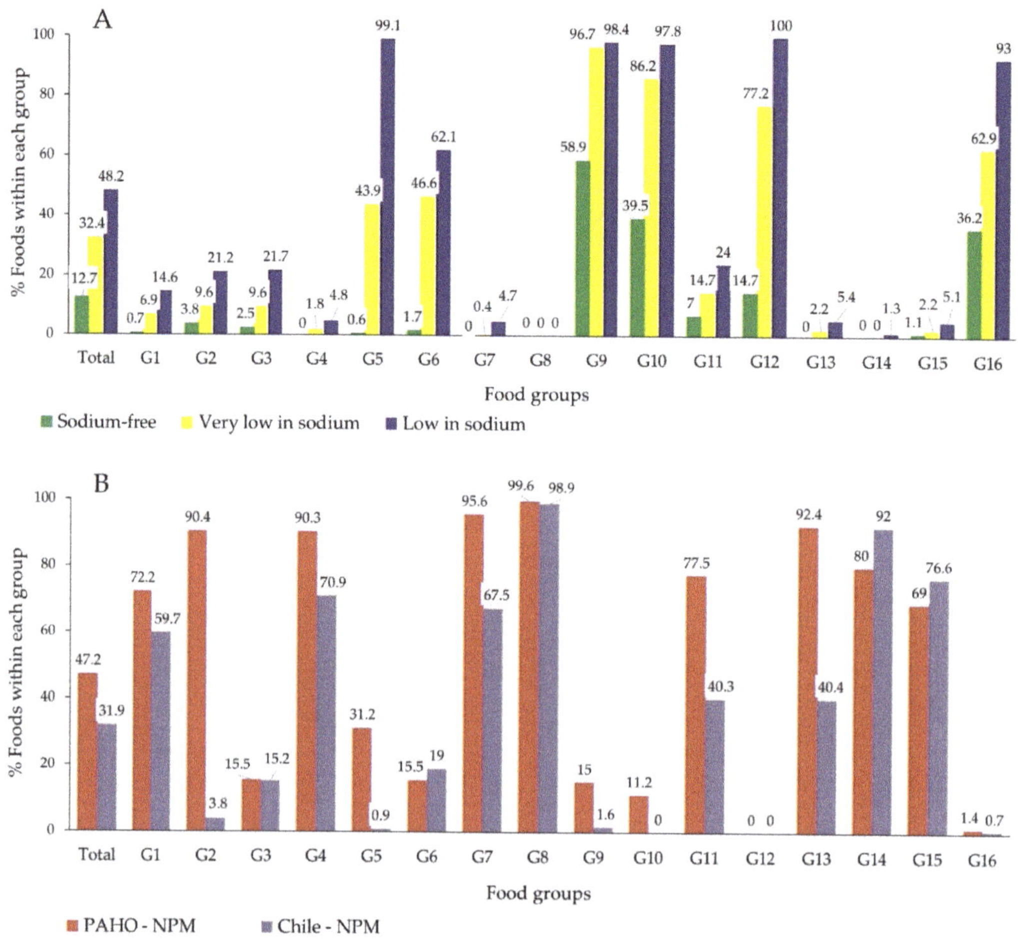

Figure 1. Foods in conformity with the nutrition claims regulated by the European Regulation No 1924/2006 [39] and the Codex Alimentarius [40] (**A**) or exceeding the NPMs thresholds (**B**) for sodium, by group.

Five groups had more than 90% of their items classified as low in sodium (G5, G9, G10, G12 and G16) (Figure 1A). Not surprisingly, none of the foods in G8 (meat) qualified for any nutrient claim on sodium. Only one could be classified as low in sodium in G14 (snacks), while less than 5% in G4 (cheese) and G7 (fish/seafood). G9 (non-alcoholic drinks) was the only group with a considerable proportion of sodium-free foods (58.9%), while for G10 (one type of ingredient) and G16 (sweets) values were below 40% (Figure 1A).

When the Nutrient Profile Models (NPMs) were applied, opposite results were obtained (Figure 1B). Near half of all foods were considered high in sodium according to the PAHO-NPM (47.2%), while around one third with the Chile-NPM (31.9%). More than 10% of all foods classified as high in sodium by both NPMs belonged to G7 (fish/seafood), G8 (meat) and G15 (snacks) each (Table S4). Only 1 and 3 foods out of 280 in the meat group (G8) was not considered high in sodium according to the PAHO-NPM and Chile-NPM respectively (Figure 1B). G4 (cheese), G7 (fish) and G14 (sauces) had also a very large proportion of foods high in sodium (Figure 1B). On the contrary, no food exceeded the maximum values in G12 (pasta), while only a few in G16 (sweets) (3 and 6 foods according to the PAHO-NPM and Chile-NPM respectively) (Figure 1B, Table S4).

It is interesting to note that the results obtained by applying both NPMs strongly differed in some groups (Figure 1B, Table S4). The κ statistics and the disagreement probability were used to compare both NPMs on foods (Table 4). The agreement was considered "substantial" for the total food database. However, important discrepancies were obtained for some food groups. On one hand, both NPMs were in accord for the G12 and had a "near perfect" agreement for G3, G6, G8 and G15. On the other hand, it was only "slight" for G2, G5, G7, G9, G13 and G14.

Table 4. Agreement between the PAHO-NPM [39] and the Chile-NPM [40].

Food Groups	κ (Confidence Interval)	Disagreement Probability [1]	Agreement [2]
Total	0.67 (0.66–0.68)	0.18	substantial
G1	0.67 (0.60–0.72)	0.15	substantial
G2	0.01 (−0.12–0.14)	0.87	slight
G3	0.86 (0.84–0.88)	0.04	near perfect
G4	0.41 (0.33–0.50)	0.19	moderate
G5	0.02 (−0.04–0.08)	0.31	slight
G6	0.88 (0.83–0.91)	0.03	near perfect
G7	0.18 (0.10–0.26)	0.27	slight
G8	-	0.01	near perfect *
G9	0.17 (0.09–0.25)	0.13	slight
G10	-	0.11	substantial *
G11	0.33 (0.22–0.43)	0.37	fair
G12	-	0	perfect *
G13	0.11 (0.02–0.19)	0.52	slight
G14	0.19 (0.04–0.34)	0.2	slight
G15	0.81 (0.78–0.83)	0.08	near perfect
G16	0.44 (0.39–0.49)	0.01	moderate

[1] Values: 0 to 1. [2] Agreement was assessed using the κ statistic as follows: 0.01–0.20 'slight'; 0.21–0.40 'fair'; 0.41–0.60 'moderate'; 0.61–0.80 'substantial'; 0.81–0.99 'near perfect'. * Agreement was assessed using the disagreement probability: >0.1 'substantial'; <0.1 'near perfect'; 0 'perfect'.

3.3. Changes in Sodium Content over the Years

When sodium content was compared over the years, some differences were observed. Foods from 2020–21 had the highest sodium content (Table 5). This may be due to different kinds of foods collected over the years. In order to minimise this confounding factor, a comparison was performed by food group. Only those with at least 30 items from three different brands were considered for this analysis (G6, G7 and G13 did not meet this requirement). Groups with no added salt (G9, G10, G12) or high heterogeneity (G11) were also discarded.

Table 5. Comparison of the sodium content (mg/100 g food) by group and year.

Food Groups	Year												p Value
	2014–16				2017–19				2020–21				
	n	25th	50th (Median)	75th	*n*	25th	50th (Median)	75th	*n*	25th	50th (Median)	75th	
Total	414	90	242 [a]	380	2455	20	100 [a]	520	1339	52	276 [b]	560	<0.001 *
G1	–	ND	ND	ND	67	140	400	600	78	400	480	560	0.881
G2	46	175	236	296	76	122	268	354	–	ND	ND	ND	0.265
G3	146	185.5	254	347.5	220	172	260	380	135	100	240	332	0.742
G4	–	ND	ND	ND	158	400	560 [a]	757	79	392	600 [b]	760	0.002 *
G5	–	ND	ND	ND	276	40	40 [a]	60	218	40	48 [b]	60	0.005 *
G8	–	ND	ND	ND	250	760	840 [a]	1110	89	720	920 [b]	1400	0.003 *
G14	–	ND	ND	ND	47	480	600 [a]	816	32	540	680 [b]	808	0.005 *
G15	34	600	700	900	123	306	640	980	151	520	680	840	0.071
G16	–	ND	ND	ND	345	0	28 [a]	64	86	16	28 [b]	51	0.005 *

* Statistically significant differences according to $p < 0.05$ by using Kruskal-Wallis test. Different lower case letters on the same line indicate significant differences in median sodium content between years. *n*: No of foods. 25th, 50th 75th: percentiles. ND: Not Determined (<30 foods or <3 brands).

As seen in Table 5, G4 (cheese), G5 (dairies and substitutes), G8 (meat) and G14 (sauces) showed statistically significant increases in sodium content in 2020–21 compared to 2017–19. No tendency to decrease sodium content was observed for any group analysed.

Further analysis was performed by comparing the same products over the years (matching products). A total of 219 foods could be studied, from 29 different brands and belonging to 12 groups. No differences were observed in sodium content (mean sodium differences = −7.1 mg/100 g; median sodium differences = 0 mg/100 g; 25th and 75th were also 0 mg/100 g).

4. Discussion

The present work analyses 3897 foods sold in the Spanish market from 2017 to 2021. Sodium content depends much on the food group. A low proportion of foods were sodium-free and almost half of foods were low in sodium. A high proportion of foods were considered high in sodium according to the two NPMs used. Both NPMs greatly disagreed in some food groups. No decrease in sodium content was observed over the years.

4.1. Sodium/Salt Content in Foods

To our knowledge, this is the first paper published in a scientific journal studying the sodium/salt content of diverse foods sold in the Spanish market. However, a previous report by the Spanish Government in 2012 showed similar results for meat, sauces, bread, cereal sweet derivatives, precooked and ready-to-eat food [35]. Our results produced higher sodium content for snacks, canned fish/seafood and cheese, while lower for canned vegetables [35].

A work on sodium content in bread in Spain was previously published in 2018 [47]. They obtained a much higher mean in bread purchased in bakeries (see Table S5) [47]. One important reason for the discrepancy may be that no bakery bread is included in the present study, but industrial bread and other similar products. In addition, the Spanish Government issued a regulation in 2019 to limit the sodium content in bread [48].

As it can be observed in Table S5, the present results are in line with preceding works [49–54]. The sodium/salt content of foods has been studied in the last five years in a number of countries (Table S5). Meat is the food group with the highest mean/median sodium content in all the countries (except for sauces in some of them) (Table S5). The values for most food groups do not vary greatly among studies (including the present one), except for sauces. Still, the discrepancies may be due to the diverse definition of the food categories (see comments in Table S5). In fact, unlike the present work, most of the publications do not describe the food groups.

4.2. Classification of Foods According to Their Sodium/Salt Content

To our knowledge, this is the first paper using the entire set of nutrient claims for salt/sodium defined by the European Commission (EC) and the Codex Alimentarius to classify foods [39,40]. However, the definition for low in sodium was previously used in a Brazilian study and it rendered 7% of the 1416 foods analysed [55]. With the same threshold, we found higher numbers (48.2%).

Our results show a great proportion of foods as high in sodium. By applying the same NPMs, a study in Honduras with 1009 foods obtained higher values: 55.8% according to PAHO-NPM (47.2% in the present work) and 68.6% when using the Chile-NPM (31.9% here) [45]. The differences could be due to the type of foods used in both studies. The work in Honduras only analysed processed and ultra-processed foods as defined by the NOVA classification [45]. However, in the present work, the NPMs were applied to all foods regardless of their level of processing.

Applying both NPMs resulted in some important differences in the present work. More foods were classified as high in sodium with the PAHO-NPM than with the Chile-NPM. The study in Honduras obtained opposite results, which may be due to the same reasons explained in the previous paragraph [45]. In addition, individual groups presented even greater discrepancies, which was also the case in the work in Honduras. This may be because, on one hand, the criteria for the PAHO-NPM is based on sodium per kcal regardless of the type of food [41]. On the other hand, the thresholds for Chile-NPM only consider the sodium content and they differ for solids and liquids [42]. Divergences are not exclusive of these two NPMs [44,45].

Recently, WHO released global sodium benchmarks for more than 50 food subcategories. Table S2 shows the thresholds for some of the food types analysed in this work. Except for breakfast cereals and salads, the median values of the rest of food types surpassed the benchmarks set by WHO [25].

4.3. Reduction in Sodium/Salt Content

In the present work, no reduction of sodium content was obtained over time either in the total food sample or in any of the nine food groups analysed. Previous papers have shown diverse outcomes. A study in Canada compared more than 6000 foods/year in 2010 vs. 2013 [32]. The authors found a significant reduction of sodium content only in 16.2% of foods categories, while no changes were observed in 81.9% of them [32]. An analysis performed in Costa Rica on more than 1000 foods/year showed decreased mean sodium content in 3 out of 18 food categories, cakes being one of them [27]. Similarly, an Indian work with 1407 products, only found a reduction in ready meals and canned vegetables, while sodium increased in 5 out of 29 food categories [30]. A paper studying salt content in sauces in the UK showed a significant reduction in median salt content in eight out of seventeen sauce categories [31].

A comparison of 219 matched products did not show a reduction in sodium content in the present paper (2017–19 vs. 2020–21). A report by the Spanish Government compared matched or similar products between 2009 and 2012 [35]. They found that sodium content decreases in breakfast cereals, soups, canned fish/seafood and industrial bread. On the contrary, sodium values increased in processed meat and sauces, which is in line with our results [35].

No changes in sodium content over time were found in a study in New Zealand comparing 182 products in 2003 vs. 2013 [33]. The same results were obtained in a food sample in Slovenia (98 foods, 2011 vs. 2015) [29]. However, an overall reduction of 23% in sodium content was obtained in a sample of 130 foods in the Australian market in 2013 vs. 1980 [34]. The comparison of 2979 matched products in the USA showed a statistically significant reduction of sodium content in 13 out of the 14 food groups analysed (2009 vs. 2015) [28].

As it seems, there is not a consensual reduction in sodium/salt content in foods in recent years. Neither the Spanish Plan for the Improvement of the Composition of Food, Beverages and Other Measures 2020 [26] has produced an effective decrease in sodium content in snacks, processed meat and sauces, according to our results. It is feasible that changes may only be detected in large samples, due to variability and bias in the collected information. However, the most probable reason is that the sodium content in foods has not really decreased over the last years.

4.4. Sodium/Salt in the Diet

As mentioned in the Introduction, daily sodium intake in the world and in Spain exceeds the recommendations [11–13]. According to the ANIBES study, processed meat and bread are the main dietary sources of sodium intake in the Spanish population [13]. Our results show that processed meat and derivatives was the food group with the highest sodium content values and that it increased over the years. In 2012, a voluntary agreement between the Spanish Agency for Food Safety and Nutrition (AESAN), the Spanish Confederation of Meat Retailers (CEDECARNE) and the Association of Manufacturer and Retailers of Food Additives and Supplements (AFCA) was signed to decrease the sodium content of these foods [56]. Whether that actually happened or not at that time, our results show that there are still reasons to be concerned about the high sodium content of processed meat.

Regarding bread, the Spanish Government released a decree in 2019 establishing a maximum sodium content for bread of 520 or 660 mg sodium/100 g, depending on the analytical method used [48]. Our data show a lower sodium content even before the decree was enforced (years 2017–19). We should mention that bread elaborated in bakeries was not included in our database while other bread-like foods were.

4.5. Strengths and Limitations

The present work has some important strengths:

- This is the first paper published in a scientific journal studying the sodium/salt content of diverse foods sold in the Spanish market;
- Foods from all groups were analysed, which provided an overview of the Spanish market;
- More than 3800 foods were analysed and the number of foods per group was significant;
- Most foods included in the database were processed, which usually have added salt.
- Data was collected following criteria completely unrelated to the aim of this study or the targeted population and, as a consequence, our results lack any bias on food choice;
- The comparative analysis of sodium content was performed at different levels in order to minimize heterogeneity in foods included in the database in different years.

The limitations are also to be mentioned:

- Data collected were reliant on the accuracy of the information provided on the manufacturer's webpage;
- Selection of brands did not follow criteria based on customer's purchase or the most popular products;
- The 3897 foods analysed may not be representative of the Spanish market due to the huge amount of foods available;
- Many of the products displayed 0 g salt/sodium, which could be wrongly rounded. The EC published a guidance document with rounding instructions, but it is not compulsory [57].
- The number of products for the matched comparative study was low, although in line with most of previous studies.

5. Conclusions

The results of our study reveal that sodium content in foods in the Spanish market is very high. Much work is still ahead of us in order to achieve the WHO challenge of reducing the sodium intake of the population by 30% by 2025. The Spanish Plan for the Improvement of the Composition of Food, Beverages and Other Measures 2020 [26] has not produced an effective decrease in sodium content in some of the food groups targeted. The voluntary nature of the agreement with food professionals is clearly insufficient to produce positive results. Unless mandatory regulations were issued, the effectiveness of such programs will be very limited. A firm compromise of all parties involved, governments, industry and consumers, is required for demanding, enforcing and monitoring a truly effective program.

Supplementary Materials: The following are available online at https://www.mdpi.com/article/10.3390/nu13103410/s1, Figure S1: Sodium content of all foods in the database, Table S1: Description of the items included in the food groups, Table S2: Sodium content by specific food type, Table S3: Sodium content per serving for the precooked and ready-to eat food group, Table S4: Foods in conformity with the nutrition claims regulated by the European Regulation No 1924/2006 and Codex Alimentarius or exceeding the NPMs thresholds for sodium, by group, Table S5: Sodium content in different studies by group.

Author Contributions: Conceptualization and coordination, A.B.R.; data collection, M.B.; data analysis, F.B. and A.B.R.; writing and reviewing the manuscript, A.B.R., M.B. and F.B.; data curation,

A.B.R. and M.B.; methodology, A.B.R. and M.B. All authors have read and agreed to the published version of the manuscript.

Funding: Publication fees were paid with a contract signed by A.B. Ropero and M. Beltrá with the Council of Ibi. The aim of this contract was to improve eating habits among pre-school and primary school students in the city.

Institutional Review Board Statement: Not applicable.

Informed Consent Statement: Not applicable.

Data Availability Statement: Not applicable.

Acknowledgments: The authors would like to thank collaborating colleagues and students for their contribution to the BADALI project. Their names are included in https://badali.umh.es/equipo-badali (accessed on 28 July 2021).

Conflicts of Interest: The authors declare no conflict of interest.

References

1. World Health Organization. Saving Lives, Spending Less. A Strategic Response to Noncommunicable Diseases. 2018. Available online: https://apps.who.int/iris/bitstream/handle/10665/272534/WHO-NMH-NVI-18.8-eng.pdf (accessed on 7 August 2021).
2. World Health Organization. Noncommunicable Diseases. 2021. Available online: https://www.who.int/news-room/fact-sheets/detail/noncommunicable-diseases (accessed on 28 July 2021).
3. GBD 2019 Diseases and Injuries Collaborators. Global burden of 369 diseases and injuries in 204 countries and territories, 1990–2019: A systematic analysis for the Global Burden of Disease Study 2019. *Lancet* **2020**, *396*, 1204–1222. [CrossRef]
4. Institute for Health Metrics and Evaluation (IHME). Country Profiles. Spain. Available online: http://www.healthdata.org/spain (accessed on 10 September 2021).
5. World Health Organization. Global Health Estimates: Leading Causes of Death. Cause-Specific Mortality, 2000–2019. Available online: https://www.who.int/data/gho/data/themes/mortality-and-global-health-estimates/ghe-leading-causes-of-death (accessed on 10 September 2021).
6. Aljuraiban, G.S.; Jose, A.P.; Gupta, P.; Shridhar, K.; Prabhakaran, D. Sodium intake, health implications, and the role of population-level strategies. *Nutr. Rev.* **2021**, *79*, 351–359. [CrossRef] [PubMed]
7. EFSA Panel on Nutrition, Novel Foods and Food Allergens (NDA). Dietary reference values for sodium. *EFSA J.* **2019**, *17*, 5778. [CrossRef]
8. World Cancer Research Fund/American Institute for Cancer Research. Diet, Nutrition, Physical Activity and Cancer: A Global Perspective. The Third Expert Report. 2018. Available online: https://www.wcrf.org/diet-and-cancer/ (accessed on 30 July 2021).
9. Robinson, A.T.; Edwards, D.G.; Farquhar, W.B. The Influence of Dietary Salt Beyond Blood Pressure. *Curr. Hypertens. Rep.* **2019**, *21*, 42. [CrossRef] [PubMed]
10. World Health Organization. Guideline: Sodium Intake for Adults and Children. 2012. Available online: https://www.who.int/nutrition/publications/guidelines/sodium_intake_printversion.pdf (accessed on 30 July 2021).
11. Powles, J.; Fahimi, S.; Micha, R.; Khatibzadeh, S.; Shi, P.; Ezzati, M.; Engell, R.E.; Lim, S.S.; Danaei, G.; Mozaffarian, D.; et al. Global, regional and national sodium intakes in 1990 and 2010: A systematic analysis of 24 h urinary sodium excretion and dietary surveys worldwide. *BMJ Open* **2013**, *3*, e003733. [CrossRef]
12. Ortega, R.M.; López-Sobaler, A.M.; Ballesteros, J.M.; Pérez-Farinós, N.; Rodríguez-Rodríguez, E.; Aparicio, A.; Perea, J.M.; Andrés, P. Estimation of salt intake by 24 h urinary sodium excretion in a representative sample of Spanish adults. *Br. J. Nutr.* **2011**, *105*, 787–794. [CrossRef]
13. Partearroyo, T.; Samaniego-Vaesken, M.L.; Ruiz, E.; Aranceta-Bartrina, J.; Gil, Á.; González-Gross, M.; Ortega, R.M.; Serra-Majem, L.; Varela-Moreiras, G. Sodium Intake from Foods Exceeds Recommended Limits in the Spanish Population: The ANIBES Study. *Nutrients* **2019**, *11*, 2451. [CrossRef]
14. He, F.J.; Tan, M.; Ma, Y.; MacGregor, G.A. Salt Reduction to Prevent Hypertension and Cardiovascular Disease. *J. Am. Coll. Cardiol.* **2020**, *75*, 632–647. [CrossRef]
15. Alonso, S.; Tan, M.; Wang, C.; Kent, S.; Cobiac, L.; MacGregor, G.A.; He, F.J.; Mihaylova, B. Impact of the 2003 to 2018 Population Salt Intake Reduction Program in England. A Modeling Study. *Hypertension* **2021**, *77*, 1086–1094. [CrossRef]
16. Brown, I.J.; Tzoulaki, I.; Candeias, V.; Elliott, P. Salt intakes around the world: Implications for public health. *Int. J. Epidemiol.* **2009**, *38*, 363–384. [CrossRef]
17. Ponzo, V.; Pellegrini, M.; Costelli, P.; Vázquez-Araújo, L.; Gayoso, L.; D'Eusebio, C.; Ghigo, E.; Bo, S. Strategies for Reducing Salt and Sugar Intakes in Individuals at Increased Cardiometabolic Risk. *Nutrients* **2021**, *13*, 279. [CrossRef]

18. Public Health England. Salt Reduction Targets for 2024. 2020. Available online: https://assets.publishing.service.gov.uk/government/uploads/system/uploads/attachment_data/file/915406/2024_salt_reduction_targets_070920-FINAL-1.pdf (accessed on 30 July 2021).
19. Government of Canada. Voluntary Sodium Reduction Targets for Processed Foods 2020–2025. Available online: https://www.canada.ca/en/health-canada/services/food-nutrition/healthy-eating/sodium/sodium-reduced-targets-2020-2025.html (accessed on 30 July 2021).
20. Boletín Oficial de la República Argentina. Ley 26.905. Consumo de sodio. Valores Máximos. 2013. Available online: https://www.boletinoficial.gob.ar/detalleAviso/primera/99389/20131216?busqueda=1 (accessed on 29 July 2021).
21. Ministério da Saúde. Brasil. Termo de Compromisso que Firmam entre si a União, por Intermédio do Ministério da Saúde, e a Associação Brasileira das Indústrias de Alimentação (ABIA), a Associação Brasileira das Indústrias de Massas Alimentícias (ABIMA), a Associação Brasileira da Indústria de Trigo (ABITRIGO) e a Associação Brasileira da Indústria de Panificação e Confeitaria (ABIP), com a Finalidade de Estabelecer Metas Nacionais para a Redução do teor de Sódio em Alimentos Processados no Brasil. 2011. Available online: http://189.28.128.100/nutricao/docs/geral/termo_abia.pdf (accessed on 29 July 2021).
22. Ministério da Saúde. Brasil. Metas de Redução sal Pactuadas nos Termos de Compromisso. Available online: http://189.28.128.100/dab/docs/portaldab/documentos/promocao/lista_categorias_metas_reducao_sodio.pdf (accessed on 29 July 2021).
23. Ministry of Health. General Directorate for Food Hygiene, Food Safety and Nutrition. Office V. Italy. Shared Objectives for Improving the Nutritional Characteristics of Food Products, with a Particular Focus on Children (3–12 years). Available online: https://www.salute.gov.it/imgs/C_17_pubblicazioni_2426_ulterioriallegati_ulterioreallegato_0_alleg.pdf (accessed on 31 July 2021).
24. Republic of South Africa. Minister of Health. Foodstuffs, Cosmetics and Disinfectants Act 54 of 1972. Regulations Relating to the Reduction of Sodium in Certain Foodstuffs and Related Matters. 2019. Available online: https://www.foodfocus.co.za/assets/documents/UPDATED_Notice%20214%20-Regulations%20relating%20reduction%20of%20sodium.pdf (accessed on 31 July 2021).
25. World Health Organization. WHO Global Sodium Benchmarks for Different Food Categories. 2021. Available online: https://www.who.int/publications/i/item/9789240025097 (accessed on 10 September 2021).
26. Gobierno de España. Ministerio de Sanidad, Consumo y Bienestar Social. Plan de Colaboración para la Mejora de la Composición de los Alimentos y Bebidas y Otras Medidas. 2020. Available online: https://www.aesan.gob.es/AECOSAN/docs/documentos/nutricion/DOSSIER_PLAN_2020.pdf (accessed on 28 July 2021).
27. Vega-Solano, J.; Blanco-Metzler, A.; Benavides-Aguilar, K.F.; Arcand, J. An Evaluation of the Sodium Content and Compliance with the National Sodium Reduction Targets among Packaged Foods Sold in Costa Rica in 2015 and 2018. *Nutrients* **2019**, *11*, 2266. [CrossRef] [PubMed]
28. Clapp, J.E.; Niederman, S.A.; Leonard, E.; Curtis, C.J. Changes in Serving Size, Calories, and Sodium Content in Processed Foods From 2009 to 2015. *Prev. Chronic Dis.* **2018**, *15*, 170265. [CrossRef]
29. Pravst, I.; Lavriša, Ž.; Kušar, A.; Miklavec, K.; Žmitek, K. Changes in Average Sodium Content of Prepacked Foods in Slovenia during 2011–2015. *Nutrients* **2017**, *9*, 952. [CrossRef] [PubMed]
30. Johnson, C.; Thout, S.; Mohan, S.; Dunford, E.; Farrand, C.; Wu, J.H.Y.; He, F.J.; Shivashankar, R.; Webster, J.; Krishnan, A.; et al. Labelling completeness and sodium content of packaged foods in India. *Public Health Nutr.* **2017**, *20*, 2839–2846. [CrossRef]
31. Tan, M.; He, F.J.; Jingmin, D.; Yuan, L.; Puhong, Z.; MacGregor, G.A. Salt content of sauces in the UK and China: Cross-sectional surveys. *BMJ Open* **2019**, *9*, e025623. [CrossRef] [PubMed]
32. Arcand, J.; Jefferson, K.; Schermel, A.; Shah, F.; Trang, S.; Kutlesa, D.; Lou, W.; L'Abbe, M.R. Examination of food industry progress in reducing the sodium content of packaged foods in Canada: 2010 to 2013. *Appl. Physiol. Nutr. Metab.* **2016**, *41*, 684–690. [CrossRef] [PubMed]
33. Monro, D.; Mhurchu, C.N.; Jiang, Y.; Gorton, D.; Eyles, H. Changes in the Sodium Content of New Zealand Processed Foods: 2003–2013. *Nutrients* **2015**, *7*, 4054–4067. [CrossRef]
34. Zganiacz, F.; Wills, R.B.H.; Mukhopadhyay, S.P.; Arcot, J.; Greenfield, H. Changes in the Sodium Content of Australian Processed Foods between 1980 and 2013 Using Analytical Data. *Nutrients* **2017**, *9*, 501. [CrossRef]
35. Ministerio de Sanidad, Servicios Sociales e Igualdad. Agencia Española de Consumo, Seguridad Alimentaria y Nutrición. Contenido de sal de los Alimentos en España 2012. 2015. Available online: https://www.aesan.gob.es/AECOSAN/docs/documentos/nutricion/estudio_contenido_sal_alimentos.pdf (accessed on 14 September 2021).
36. Base de Datos de Alimentos, BADALI. Available online: https://badali.umh.es (accessed on 28 July 2021).
37. Ropero, A.B.; Marquina, E.; Sarmiento, V.M.; Beltrá, M. BADALI: Una herramienta de promoción de la salud. *Rev. Esp. Nutr. Hum. Diet.* **2017**, *21*, 335–350. [CrossRef]
38. Ropero, A.B.; Blain, N.; Beltrá, M. Nutrition Claims Frequency and Compliance in a Food Sample of the Spanish Market: The BADALI Study. *Nutrients* **2020**, *12*, 2943. [CrossRef]
39. Regulation (EC) No 1924/2006 of the European Parliament and of the Council of 20 December 2006 on Nutrition and Health Claims Made on Foods. 2006. Available online: https://eur-lex.europa.eu/legal-content/EN/TXT/PDF/?uri=CELEX:32006R1924&from=en (accessed on 28 July 2021).
40. World Health Organization and Food and Agriculture Organization of the United Nations. Codex Alimentarius. Food Labelling. Fifth Edition. Available online: http://www.fao.org/3/a1390e/a1390e.pdf (accessed on 10 September 2021).

41. Pan American Health Organization. Nutrient Profile Model. 2016. Available online: https://iris.paho.org/bitstream/handle/10665.2/18621/9789275118733_eng.pdf?sequence=9&isAllowed=y (accessed on 28 July 2021).
42. Diario Oficial de la República de Chile. Cuerpo I—12. 26 de Junio de 2015. No 41.193. (IdDO 918813) Modifica Decreto Supremo No 977, de 1996, Reglamento Sanitario de los Alimentos Núm. 13. 2015. Available online: https://www.minsal.cl/wp-content/uploads/2015/08/decreto_etiquetado_alimentos_2015.pdf (accessed on 30 July 2021).
43. Duran, A.C.; Ricardo, C.Z.; Mais, L.A.; Martins, A.P.B.; Taillie, L.S. Conflicting Messages on Food and Beverage Packages: Front-of-Package Nutritional Labeling, Health and Nutrition Claims in Brazil. *Nutrients* **2019**, *11*, 2967. [CrossRef] [PubMed]
44. Poon, T.; Labonté, M.; Mulligan, C.; Ahmed, M.; Dickinson, K.M.; L'Abbé, M.R. Comparison of nutrient profiling models for assessing the nutritional quality of foods: A validation study. *Br. J. Nutr.* **2018**, *120*, 567–582. [CrossRef]
45. Hernandez, A.; Waleska, S.; Rojas, D.; Enríquez, J.P.; Di lorio, A.B. Evaluation of the Nutritional Quality of Processed Foods in Honduras: Comparison of Three Nutrient Profiles. *Int. J. Environ. Res. Public Health* **2020**, *17*, 7060. [CrossRef] [PubMed]
46. Beltrá, M.; Soares-Micoanski, K.; Navarrete-Muñoz, E.-M.; Ropero, A.B. Nutrient Composition of Foods Marketed to Children or Adolescents Sold in the Spanish Market: Are They Any Better? *Int. J. Environ. Res. Public Health* **2020**, *17*, 7699. [CrossRef] [PubMed]
47. Pérez Farinós, N.; Santos Sanz, S.; Dal Re, M.-A.; Yusta Boyo, M.J.; Robledo, T.; Castrodeza, J.J.; Campos Amado, J.; Villar, C. Salt content in bread in Spain, 2014. *Nutr. Hosp.* **2018**, *35*, 650–654. [CrossRef]
48. España. Ministerio de la Presidencia, Relaciones con las Cortes e Igualdad. Real Decreto 308/2019, de 26 de Abril, por el que se Aprueba la Norma de Calidad para el pan. BOE-A-2019-6994. Available online: https://www.boe.es/eli/es/rd/2019/04/26/308 (accessed on 31 July 2021).
49. Allemandi, L.; Tiscornia, M.V.; Guarnieri, L.; Castronuovo, L.; Martins, E. Monitoring Sodium Content in Processed Foods in Argentina 2017-2018: Compliance with National Legislation and Regional Targets. *Nutrients* **2019**, *11*, 1474. [CrossRef]
50. Arcand, J.; Blanco-Metzler, A.; Benavides Aguilar, K.; L'Abbe, M.R.; Legetic, B. Sodium Levels in Packaged Foods Sold in 14 Latin American and Caribbean Countries: A Food Label Analysis. *Nutrients* **2019**, *11*, 369. [CrossRef]
51. Carcea, M.; Narducci, V.; Turfani, V.; Aguzzi, A. A Survey of Sodium Chloride Content in Italian Artisanal and Industrial Bread. *Foods* **2018**, *7*, 181. [CrossRef] [PubMed]
52. Peters, S.A.E.; Dunford, E.; Ware, L.J.; Harris, T.; Walker, A.; Wicks, M.; Van Zyl, T.; Swanepoel, B.; Charlton, K.E.; Woodward, M.; et al. The Sodium Content of Processed Foods in South Africa during the Introduction of Mandatory Sodium Limits. *Nutrients* **2017**, *9*, 404. [CrossRef]
53. Coyne, K.J.; Baldridge, A.S.; Huffman, M.D.; Jenner, K.; Xavier, D.; Dunford, E.K. Differences in the sodium content of bread products in the USA and UK: Implications for policy. *Public Health Nutr.* **2018**, *21*, 632–636. [CrossRef] [PubMed]
54. Shahar, S.; You, Y.X.; Zainuddin, N.S.; Michael, V.; Ambak, R.; Haron, H.; He, F.J.; MacGregor, G.A. Sodium content in sauces—A major contributor of sodium intake in Malaysia: A cross-sectional survey. *BMJ Open* **2019**, *9*, e025068. [CrossRef] [PubMed]
55. Martins, C.; De Sousa, A.; Veiros, M.; González-Chica, D.; Proença, R. Sodium content and labelling of processed and ultra-processed food products marketed in Brazil. *Public Health Nutr.* **2015**, *18*, 1206–1214. [CrossRef] [PubMed]
56. España. Agencia Española de Consumo Seguridad Alimentaria y Nutrición. Convenio AESAN-CEDECARNE-AFCA para Reducir la sal y la grasa en los Productos de Carnicería-Charcutería (2012). Available online: https://www.aesan.gob.es/AECOSAN/web/noticias_y_actualizaciones/noticias/2012/reducir_sal.htm (accessed on 30 July 2021).
57. European Commission. Health and Consumers Directorate-General. Guidance Document for Competent Authorities for the Control of Compliance with EU Legislation on: Regulation (EU) No 1169/2011 of the European Parliament and of the Council of 25 October 2011 on the Provision of Food Information to Consumers, Amending Regulations (EC) No 1924/2006 and (EC) No 1925/2006 of the European Parliament and of the Council, and Repealing Commission Directive 87/250/EEC, Council Directive 90/496/EEC, Commission Directive 1999/10/EC, Directive 2000/13/EC of the European Parliament and of the Council, Commission Directives 2002/67/EC and 2008/5/EC and Commission Regulation (EC) No 608/2004 and Council Directive 90/496/EEC of 24 September 1990 on Nutrition Labelling of Foodstuffs and Directive 2002/46/EC of the European Parliament and of the Council of 10 June 2002 on the Approximation of the Laws of the Member States Relating to Food Supplements with Regard to the Setting of Tolerances for Nutrient Values Declared on a Label. 2012. Available online: https://www.fsai.ie/uploadedFiles/guidance_tolerances_december_2012.pdf (accessed on 13 August 2021).

nutrients

MDPI

Article

Non-Conjugated-Industrially-Produced-Trans Fatty in Lebanese Foods: The Case of Elaidic and Linolelaidic Acids

Maha Hoteit [1], Edwina Zoghbi [2], Alissar Rady [2], Iman Shankiti [2], Carla Ibrahim [1,3] and Ayoub Al-Jawaldeh [4,*]

1 PHENOL Research Group (Public Health Nutrition Program-Lebanon), Faculty of Public Health, Lebanese University, Beirut 6573, Lebanon; m.hoteit@ul.edu.lb (M.H.); carla.t.ibrahim@outlook.com (C.I.)
2 Country Office for Lebanon, World Health Organization, Beirut 5391, Lebanon; zoghbi@who.int (E.Z.); Radya@who.int (A.R.); shankitii@who.int (I.S.)
3 Department of Nutrition and Food Sciences, Faculty of Arts and Sciences, Holy Spirit University of Kaslik, Jounieh P.O. Box 446, Lebanon
4 World Health Organization Regional Office for the Eastern Mediterranean, Cairo 11371, Egypt
* Correspondence: aljawaldeha@who.int

Abstract: To determine Industrially-Produced Trans fatty acids (IP-TFAs) distribution of Lebanese traditional foods, especially regarding Elaidic acid (EA; 9t18:1) and Linolelaidic acid (LEA; 9t12t18:2), a mapping exercise was enrolled between January 2019 and April 2021 in which 145 food samples of three categories (traditional dishes, Arabic sweets, and market food products) were analyzed using Gas chromatography methods. Results showed that about 93% of the products tested in Lebanon, between 2019 and 2021, met the World Health Organization recommendations, while about 7% exceeded the limit. The mean level of the IP-TFAs Elaidic and Linolelaidic acid in most Traditional dishes (0.9%), Arabic sweets (0.6%), butter and margarine (1.6%), and market foods (0.52%) were relatively low compared with other countries. Despite that, the relative impact of IP-TFAs on heart diseases mortality in Lebanon is limited but unambiguously still substantial. The persistence of food products with high IP-TFAs levels threatens the health of Lebanese people. Fortunately, this problem is fairly easy to solve in Lebanon via proper legislation.

Keywords: industrially-produced trans fatty acids; Elaidic acid; Linolelaidic acid; traditional dishes; Arabic sweets; market foods; Lebanon

Citation: Hoteit, M.; Zoghbi, E.; Rady, A.; Shankiti, I.; Ibrahim, C.; Al-Jawaldeh, A. Non-Conjugated-Industrially-Produced-Trans Fatty in Lebanese Foods: The Case of Elaidic and Linolelaidic Acids. *Nutrients* **2021**, *13*, 3664. https://doi.org/10.3390/nu13103664

Academic Editors: Alessandra Durazzo, Massimo Lucarini and Laura den Hartigh

Received: 11 August 2021
Accepted: 13 October 2021
Published: 19 October 2021

1. Introduction

The intake of Industrially-produced-Trans fatty acids (IP-TFAs) is associated with an increased risk of heart attacks and death from coronary heart disease (CHD) [1]. A 2% absolute increase in energy intake from IP-TFAs has been associated with a 23% increase in cardiovascular risk [2]. In 2018, IP-TFAs elimination was identified as one of the priority targets in the World Health Organization (WHO) 13th General Programme of Work, which guides the five-year work of WHO in 2019–2023 [3]. Also, in 2018, the REPLACE action package was launched to help countries removing IP-TFAs from their food supplies [4]. In addition, WHO released additional resources in 2019 to support country actions, including six implementation modules and a live policy tracking map—the TFAs Country Score Card 1—to monitor global progress towards the 2023 target [3]. In 2020, WHO established an indicator that records whether countries have adopted WHO best-practice policies for eliminating IP-TFAs [5]. Around fifty-eight countries have introduced laws to date that will protect more than 3 billion people from TFAs by the end of 2021 [3]. However, more than 100 countries have yet to act to eliminate TFAs from their national food supply and make the world TFAs free by 2023 [3]. The European Region has the largest number of mandatory TFA limits in place and has had the most policy progress of all WHO regions since 2019. Since Denmark's effort (2004), Austria (2009), Iceland (2011), Hungary (2014), Norway (2014), Latvia (2018), Slovenia (2018) [6], and New Zealand (2008) have passed similar best-practice regulations [7]. Switzerland, one of the first countries in Europe to take legal action to restrict TFA, has a TFA limit in oils and fats (2008) [6]. The Eastern Mediterranean Region (EMR), as well as Lebanon, have witnessed rapid modernization in the last thirty years that has led to a dramatic transformation affecting people's lifestyles and diets. The average intake of saturated fatty acids (SFAs) and IP-TFAs in EMR exceeded the WHO upper limits and was estimated to be 10.3% and 1.9% of total energy intake (EI), respectively [8]. The highest SFAs intake was reported in Djibouti, Kuwait, Saudi Arabia, Lebanon, and Yemen, while the highest intake of IP-TFAs was reported in Egypt and Pakistan [8]. According to recent national data, the proportion of coronary heart diseases (CHD) death due to IP-TFA intake is 9.4% (>0.5% energy) [5] and a high burden of NCDs, accounting for 91% of total annual deaths with CVDs responsible for 47% of total deaths [9] was observed in Lebanon. As a result, the urgent need for policy measures to protect cardiovascular health is more apparent than ever and presents a historic imperative to prioritize and invest in public health by adopting health-promoting policy measures, including industrially produced Trans fatty acids (IP-TFAs) elimination. Although limited data are available on IP-TFAs intake globally, a recent report estimated that the 2017 global market volume of partially hydrogenated vegetable oils (PHVO)—the main source of IP-TFAs in food—was approximately 13.6 million tones [10]. PHVO constitutes 25% to 45% of total fat [6]. Their removal from the global food supply could prevent up to 17 million deaths by 2040 and would be the first time an NCDs risk factor has been eliminated [11]. The most common non-conjugated IP-TFA in the human's daily diet are 18-carbon fatty acids with one double bond in the 9-carbon transposition or two double bonds in the 9 and 12 carbon, called Elaidic acid (EA; 9t18:1) and Linolelaidic acid (LEA; 9t12t18:2) respectively [12]. EA and LEA were associated with various health problems [13]. EA, which is the *trans* form of oleic acid (OA, C18:1 *cis*), is the principal IP-TFA found in PHVO and margarine. EA intake resulted in significant hyperlipidemia, inflammation, and fatty liver alterations [14]. LEA is an omega-6 TFA (9E,12E-9t12t18:2), principally discovered in foods with fried or high-heat cooking or PHVO [15]. It was suspected to enhance the adipogenic differentiation favoring obesity [15]. Moreover, LEA appeared to be potentially more detrimental than EA and LEA contributed to higher risks of sudden cardiac death compared with other TFAs [16]. Because IP-TFAs increases the risk of heart disease and are estimated to cause more than 500,000 deaths per year [3] and based on the WHO recommendation that IP-TFAs intake should not exceed 1% of total daily energy intake (equivalent to less than 2.2 g/day in a 2000-calorie diet), providing baseline information on dietary sources of IP-TFAs in Lebanon is a crucial stepstone to reduce the risk of death and hospitalization by CVDs and is one of the strategic interventions under the area of prevention and reduction of risk factors in the Regional Framework for Action on NCDs [17]. To our knowledge, this is the first national study that assesses the content of EA and LEA in food. The main objectives of this article are to:

Assess IP-TFAs levels, mainly EA and LEA in frequently consumed traditional dishes, Arabic sweets, processed foods, butter, and margarines in Lebanon.

Review of the findings retrieved from online databases on dietary sources of IP-TFAs in Lebanon and compare them with other countries.

Establish a steppingstone for required policies and regulations to mandate limits of IP-TFAs levels in foods imported or produced locally.

2. Materials and Methods

2.1. Food Sampling

A series of samples collections were conducted over the last two years. The 2019 samples collection, conducted in November 2019, was not centrally coordinated at the capital city Beirut but instead pooled data from five separate sources from the five main governorates in Lebanon (Beirut, Beqaa, Tripoli, Saida, and Mount Lebanon). In this sample collection, we collected thirty types of traditional dishes. Traditional composite dishes are defined as dishes consumed at main meals (i.e., lunch or dinner), containing ingredients from at least three of the five main food groups and requiring preparation using culinary skills [18–20]. A total of 30 traditional composite dishes were identified as most frequently consumed and hence were included for analysis. The names of the food dishes were reported in the current analysis considering the most familiar name used for the dish at a national level with respect to its ingredients. The ingredients of these traditional dishes were described in Hoteit et al. [18–20], and the food samples were collected from five different central kitchens in the 5 governorates listed above. The central kitchens were randomly chosen based on (1) their specialties in cooking homemade dishes, (2) their popularity in the area, (3) their implications in social entrepreneurship and women empowerment (e.g., household women who cook for these central kitchens). Consequently, the food samples were classified into 5 strata, per governorate area [20]. The samples were identified according to their frequency of consumption [21,22] and selected for IP-TFAs analysis mainly for two non-conjugated fatty acids (EA and LEA). In contrast, the subsequent samples collections 2020, conducted in April, were centrally coordinated at Beirut having the broadest coverage in terms of products selected and had a sample of thirty-five types of Arabic sweets and forty-six types of market food products. The full methodology of food list identifications and food sampling is described elsewhere [18–20]. On the other hand, the 2021 sample collections, conducted in March, were nationally coordinated, with a coverage of 34 available types of butter and margarines purchased from all the Lebanese markets. Lot numbers were checked to ensure that each unit belonged to a different lot. The samples were stored, labeled, and analyzed before expiry dates. Samples were selected to include all types of butter and margarines in Lebanon. The analyses were carried out in duplicate for each sample. Thus, a composite sample from each type of food, according to each governorate, was prepared and analyzed. To further interpret current levels of IP-TFAs in Lebanese foods, product categories were compared with similar products found in other countries. A graphical scheme for the whole study is shown in Figure 1.

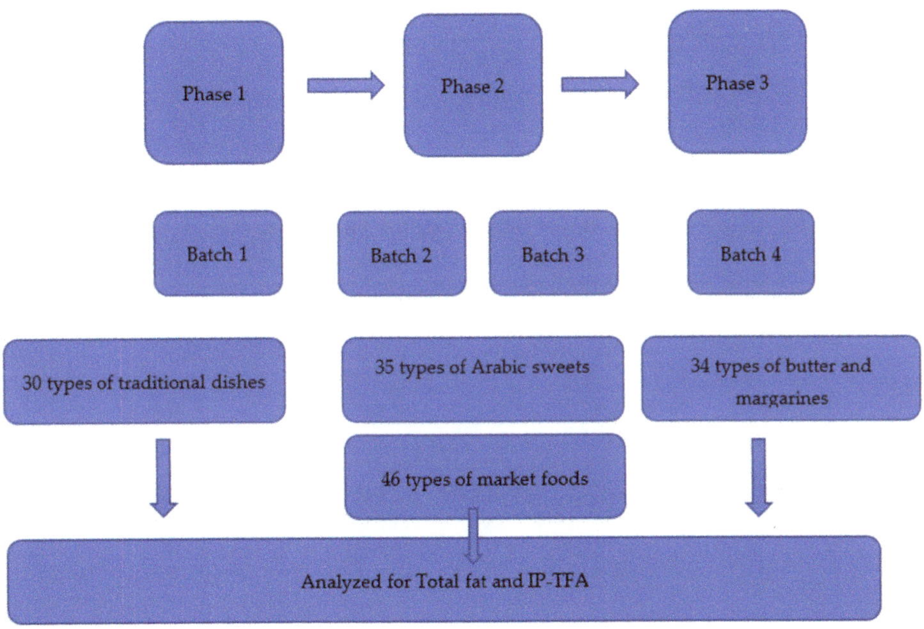

Figure 1. Graphical scheme of the current study.

2.2. Laboratory Analysis Protocol

Around 500 g of each sample was mashed, then analyzed, and the remaining samples were kept frozen at −18 °C for further analysis. The fatty acid profile was measured using gas chromatography. The IP-TFA analysis method was selected considering guidance from the technical committee at the Industrial Research Institute laboratories in Beirut and following standardized protocols. The Association of Official Analytical Chemists (AOAC) methods were used for the analysis of nutrients in food matrices [23].

Soxhlet extraction (Total fatty acids extraction):

The Roese–Gotlieb method was used in the investigation of the fat content [24]. Between 1–2 g of the dried food sample was filtered using a piece of filter paper. Later, it was wrapped and introduced into the soxhlet thimble. To avoid sample spilling, a cotton plug was placed at the top of the thimble. The soxhlet apparatus was assembled, and light petroleum, hexane, heptane, diethyl ether, or cyclohexane was used. The extraction was performed overnight. If solids from the thimble or sample were found in the solvent extract, a filtration before evaporation into another tarred flask or beaker was performed. Then it was dried to a constant weight, which was attained when successive 1 h drying periods showed additional loss of less than 0.05 fat. The percentage (%) fat was equal to (g) fat × 100 g sample.

Fatty Acid Profile (saturated, unsaturated, trans):

The extracted fat of the sample that was obtained during fat determination was used to analyze the fatty acid profile. Between 200–500 mg of the lipid sample was placed in a boiling flask with chips; then, 5 mL of 0.5 M methanolic KOH was added. Esterification was performed by boiling under reflux for 3–5 min. An addition of 15 mL of esterification reagent (2 g NaNH4 + 60 mL methanol + 3 mL conc sulfuric acid) through the condenser was performed, and then the sample was boiled for 15 min. After cooling, there was an addition of 50 mL of distilled water and 25 mL of the solvent. The organic layer was isolated by means of a separatory funnel. Finally, the solvent layer was washed twice with distilled water.

Chromatographic analysis:

- Column: fatty acid methyl esters (FAME) length: 30 m, 0.32 ID
- Injection volume: 1 µL
- Injector temperature (PTV): injection: 60 °C for 0.1 min
- Transfer: ramp 10 °C/min to 270 °C, 1 min hold
- Carrier flow (He): 2.0 mL/min
- Split flow: 20 mL/min (split ratio: 20)
- Detector temperature (flame ionization detection (FID) 280 °C
- Detector gases flows: Air 350 mL/min, Hydrogen 32 mL/min, Make-up (N2) 30 mL/min

Oven Program:

- Initial temperature: 100 °C, hold 3 min
- -Ramp 1: 10 °C/min to 200 °C, hold 3 min
- -Ramp 2: 10 °C/min to 250 °C, hold 5 min

Using the chromatograph software Chromquest 4.2.34, USA, an integration of the areas under the peak was detected under the standards peaks. A calculation as percentage areas was done. The sum of trans fatty acids was calculated accordingly [24]. TFAs isomers were later on detected through SP-2560 100 m capillary column (180 °C isothermal, H2 at 1.0 mL/min) [23].

Statistical tests:

The study variables were presented as continuous variables and listed as reported values per 100g. Means and standard deviations were calculated for the group of traditional dishes, Arabic sweets, and other market products. T-test was used to compare the mean content of the group of foods in terms of EA and LEA. A p-value less than 0.05 indicates a statistically significant difference. Statistical analysis was conducted on IBM SPSS Statistics for Mac, Version 24, USA.

3. Results

3.1. Trans Fatty Acid Acids in Frequently Consumed Traditional Dishes

The mean levels of total IP-TFAs (total of Elaidic acid and Linolelaidic acid) in the tested traditional dishes was equal to 0.9% ± 0.62 and ranged from less than 0.1 to 2.8 g/100 g of total fat except for the dishes *Riz a dajaj* and *Shawarma Lahma* in which total IP-TFA exceeded 2% of the total fat (Table 1). The comparison between the mean values of the IP-TFA (EA and LEA) in the traditional dishes tested shows that EA was significantly higher than LEA in all traditional dishes (p-value = 0.00). Per each governorate, the mean level of total IP-TFAs in all dishes was 0.78 ± 0.65% in Mount Lebanon (range: <0.1–2.1%), 1.1% ± 0.6 in Beqaa (range: <0.1–3%), 1 ± 2.4% in Beirut (range: <0.1–11.2%), 1.2 ± 0.9% in Tripoli (range: <0.1–3.2%) and 0.8 ± 0.4% in Saida (range: <0.1–2%) (Data not shown).

Table 1. Total fat in 100 g of edible portions, total IP-TFAs and IP-TFAs (EA and LEA) in 100 g of fat of frequently consumed traditional dishes among all Lebanese governorates.

			IP-TFAs in 100 g of Total Fat	
Dish	**Total Fat (g) in 100 g**	**Total IP-TFA * per 100 g of Total Fat**	**Trans-C18:1n9t (Elaidic Acid)**	**Trans-C18:2n6t (Linolelaidic Acid)**
Baba ghanouj	9.44	0.74	0.66	0.08
Batata mahchi	1.24	1.98	1.92	0.06
Borgul bil banadoura	5.02	0.48	0.34	0.14
Chichbarak	4.62	0.98	0.86	0.12
Falafel	11.70	0.36	0.32	0.04
Fatayer sabanikh	11.16	0.18	0.12	0.04
Fattat Hommos	7.04	0.66	0.58	0.08
Fattoush	2.94	0.8	0.5	0.3
Foul moudamas	3.48	0.46	0.38	0.08
Hindbe bil zet	10.70	0.18	0.18	0
Hommos bi tahini	6.44	0.38	0.24	0.14
Kafta wa batata	6.32	1.28	1.18	0.1
Kebba bil sayniya	6.40	0.86	0.74	0.12
Koussa mahchi	2.42	1.26	1.1	0.16
Lahm bil ajin	8.96	0.34	0.22	0.12
Loubia bil zet	5.68	0.52	0.46	0.06
Malfouf mahchi	2.12	1.1	1.02	0.1
Moujadara	5.80	0.36	0.36	0
Moghrabia	3.94	0.86	0.76	0.1
Mousaka batinjan	6.58	0.5	0.34	0.16
Riz a dajaj	5.42	2.82	2.66	0.16
Riz bi lahma	6.52	0.82	0.78	0.04
Sayadia	6.48	0.22	0.18	0.04
Shawarma dajaj	6.94	0.24	0.16	0.08
Shawarma lahma	8.28	2.24	2.08	0.16
Tabboula	4.24	0.38	0.26	0.12
Warak enab	3.98	1.24	1.06	0.18
Yakhnat Bamia	5.42	1.24	1.02	0.22
Yakhnat Fassoulia	3.90	0.76	0.64	0.12
Yakhnat Mouloukhia	4.28	1	0.8	0.2

* This represents the sum of EA and LEA only.

3.2. *Trans Fatty Acid Acids in Frequently Consumed Arabic Sweets*

The average of the total IP-TFAs in all samples of Arabic sweets was 0.6 ± 0.3%, predominantly from the EA type. (Table 2). Among 35 samples of Arabic sweets, none exceeded 2% as total IP-TFA in 100 g of total fat. The comparison between the mean values of the IP-TFA (EA and LEA) in the Arabic sweets tested shows that EA was significantly higher than LEA in Arabic sweets (p-value = 0.00).

Table 2. Total fat in 100 g of edible portions, total IP-TFAs and IP-TFAs (EA and LEA) in 100 g of total fat of frequently consumed Arabic sweets.

			IP-TFAs in 100 g of Total Fat	
Name	Total Fat (g) in 100 g	Total IP-TFAs per 100 g of Total Fat *	Trans-C18:1n9t (Elaidic Acid)	Trans-C18:2n6t (Linolelaidic Acid)
Baklava Mixed	23.45	0.25	0.2	0.05
Baklava Mixed Light	20.5	0.3	0.3	0
Halawat El Jiben	8.95	1.2	1.05	0.15
katayef Kashta	6.65	0.9	0.65	0.25
Kounafa bil jiben	12.25	0.4	0.25	0.15
Maakaroun	12	0.1	0.1	0
Maamoul Tamer	17.4	0.4	0.25	0.15
Maamoul mad Kashta	10.65	0.4	0.25	0.15
Maamoul mad joz	19.2	0.45	0.4	0.05
Maamoul joz	21.5	0.85	0.75	0.1
Mafrouka Kashta	13.25	0.4	0.2	0.2
Mafroukeh fostok	10.6	0.6	0.5	0.1
Moushabak	20.1	0.4	0.4	0
Nammoura	5.9	1.5	1.3	0.2
Osmaliya	16.25	0.5	0.4	0.1
Saniora	23.8	1.15	0.85	0.3
Sfouf	12.45	1.45	1.2	0.25
Barazik	16.5	0.5	0.5	0
Boundoukia	19.5	0.3	0.3	0
Daoukia	14.8	0.4	0.3	0.1
Foustoukia	20.4	0.4	0.4	0
Ghourayba	0.325.8	0.6	0.4	0.2
Ish el bulbul	25.1	0.2	0.2	0
kallaj kashta	9.6	<0.1	-	-
Karabij joz maa crema	18.8	0.4	0.2	0.2
kounafa kashta maa kaak	10	0.4	0.1	0.3
Maakroun wa Moushabak	13.7	<0.1	-	-
Maamoul fostok	19.1	0.7	0.5	0.2
Madlouka	11.5	0.6	0.5	0.1
Moufattaka	20.7	<0.1	-	-
Mouhallabiya	4	0.5	0.1	0.4
Riz bil Halib	4.4	<0.1	-	-
Shaaybiyat	16.1	<0.1	-	-
Ward el sham	14.2	0.5	0.5	0
Znoud El sitt	12.3	<0.1	-	-

* This represents the sum of EA and LEA only.

3.3. Trans Fatty Acid in Market Foods

3.3.1. Cereals and Breads Group

In the group of cereals and breads, the mean level of total IP-TFAs was less than 2% of total fat except for *pain au lait (total IP-TFAs: 3.8%)*, which is usually prepared from wheat, milk, and butter or ghee to be consumed frequently by children as a sandwich (Tables 3 and 4).

Table 3. Total fat in 100 g of edible portions, total IP-TFAs and IP-TFAS (EA and LEA) in 100 g of total fat of Market food products collected from Lebanese markets.

			IP-TFAs in 100 g of Total Fat	
Product	**Total Fat (g) in 100 g**	**Total IP-TFA per 100 g of Total Fat**	**Trans-C18:1n9t (Elaidic Acid)**	**Trans-C18:2n6t (Linolelaidic Acid)**
Arabic Bread-White	2.3	<0.1	-	-
Arabic Bread-Whole wheat	4	<0.1	-	-
Baguette	0.5	<0.1	-	-
Biscuits Chocolate Quinoa	13.4	0.1	0.1	-
Biscuits Digestive	17.1	0.3	0.1	0.2
Biscuits Digestive Light	13.8	0.3	0.1	0.2
Biscuits with cream	15.5	<0.1	-	-
Breakfast Cereals	2.1	<0.1	-	-
Breakfast Cereals-Chocolate	2.4	0.3	0.1	0.2
Butter (n = 17 samples)	80	0.8	0.6	0.2
Cake with Cream	16.1	<0.1	-	-
Chocolate Dark	33.6	<0.1	-	-
Chocolate Milk-1	36.6	0.1	0.1	-
Chocolate Milk-2	35	<0.1	-	-
Coffee without cardamon	16.8	0.2	0.1	0.1
Coffee with cardamon	17.7	0.3	0.1	0.2
Corn Oil	100	<0.1	-	-
Croissant Zaatar-1 (cheap)	16.1	0.7	0.4	0.3
Croissant zaatar-2 (expensive)	22.5	0.1	0.1	
De-hulled Pumpkin Seeds	50.6	0.6	0.3	0.3
De-hulled Sunflower Seeds	52.5	0.7	0.3	0.4
Doughnuts	19.6	0.5	0.5	-
English Cake-Chocolate	18.6	2.6	2.6	-
Margarines (n = 18)	100	2.4	2.2	0.2
Halawa	25.5	0.4	0.4	-
Halawa Light	29.9	1.3	1.1	0.2
Hot Chocolate Powder	5.4	0.3	0.3	-
Instant Coffee	10.8	0.2	0.2	-
Kaak asrouni **	1.5	Tr	-	-
Kaak debes and Cacao ***	11.9	0.3	0.2	0.1
Kaak korshalli ****	6.9	0.5	0.5	-

Table 3. *Cont.*

Product	Total Fat (g) in 100 g	Total IP-TFA per 100 g of Total Fat	IP-TFAs in 100 g of Total Fat	
			Trans-C18:1n9t (Elaidic Acid)	Trans-C18:2n6t (Linolelaidic Acid)
Mixed Kernels	53.6	<0.1		-
Mixed Nuts	25.7	0.3	0.2	0.1
Olive Oil	100	<0.1	-	-
Pain au Lait	3.8	2.7	2.7	-
Petit Fours-1 (cheap)	25.6	0.2	0.2	-
Petit Fours-2 (expensive)	29.6	0.2	-	0.2
Potato Chips-1	29.9	0.1	0.1	-
Potato Chips-2	15.4	0.3	0.2	0.1
Potato Chips Light-1	26.9	0.1	0.1	-
Potato Chips light-2	22.9	0.3	0.3	-
Sunflower Oil	100	<0.1	-	-
Tahina	59.4	0.1	-	0.1
Tuna Packed in Oil	6.8	0.3	0.1	0.2
Tuna Packed in Water	0.5	0.6	0.6	-
Wafer-Chocolate-1	21.7	<0.1	-	-
Wafer-Chocolate-2 (manufactured in Lebanon)	24.2	6.5	6.2	0.3

** Kaak Asrouni: type of Lebanese street bread. *** Kaak Debes and Cacao: Cacao cookies with molasses. **** Kaak korshalli: toast bagel (elongated shape).

Table 4. Industrially-Produced trans fatty acids (EA and LEA) in 100 g of total fat per food groups among different countries.

Countries	IP-TFAs	
	Trans-C18:1n9t (Elaidic Acid) (%)	Trans-C18:2n6t (Linolelaidic Acid) (%)
France	Cake: 24.43 Cereals: 28.9 Roasted bread: 33.1 Toasted bread: 25.8 Bread: 30.3 Cookies 38.9	-
France	Cake: 18.5	-
New Zealand	Margarines and table spreads (low trans): 0.1 Margarines and table spreads: 12.3 Margarine/butter blends: 8.3 Butters: 5.2	Margarines and table spreads (low trans): 0.1 Margarines and table spreads: 1.3 Margarine/butter blends: 1.6 Butters: 1.7
Spain	Spanish margarines: 8.17	Spanish margarines: 0.49
Bulgaria	Imported margarines: 8.4 Bulgarian margarines: 1.12	-
Turkey	Margarine tub: 3.85 Margarine stick: 16.88	Margarine tub: 0 Margarine stick: 2.09
Korea	Breakfast cereal: 6.75 Cream-filled biscuit: 15.57 Cream-stuffed cake: 20.96 Canned coffee: 2.3	Breakfast cereal: 0.25 Cream-filled biscuit: 0.43 Cream-stuffed cake: 0.66 Canned coffee: 0.3

Table 4. *Cont.*

	IP-TFAs	
Countries	**Trans-C18:1n9t (Elaidic Acid) (%)**	**Trans-C18:2n6t (Linolelaidic Acid) (%)**
New Zealand	Biscuits and cakes: 0.9 Margarines/spreads: 4.9 Chocolate: 1.1 Snack bars: 0.4 Pies and pastry: 3.7 Partially cooked chips/wedges: 2.5	Biscuits and cakes: 0 Margarines/spreads: 0.1 Chocolate: 0 Snack bars: 0.1 Pies and pastry: 0.4 Partially cooked chips/wedges: 0.4
Pakistan	Margarines: 7.89 Butter: 3.82	Margarines: 0.45
Turkey	Margarines and shortenings: 10.55	-
Canada	Tub margarines: 3.4 Print margarines: 5.5	Tub margarines: 0.1 Print margarines: 0.3
Costa Rica	Corn oil: 0.35 Sunflower oil: 0.28 Olive oil: 0.26 Margarines: 10.15 Butter: 5.1 Mixed nuts: 0.2 Canned tuna (oil): 0.54 Canned tuna (water): 1.07 Nondairy coffee creamer: 30.84	Corn oil: 0.07 Sunflower oil: 0.09 Olive oil: 0 Margarines: 0.35 Butter: 0.23 Mixed nuts: 0 Mayonnaise: 0.02 Canned tuna (oil): 0.08 Canned tuna (water): 0 Nondairy coffee creamer: 1.15
Korea	Breakfast cereal: 0.5 Cream-filled biscuit: 2.4 Cream-stuffed cake: 1.36 Canned coffee: 2.3	Breakfast cereal: 0.3 Cream-filled biscuit: 0.25 Cream-stuffed cake: 0.26 Canned coffee: 0.7
Pakistan	Margarines: 19.48	Margarines: 0.49
Brazil	Regular dark Chocolate: 0.078 Regular chocolate: 0.075	-
Germany	Margarines/spreads: 0.2 Shortenings/cooking fats: 0.51 Doughnuts: 2.07 Chocolate products: 0.44 Biscuits: 0.18 Instant coffee products: 0.36 Butter: 0.23	-
Mexico	Spreadable margarines: 4.73 Stick margarines: 7.4	Spreadable margarines: 0.39 Stick margarines: 0.94
Turkey	Potato crisps: 0.13 Corn crisps: 0.24 Cocoa cakes: 0.37 Chocolate cakes: 0.55 Cream cakes: 0.78 Fruity cakes: 1	Potato crisps: 0.15 Corn crisps: 0.16 Cocoa cakes: 0.11 Mosaic cakes: 0.05 Chocolate cakes: 0.08 Cream cakes: 0.24 Hazelnut-cocoa cakes: 0.14 Fruity cakes: 0.09
India	Biscuit: 0.01 Pastry: 0.85 Cake: 1.92 Bread: 0.18 Bun: 1.31	Biscuit: 0 Pastry: 0 Cake: 0.04 Bread: 0.007 Bun: 0.03

Table 4. *Cont.*

	IP-TFAs	
Countries	**Trans-C18:1n9t (Elaidic Acid) (%)**	**Trans-C18:2n6t (Linolelaidic Acid) (%)**
Iran	Cakes: 18 Cream biscuits: 12 Simple biscuits: 9 Simple chocolates: 5 Potato chips: 10 Margarine: 3.2	Cakes: 0 Cream biscuits: 2 Simple biscuits: 2 Simple chocolates: 0 Potato chips: 4 Margarine: 0.9
UK	Breakfast cereal products: 0.03 Margarine, hard block: 0.05 Potato chips, takeaway: 0.97 Potato chips, fine cut, takeaway: 0.08 Potato chips, oven baked: <0.02 Potato snacks and corn snacks: 0.08 Confectionery, non-chocolate: 0.05 Confectionery, chocolate: 0.08 Butter, spreadable: 0.22	-
Malaysia	Cakes: <0.001 Doughnuts: <0.001 Croissants: <0.001–0.02 White bread: <0.001 Whole grain bread: <0.001 Buns: <0.001 Cream crackers: <0.001–0.33 Chocolate biscuits: <0.001 Potato chips: <0.001–0.87 Chocolate bars: <0.001 Chocolate wafers: <0.001–0.38 Olive oil: 0.79 Blended oil (canola, soybean and olive): 0.82 Soybean oil: 1.76 Palm oil: 1.79 Corn oil: <0.001 Coco-coated cereal: 1.57 Corn cereal: <0.001 Cereal beverages: <0.001	Cakes: <0.001 Doughnuts: <0.001 Croissants: <0.001 White bread: 3.12 Whole grain bread: <0.001 Buns: <0.001–1.21 Cream crackers: <0.001 Chocolate biscuits: <0.001–0.02 Potato chips: <0.001–1.02 Chocolate bars: <0.001–0.54 Chocolate wafers: <0.001 Olive oil: <0.001 Blended oil (canola, soybean and olive): 3.24 Soybean oil: 4.06 Palm oil: <0.001 Corn oil: 2.13 Coco-coated cereal: <0.001 Corn cereal: 4.82 Cereal beverages: <0.001–6.60
Iran	Liquid frying oils: 0.08 Solid frying oils: 1.26	Liquid frying oils: 0.01 Solid frying oils: 0.03
Saudi Arabia	Margarines and shortenings: 5.43	Margarines and shortenings: 1.49
Iran	Margarines: 5.99	Margarines: 0.66
Iran	Biscuit: 12.86 Cake: 6.95 Shortcake: 3.38 Donuts: 3.29 Bread tan: 2.99 Baklava: 2.5 Chocolate: 1.24 Chips: 0.61 Snack: 0.52	-
Iran	Edible oils: 0.07 Margarines: 5.3	-
India	Cakes: 3.93	Cakes: 2.82

Table 4. *Cont.*

Countries	IP-TFAs	
	Trans-C18:1n9t (Elaidic Acid) (%)	**Trans-C18:2n6t (Linolelaidic Acid) (%)**
Lebanon	Cakes: 1.7 Biscuits: 3.7 Croissant: 2.7 Wafers: 5.6	Cakes: 0.1 Biscuits: 0.1 Croissant: 0.1 Wafers: 0.1
Slovenia	Margarines and shortenings: 34.63	Margarines and shortenings: 21.38
Serbia	Crackers: 0.9 Chips and flips: 5.34 Fried corn nuts: 1.7	Crackers: 0.5 Chips and flips: 0.152 Fried corn nuts: 0.1
Poland	Biscuits: 2.81 French pastry cookies: 1.65	Biscuits: 0.21 French pastry cookies: 0.275
Tunisia	Margarines: 4.47 Frying oil: 0.14	Margarines: 4.47 Frying oil: 0.24
Lebanon 2021 (current study)	Traditional dishes: 0.7 Arabic sweets: 0.5 Butter and margarines: 1.4 Biscuits, doughnuts, cake: 0.4 Cereals and breads group: 0.3 Tuna: 0.35 Chocolate and chocolate wafers: 1.26 Cooking oils: 0 Coffee and instant coffee: 0.2 Chips, nuts and seeds: 0.2 Tahina and Halawa: 0.5	Traditional dishes: 0.9 Arabic sweets: 0.6 Butter and margarines: 1.6 Biscuits, doughnuts, cake: 0.5 Cereals and breads group: 0.3 Tuna: 0.45 Chocolate and chocolate wafers: 1.3 Cooking oils: 0 Coffee and instant coffee: 0.25 Chips, nuts and seeds: 0.3 Tahina and Halawa: 0.6

3.3.2. Butter and Margarines

Particular attention was given to the margarine group as it is used as an ingredient and therefore amongst the main sources of IP-TFAs in processed foods. The average of total IP-TFAs in 18 margarines used frequently in Lebanon was 2.4 ± 0.4% (Table 3) with a range between <0.1% and 11.8% (Data not shown). The dominant IP-TFA was EA in almost all these products (Table 3). Within the group of butter, none of the samples exceeded 2% of total fat. The average of total IP-TFAs in the butter and margarines group was 1.6 ± 0.6% of total fat in which EA predominates in these products. Generally, the level of total IP-TFAs in cooking oils, Halawa and Tahina was negligible (Tables 3 and 4).

3.3.3. Snacks and Processed Foods

As for the group of biscuits, doughnuts, and cakes group, negligible amounts of IP-TFAs were found in these products (Average: 0.5% ± 0.2) (Table 4). On the other hand, the unlabeled English cake (chocolate flavor) had an apparently high amount of total IP-TFAs (2.6% in the total fat) in which EA was dominantly available (Table 3). Despite being unable to discuss the fat type used in unlabeled samples, based on this data, partially hydrogenated fats were certainly present in high amounts.

The data on chocolate products presented an amount of 1.3% ± 0.3 as total IP-TFAs (Figure 1), except for the case of wafer-coated chocolate originally manufactured in Lebanon which contains a level of 6.5% (Table 3).

According to Tables 3 and 4, it appears that all samples of potato chips, nuts, seeds, coffee, instant coffee, and packed tuna contained low amounts of total IP-TFAs that are below 2% of total fat. When comparing the mean values of the IP-TFA (EA and LEA) in the market foods, EA and LEA didn't show any significant difference (p-value = 0.16).

4. Discussion

Industrially-Produced Trans fatty acid content in frequently consumed foods in Lebanon compared with different countries.

The available data, the first of its kind in Lebanon, demonstrate that categories with the highest IP-TFAs levels included *Riz a dajaj, Shawarma Lahma, Pain au lait, English cake, Chocolate wafers, and margarines.* About 93% of the products tested in Lebanon, between 2019 and 2021, met the WHO recommendations (less than 2% of Trans fatty acid in total fat), while about 7% exceeded the limit. As per Tables 1–3, all in all, EA was dominant in almost all the analyzed samples and its higher amount indicates that hydrogenated oils were a major contributor in the processing of food products or baking and cooking meals. In comparison to other countries all over the globe, a broad range of EA was observed in many food products (Table 4). For instance, the mean level of EA in Baklava (0.2%) was relatively low in our study in comparison with the content of EA in Baklava in Iran (2.5%) [25]. Furthermore, our findings showed that the mean levels of EA in cakes (2.6%) was much lower than the content of EA found in cakes in France (18.5–25.6%) [26], Iran (6.95–18%) [25–27], Poland (7.95%) [28], India (1.92–3.93%) [29], and higher than EA cake content tested in Lebanon in 2015 (1.7%) [30], Korea (1.36%) [31], Turkey (0.37–1.43%) [32], New Zealand (0.9%) [7], and Malaysia (<0.001%) [33] (Table 4). In addition, the mean levels of EA in biscuits in Iran (9–12.86%) [27], Lebanon 2015 (3.7%) [30], Poland (2.81%) [28], Korea (2.4%) [31], New Zealand (0.9%) [7], and Germany (0.18%) [34] were higher than our results (0.1%), except for Malaysia (<0.001%) [33] and India (0.01%) [29] (Table 4). As for the breakfast cereals, the mean level of EA in our study (0.1%) was much lower than in France (28.9–32.4%) [26] and Korea (0.5–6.75%) [31], and higher than in the UK (0.03%) [35] and Malaysia (<0.001%) [33] (Table 4). Moreover, our findings showed that the mean level of EA in chocolate wafers were six times more than EA content in chocolate wafers in Malaysia [33]. As for the butter, the New Zealand [7] and Costa Rican butter [36] contained five times more EA, and the Pakistani butter [37] contained three more times EA, compared with our results (Table 4). However, the butter in UK, Germany, and Iran contained 0.22% [35], 0.23% [34], and 0.3% [27] EA respectively; this is lower than the content of EA tested in our study (0.6%) (Table 4). Also, Table 4 showed that the margarines in Slovenia contained the highest content of EA (34.63%) [38] compared to our findings (2.2%) and other countries. As for the EA content in chips, Iranian chips showed the highest level of EA (10%) compared to our results (0.1–0.3%) and other countries [27] (Table 4). On the other hand, the results of LEA in the food products tested in our study and those in other countries are available in Table 4.

4.1. Comparison between Lebanese Market Basket Investigation and Other Global and Regional Market Investigations

According to many studies, there was an impact of TFAs labeling on reducing the burden of CVDs due to TFAs [39]. According to an unpublished study conducted by our team, 32% only of the products available in the Lebanese markets reported TFAs on their labels (Data not shown). Our finding came to hand by hand with Kamel et al. [40], in which 181 food products were sampled from local supermarkets in Saudi Arabia and showed that one-third of the products mentioned TFAs on the nutrition label. Moreover, while the majority of the investigated samples in our project had low levels of TFAs, up to 14 g of TFAs per 100 g of food was observed in certain oils and fats sold at the Lebanese markets. Our findings, concerning the range of TFAs in-market products, were relatively low compared with the market investigations published in Stender et al. (2019–2020) [41,42].

4.2. Investigation of the Country of Origin of Imported Food Products in Lebanon

Lebanon imported its food products from France ($107,957), Germany ($98,250), Turkey ($97,015), United Kingdome ($75,571), Italy ($70, 571), Argentina ($69,989), Saudi Arabia ($64, 332) and United States ($57,785). In addition, the main importation sources of butter, oils, and fats are Denmark, Netherlands, France, Belgium, Ukraine, New Zealand, United Kingdom, and Argentina [43]. According to the nutrition labels of tested butter and margarines, the country of origin from which all the butter and margarines were imported to Lebanon were Turkey (n = 5), Egypt (n = 4), Malaysia (n = 3), Saudi Arabia (n = 1), Sri Lanka (n = 3), UAE (n = 1), Netherland (n = 2), Belgium (n = 3), France (n = 4), Italy (n = 1), Ukraine (n = 1), Germany (n = 2), and Denmark (n = 2). Among all these countries, 33 percent (five countries over 15) are implementing mandatory national limits and adopting monitoring mechanisms for mandatory of TFAs limits. On the other hand, in the remaining countries, the best-practice TFAs policy passed but was not yet in effect [5]. Lebanon, long considered a middle-income country, is rapidly sinking into poverty as it faces a triple shock from the unprecedented economic crisis, the impact of COVID-19 on employment and public health, and the consequences of Beirut port explosions. Despite that, the actual relative impact of IP-TFAs exposure on heart disease mortality in Lebanon is limited, but unambiguously still considerable. The findings in our report highlight the importance of controlling the importation of food products from countries controlling IP-TFAs levels in food to avoid sinking Lebanese markets with IP-TFAs rich food products [44], both of which are often ultra-processed, unhealthy, and rich in IP-TFAs. Therefore, this population group is at higher risk of IP-TFAs-attributable CVDs.

4.3. Limits, Advantages, and Future Directions

This study presents some limitations. First, there are many challenges facing the laboratories in Lebanon concerning the testing of IP-TFA, and the lack of standards limits testing other forms of isomers. Second, the food products compared between regions were compared in terms of food groups and not in terms of brands. Moreover, the comparison between traditional dishes or Arabic sweets omits the cooking preparations and ingredients. Third, in the current study, the WHO technique was followed to test the IP-TFA levels in foods tested, however, this was not always reported in many other countries.

Despite these limitations, this study, the first of its kind in Lebanon, should provide the impetus for continuous comprehensive analysis of IP-TFA levels in foods in the regional and national kitchens and markets and the adaptation of the approaches for curbing the health hazards associated with IP-TFA consumption.

5. Conclusions

For the first time in Lebanon, a database on IP-TFA, mainly EA and LEA content in traditional dishes and market products is available and ready to be used by health care providers. There is more than enough convincing evidence that a high IP-TFAs, mainly EA and LEA intake is detrimental to cardiovascular health. Fortunately, this problem in Lebanon is fairly easy to solve via proper legislation. Despite the poorness of Lebanese dishes in IP-TFAs, however, the persistence of food products with high IP-TFAs levels in Lebanon means that subgroups of the Lebanese population, mainly vulnerable and food-insecure people, are threatened by high levels of IP-TFAs due to frequent consumption of risky products. The inauguration and implementation of policies to curtail IP-TFAs in Lebanon may therefore be legitimized, and such efforts should underline added fats and packaged foods. The economic crises in Lebanon pushed the Lebanese people to select cheap oils, including butter and margarines instead of vegetable oils. Thus, it appears reasonable that the Lebanese government and ministries should strive to raise public awareness about the issue and lobby for implementing anti-IP-TFAs laws either on the level of national industries or, on the level of food products importation.

Author Contributions: Conceptualization, M.H. and E.Z.; methodology, M.H. and E.Z.; validation, A.A.-J. and I.S.; formal analysis, M.H. and C.I.; investigation, M.H., E.Z., C.I.; resources, E.Z., A.A.-J.; data curation, M.H., E.Z., C.I.; writing—original draft preparation, M.H., E.Z., C.I.; writing—review and editing, All the others; supervision, A.A.-J., A.R. and I.S. All authors have read and agreed to the published version of the manuscript.

Funding: This project was self-funded.

Institutional Review Board Statement: Not applicable.

Informed Consent Statement: Not applicable.

Data Availability Statement: All data underlying the results are available as part of the article and no additional source data are required.

Acknowledgments: The authors would like to thank the head of the Chemical department Joseph Matta and the laboratory technician Halim El Bayeh at the Industrial Research Institute, the Research assistants. Nadia Hallak and Iman Kheir and the students Jessica Khalifeh and Layla Abadi at the Lebanese University. The authors confirm the approval of the acknowledged persons to be acknowledged.

Conflicts of Interest: The authors declare no conflict of interest.

Disclaimer: The authors alone are responsible for the views expressed in this article and they do not necessarily represent the views, decisions or policies of WHO or the other institutions with which the authors are affiliated.

References

1. World Health Organization-Noncommunicable Diseases (NCD) Country Profiles 2018. Available online: https://www.who.int/publications/i/item/ncd-country-profiles-2018 (accessed on 2 June 2021).
2. Remig, V.; Franklin, B.; Margolis, S.; Kostas, G.; Nece, T.; Street, J.C. Trans fats in America: A review of their use, consumption, health implications, and regulation. *J. Am. Diet. Assoc.* **2010**, *110*, 585–592. [CrossRef] [PubMed]
3. World Health Organization. *Countdown to 2023: WHO Report on Global Trans-Fat Elimination 2019*; World Health Organization: Geneva, Switzerland, 2019.
4. World Health Organization. *An Action Package to Eliminate Industrially-Produced Trans Fat from the Global Food Supply*; World Health Organization: Geneva, Switzerland, 2018; Available online: http://www.who.int/nutrition/topics/replace-transfat (accessed on 15 March 2021).
5. World Health Organization. *Countdown to 2023: WHO Report on Global Trans-Fat Elimination 2020*; World Health Organization: Geneva, Switzerland, 2020.
6. World Health Organization. REPLACE Trans Fat: An Action Package to Eliminate Industrially Produced Trans-Fatty Acids. In *Module 3: Legislate or Regulate. How-to Guide for Trans Fat Policy Action*; World Health Organization: Geneva, Switzerland, 2020; Licence: CC BY-NC-SA 3.0 IGO.
7. Saunders, D.; Jones, S.; Devane, G.J.; Scholes, P.; Lake, R.J.; Paulin, S.M. Trans fatty acids in the New Zealand food supply. *J. Food Comp. Anal.* **2008**, *21*, 320–325. [CrossRef]
8. Jawaldeh, A.A.; Al-Jawaldeh, H. Fat Intake Reduction Strategies among Children and Adults to Eliminate Obesity and Non-Communicable Diseases in the Eastern Mediterranean Region. *Children* **2018**, *5*, 89. [CrossRef] [PubMed]
9. World Health Organization. Noncommunicable Diseases: Key Facts. Geneva. 2018. Available online: https://www.who.int/news-room/fact-sheets/detail/noncommunicable-diseases (accessed on 16 March 2021).
10. Persistence Market Research. PHO and non-PHO Based Oils and Fats Market: Global Industry Analysis 2013–2017 and Forecast 2018–2026. Available online: https://www.persistencemarketresearch.com/market-research/pho-and-nonpho-based-oils-and-fats-market.asp (accessed on 17 June 2021).
11. Bösch, S.; Westerman, L.; Renshaw, N.; Pravst, I. Trans Fat Free by 2023—A Building Block of the COVID-19 Response. *Front. Nutr.* **2021**, *8*, 645750. [CrossRef]
12. Ohmori, H.; Fujii, K.; Kadochi, Y.; Mori, S.; Nishiguchi, Y.; Fujiwara, R.; Kishi, S.; Sasaki, T.; Kuniyasu, H. Elaidic Acid, a Trans-Fatty Acid, Enhances the Metastasis of Colorectal Cancer Cells. *Pathobiol. J. Immunopathol. Mol. Cell. Biol.* **2017**, *84*, 144–151. [CrossRef] [PubMed]
13. Li, C.; Cobb, L.K.; Vesper, H.W.; Asma, S. Global Surveillance of trans-Fatty Acids. *Prev. Chronic Dis.* **2019**, *16*, E147. [CrossRef]
14. Park, K.-H.; Kim, J.-M.; Cho, K.-H. Elaidic acid (EA) generates dysfunctional high-density lipoproteins and consumption of EA exacerbates hyperlipidemia and fatty liver change in zebrafish. *Mol. Nutr. Food Res.* **2014**, *58*, 1537–1545. [CrossRef]
15. Wang, J.; Liang, Y.; Jian, L.; Zhang, J.; Liang, S.; Xiao, S.; Liu, B.; Wang, H. Linoelaidic acid enhances adipogenic differentiation in adipose tissue-derived stromal cells through suppression of Wnt/β-catenin signaling pathway in vitro. *Prostaglandins Leukot. Essent. Fat. Acids* **2016**, *110*, 1–7. [CrossRef]
16. Li, J.; Rao, H.; Bin, Q.; Fan, Y.-W.; Li, H.-Y.; Deng, Z.-Y. Linolelaidic acid induces apoptosis, cell cycle arrest and inflammation stronger than elaidic acid in human umbilical vein endothelial cells through lipid rafts. *Eur. J. Lipid Sci. Technol.* **2016**, *119*, 1600374. [CrossRef]
17. World Health Organization. Regional framework for action. In *WHO/Noncommunicable Diseases*; WHO Regional Office for Eastern Mediterranean: Cairo, Egypt, 2015.
18. Hoteit, M.; Zoghbi, E.; Al Iskandarani, M.; Rady, A.; Shankiti, I.; Matta, J.; Al-Jawaldeh, A. Nutritional value of the Middle Eastern diet: Analysis of total sugar, salt, and iron in Lebanese traditional dishes. *F1000Research* **2020**, *9*, 1254. [CrossRef]

19. Hoteit, M.; Zoghbi, E.; Rady, A.; Shankiti, I.; Al-Jawaldeh, A. Development of a Lebanese food exchange system based on frequently consumed Eastern Mediterranean traditional dishes and Arabic sweets. *F1000Research* **2021**, *10*, 12. [CrossRef]
20. Food Composition Data: Traditional Dishes, Arabic Sweets and Market Foods. 2021. Available online: https://ul.edu.lb/files/ann/20210422-LU-RePa-Report.pdf (accessed on 29 April 2021).
21. Batal, M.; Hunter, E. Traditional Lebanese recipes based on wild plants: An answer to diet simplification? *Food Nutr. Bull.* **2007**, *28* (Suppl. 2), S303–S311. [CrossRef]
22. Issa, C.; Salameh, P.; Batal, M.; Vieux, F.; Lairon, D.; Darmon, N. The nutrient profile of traditional Lebanese composite dishes: Comparison with composite dishes consumed in France. *Int. J. Food Sci. Nutr.* **2009**, *60* (Suppl. 4), 285–295. [CrossRef]
23. Association of Official Analytical Chemists. *Official Methods of Analysis*; Washington Ce 1h-05; 23; Association of Official Analytical Chemists: Washington, DC, USA, 2017; (Revised 2017).
24. Association of Official Analytical Chemists. *Official Methods of Analysis*, 21st ed.; Association of Official Analytical Chemists: Washington, DC, USA, 2019.
25. Bahrami, G.; Bahrami, S.; Bahrami, M.; Pasdar, Y.; Darbandi, M.; Rezaei, M. The Trans Fatty Acid Content of Snacks Offered in Kermanshah. *Int. J. Health Life Sci.* **2015**, *1*, e69803.
26. Wolff, R.L.; Combe, N.A.; Destaillats, F.; Boué, C.; Precht, D.; Molkentin, J.; Entressangles, B. Follow-up of the Δ4 to Δ16 trans-18:1 isomer profile and content in French processed foods containing partially hydrogenated vegetable oils during the period 1995–1999. Analytical and nutritional implications. *Lipids* **2000**, *35*, 815–825. [CrossRef]
27. Nazari, B.; Asgary, S.; Azadbakht, L. Fatty acid analysis of Iranian junk food, dairy, and bakery products: Special attention to trans-fats. *J. Res. Med. Sci. Off. J. Isfahan Univ. Med. Sci.* **2012**, *17*, 952–957.
28. Żbikowska, A.; Onacik-Gür, S.; Kowalska, M.; Rutkowska, J. Trans fatty acids in polish pastry. *J. Food Prot.* **2019**, *82*, 1028–1033.
29. Reshma, M.V.; Kiran, R.C.; Nisha, P.; SobanKumar, D.R.; Sundaresan, A.; Jayamurthy, P. Trans fat content in labeled and unlabelled indian bakery products including fried snacks. *Int. Food Res. J.* **2012**, *19*, 1609–1614. Available online: http://ezproxy.usek.edu.lb/login?url=https://www-proquest-com.ezproxy.usek.edu.lb/scholarly-journals/trans-fat-content-labeled-unlabelled-indian/docview/1318005982/se-2?accountid=43544 (accessed on 2 July 2021).
30. Saadeh, C.; Toufeili, I.; Habbal, M.Z.; Nasreddine, L. Fatty acid composition including trans-fatty acids in selected cereal-based baked snacks from Lebanon. *J. Food Compos. Anal.* **2015**, *41*, 81–85. [CrossRef]
31. Lee, J.H.; Adhikari, P.; Kim, S.-A.; Yoon, T.; Kim, I.-H.; Lee, K.-T. Trans Fatty Acids Content and Fatty Acid Profiles in the Selected Food Products from Korea between 2005 and 2008. *J. Food Sci.* **2010**, *75*, C647–C652. [CrossRef]
32. Cakmak, Y.S.; Guler, G.O.; Yigit, S.; Caglav, G.; Aktumsek, A. Fatty Acid Composition and TransFatty Acids in Crisps and Cakes in Turkey's Markets. *Int. J. Food Prop.* **2011**, *14*, 822–829. [CrossRef]
33. Akmar, Z.D.; Norhaizan, M.E.; Azimah, R.; Azrina, A.; Chan, Y.M. The trans fatty acids content of selected foods in Malaysia. *Malays. J. Nutr.* **2013**, *19*, 87–98.
34. Kuhnt, K.; Baehr, M.; Rohrer, C.; Jahreis, G. Trans fatty acid isomers and the trans-9/trans11 index in fat containing foods. *Eur. J. Lipid Sci. Technol.* **2011**, *113*, 1281–1292. [CrossRef] [PubMed]
35. Roe, M.; Pinchen, H.; Church, S.; Elahi, S.; Walker, M.; Farron-Wilson, M.; Buttriss, J.; Finglas, P. Trans fatty acids in a range of UK processed foods. *Food Chem.* **2013**, *140*, 427–431. [CrossRef]
36. Baylin, A.; Siles, X.; Donovan-Palmer, A.; Fernandez, X.; Campos, H. Fatty acid composition of Costa Rican foods including trans fatty acid content. *J. Food Compos. Anal.* **2007**, *20*, 182–192. [CrossRef]
37. Anwar, F.; Bhanger, M.I.; Iqbal, S.; Sultana, B. Fatty acid composition of different margarines and butters from pakistan with special emphasis on trans unsaturated contents. *J. Food Qual.* **2006**, *29*, 87–96. [CrossRef]
38. Abramovič, H.; Vidrih, R.; Zlatić, E.; Kokalj, D.; Schreiner, M.; Žmitek, K.; Kušar, A.; Pravst, I. Trans fatty acids in margarines and shortenings in the food supply in Slovenia. *J. Food Compos. Anal.* **2018**, *74*, 53–61. [CrossRef]
39. Satia, J.A.; Galanko, J.A.; Neuhouser, M.L. Food nutrition label use is associated with demographic, behavioral, and psychosocial factors and dietary intake among African Americans in North Carolina. *J. Am. Diet. Assoc.* **2005**, *105*, 392–403. [CrossRef] [PubMed]
40. Kamel, S.; Otaibi, H.A. Trans-Fats Declaration, Awareness and Consumption in Saudi Arabia. *Curr. Res. Nutr. Food Sci.* **2018**, *6*, 748–756. [CrossRef]
41. Stender, S. Industrially produced trans fat in popular foods in 15 countries of the former Soviet Union from 2015 to 2016: A market basket investigation. *BMJ Open* **2019**, *9*, e023184. [CrossRef]
42. Stender, S. Trans fat in foods in Iran, South-Eastern Europe, Caucasia and Central Asia: A market basket investigation. *Food Policy* **2020**, in press. [CrossRef]
43. Lebanon Dairy Produce; Butter and Other Fats and Oils Derived from Milk Imports by Country/2018/Data (worldbank.org). Available online: https://wits.worldbank.org/trade/comtrade/en/country/LBN/year/2018/tradeflow/Exports/partner/ALL/product/040500 (accessed on 1 July 2021).
44. World Food Program. WFP Lebanon Country Brief February 2021. Available online: www.wfp.org/countries/Lebanon (accessed on 2 July 2021).

nutrients

Article

Energy, Sodium, Sugar and Saturated Fat Content of New Zealand Fast-Food Products and Meal Combos in 2020

Sally Mackay [1,*], Teresa Gontijo de Castro [1,2], Leanne Young [3], Grace Shaw [1], Cliona Ni Mhurchu [3,4] and Helen Eyles [1,3]

1 Department of Epidemiology and Biostatistics, The University of Auckland, Auckland 1010, New Zealand; t.castro@auckland.ac.nz (T.G.d.C.); grace.shaw@auckland.ac.nz (G.S.); h.eyles@auckland.ac.nz (H.E.)
2 Nutrition Section, The University of Auckland, Auckland 1010, New Zealand
3 National Institute for Health Innovation, The University of Auckland, Auckland 1010, New Zealand; leanne.young@auckland.ac.nz (L.Y.); c.nimhurchu@auckland.ac.nz (C.N.M.)
4 The George Institute for Global Health, Newtown, NSW 2042, Australia
* Correspondence: sally.mackay@auckland.ac.nz

Abstract: This study aimed to benchmark the healthiness of the New Zealand (NZ) fast-food supply in 2020. There are currently no actions or policies in NZ regarding the composition, serving size and labeling of fast food. Data on serving size and nutrient content of products was collected from company websites and in-store visits to 27 fast-food chains. For each fast-food category and type of combo meal, medians and interquartile ranges were calculated for serving size and energy, sodium, total sugar, and saturated fat per serving. Nutrient contents/serving were benchmarked against the United Kingdom (UK) soft drinks levy sugar thresholds and targets for salt for away from home foods, the NZ daily intake guidelines for energy, sodium, and saturated fat, and the World Health Organization (WHO) recommendation for free sugars. Analyses were conducted for the 30.3% (n = 1772) of products with available nutrition information and for 176 meal combos. Most (n = 67; 91.8%) sugar-sweetened drinks would qualify for a UK soft drink industry levy and 47% (n = 1072) of products exceeded the relevant UK sodium target. Half of the meal combos provided at least 50.3% of the daily energy requirements and at least 88.6% of the maximum recommended intake of sodium. Fast-food products and combo meals in NZ contribute far more energy and negative nutrients to recommended daily intake targets than is optimal for good health. The NZ Government should set reformulation targets and serving size guidance to reduce the potential impact of fast-food consumption on the health of New Zealanders.

Keywords: fast food; sodium; total sugars; population health; food environments; meal combos

Citation: Mackay, S.; Gontijo de Castro, T.; Young, L.; Shaw, G.; Ni Mhurchu, C.; Eyles, H. Energy, Sodium, Sugar and Saturated Fat Content of New Zealand Fast-Food Products and Meal Combos in 2020. *Nutrients* **2021**, *13*, 4010. https://doi.org/10.3390/nu13114010

Academic Editors: Alessandra Durazzo and Massimo Lucarini

Received: 8 October 2021
Accepted: 8 November 2021
Published: 10 November 2021

1. Introduction

New Zealand (NZ) has high rates of non-communicable disease with two of three adults and one in four children overweight or obese [1], and in 2017 unhealthy diets accounted for nearly 20% of illness and early death [2]. The average NZ diet is low in fruit, vegetables, wholegrains, legumes, nuts and seeds and contains an excess of foods high in sugar and sodium [1,2].

Fast-food is heavily marketed, cheap in comparison to other restaurant foods, convenient, accessible and palatable [3,4]. Fast-food has been independently associated with increased energy intakes and accelerating rates of weight gain and obesity [5,6]. In the United States, fast-food consumption has been associated with an additional 814 kJ of dietary energy per day and higher intakes of saturated fat and sodium [7]. Fast-food meals are generally characterized by large portion sizes, low levels of health-promoting nutrients such as fibre, and high levels of energy and adverse nutrients including saturated fat, added sugar and sodium [8]. Many chains offer combination meals (meal combos) in addition to individual items. These bundle unhealthy food options for a cheaper price and are a common tool used by the fast food industry to increase consumption [9].

In NZ, the percentage of the household food budget spent on restaurants and take-aways increased from 22% in 2000 to 27% in 2020 [10], and in 2019, 53% of adults bought a takeout meal from a fast-food or takeaway shop at least once a week [11]. Euromonitor trends for NZ from 2015 to 2020 show a 3.5% increase in foodservice value growth for limited-service restaurants with sales, and the number of fast-food outlets is forecast to continue to grow [12]. A previous assessment of the energy, serving size and sodium content of the NZ fast-food supply [13] reported an increase in the serving size of fast-food items from 2012 to 2016 and an increase in sodium and energy per serving, although there was no increase in product sodium or energy density.

Internationally there is a lack of agreed guidelines for portion or serving sizes or nutrients for fast-foods, and there is wide variation in serving sizes and nutrient contents of fast foods within and across countries [14]. However, the United Kingdom (UK) has government-led programs to reduce the energy, salt and sugar content in the "out-of-home" food sector [15]. The voluntary targets aim to reduce the levels of salt and sugar in the foods that contribute most to dietary intakes for UK adults. The mandatory UK "Soft Drinks Industry Levy" was introduced in 2018 to incentivize the industry to reduce the sugar content of soft drinks or pay a variable levy depending on the sugar content of the drink [16].

There are no government regulations in NZ related to fast-food composition targets, menu labeling of energy or nutrients [17], or to limit density of outlets. There is little nutrition information provided at the point of purchase to assist customers to purchase healthier options [18]. The NZ Government has not set food composition targets for any foods. The Heart Foundation has a voluntary HeartSAFE program focused on reducing sodium and sugar in low-cost, high-volume processed foods [19] but only four of 38 food categories have targets for foods that are consumed away from the home. In 2018, the NZ Ministers of Health and Primary Industries requested that the food industry convene a Food Industry Taskforce to show how they could contribute to obesity reduction. The resulting voluntary recommendations of the Taskforce related to fast food were to develop serving size ranges, best-practice portion guidance, education, providing nutrition information and to consider voluntary menu labeling [18] but there has been no indication of implementation. Annual cross-sectional surveys of all food and beverage products available for sale at fast-food chains in NZ are undertaken as part of data collection for the Nutritrack database [20]. Fast-food chains were defined according to Fleischhacker et al. [21] as restaurants providing food which is generally cheap, requires minimal preparation, and where no table service is provided.

This paper aims to benchmark the healthiness of products and combo meals available in the NZ fast-food supply in 2020 to provide recent evidence to inform effective policies and actions regarding reformulation and consumer information. To achieve this we: (i) assessed the sugar, salt, saturated fat and energy content of key fast-food product categories and meal combos and compared their nutrient content to national and international daily recommended intakes and; (ii) benchmarked selected fast-food product categories against accepted sodium and sugar targets. This is also the first study in NZ to assess the healthiness of meal combos provided by fast-food chains.

2. Materials and Methods

2.1. Data Sources

2.1.1. Fast-Food Categories and Products

We conducted a cross-sectional survey of all food and beverage products available for sale at fast-food chains in NZ in 2020 [20]. All chains with ≥20 stores nationwide were selected (n = 27). Of these, twenty-two chains provided nutrition information for some or all products. Eleven chains were international chains and eleven were national chains. The following data were collected by trained fieldworkers between February and March in 2020, the year used for the current analysis: product name, serving and/or pack size, and nutrient information. Data were recorded directly from company websites. Visits to one large store representing each fast-food chain were also completed to capture any additional information not available on-line e.g., that on menu boards. Stores selected for visits were in Auckland, New Zealand's largest city, and chosen based on size and location to reflect the largest product range possible. It is not mandatory for nutrition information to be available or displayed under the Australian and NZ Food Standards Code [17], and thus serving size and nutrient data were missing for some products.

2.1.2. Fast-Food Combo Meals

We also created a database containing the nutrient composition of meal combos. This database used the data collected for the fast-food products (Section 2.2.1). We calculated the nutritional composition of meal combos by summing nutritional contents of individual products from this database. A combination of products was considered a meal combo if: (i) it offered two or more products, one of which may be a beverage; (ii) the meals were considered a main meal (e.g., burger, fries, soft drink) rather than a snack (e.g., muffin and coffee) and; (iii) the combo was promoted as a meal to be shared between a specific number of people (e.g., contained four burgers and four beverages). Some 'party packs' and pizza deals were not considered meals as it was not obvious how many people they would feed (e.g., three large pizzas, two large fries and 1.5 L soft drink).

2.2. Data Preparation

2.2.1. Fast-Food Categories and Products

Of the 5840 products in the 2020 database, 30.4% were included in the analysis (n = 1772) and 59.8% (n = 3492) products were removed from the dataset as no information was available on serving size, package size and nutrients, with an additional 9.9% removed due to missing key nutrient data, implausible nutrient values, and/or no information on serving or package size. Food categories with less than 30 products were excluded except for 'Fries' (n = 25 products) because this is one of the most frequently consumed fast-food items [22], and there is little variation of products within the category. Those 'Sides' not available for sale separately were excluded (n = 118, 2.0%) from the analysis (Figure 1). Identical products with different serving sizes were retained as some nutrient targets [23] are set according to serving size. For products where the nutritional information was provided only per 100 g and the serving size was also available, the products' nutrition composition per serving was calculated.

Where applicable, we assessed the proportion of food items within the fast-food categories that exceeded the UK targets for sodium [23] and sugar [16]. This required further categorisation of some of the specific food items as the target to apply depended on certain ingredients (such as containing processed meat) or product size (fries < 8 mm or ≥8 mm in width).

2.2.2. Fast-Food Combo Meals

The nutrient composition of meal combos in the database was calculated by combining per product nutrient values for existing individual products. A combination of products was considered a meal combo if: (i) it offered two or more products, one of which may be a beverage; (ii) the meals were considered a main meal (e.g., burger, fries, soft drink) rather than a snack (e.g., muffin and coffee) and; (iii) the combo was promoted as a meal to be shared between a specific number of people (e.g., contained four burgers and four beverages). Some 'party packs' and pizza deals were not considered meals as it was not obvious how many people they would feed (e.g., three large pizzas, two large fries and 1.5 L soft drink).

Meal combos were then categorized by type based on the presence of key common products e.g., burger, chicken, pie, sandwich, etc. Many of the combos offered the consumer a choice for one or more of the products within. When this occurred, an option offered with the lowest energy choice (e.g., artificially sweetened beverage (ASB), lowest calorie sandwich filling) and an option with the highest energy choice (e.g., sugar sweetened beverage (SSB), highest calorie sandwich filling) were created.

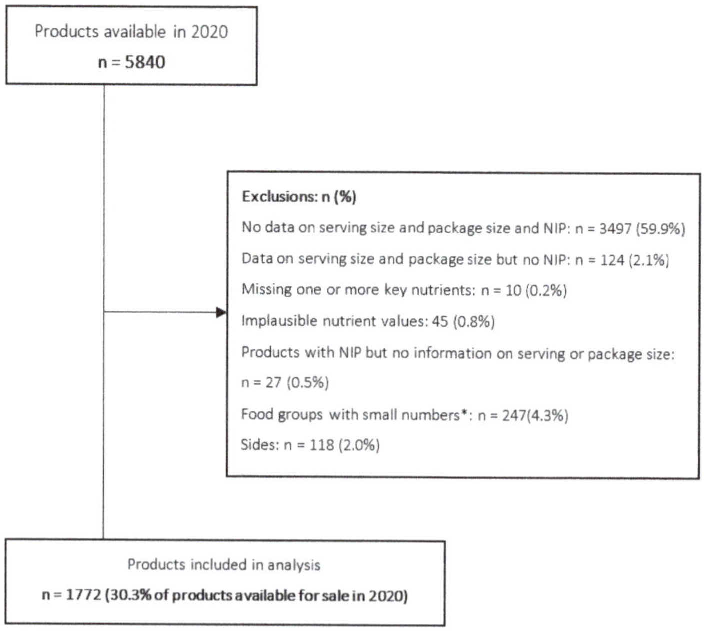

Figure 1. Flow-chart indicating the reasons and number of fast-food products excluded and included in the analyses, New Zealand, 2020. * Asian—Chinese, juice, beverages other, tea/coffee/hot chocolate, water, breakfast sweet, dressings/condiments sweet, other, seafood, soups.

In total, 176 meal combos from nine fast-food chains were identified, which were allocated to one of 20 combo categories. The number of combos within a category ranged from three (5 categories) to 20 (4 categories).

Outcomes

A range of indicators were chosen to benchmark the healthiness of fast-food items and meal combos in relation to existing relevant targets and daily population recommendations for energy, sodium, added sugar and saturated fat intakes [24,25]

Serving size and nutritional composition/serving: The average amount of energy (kilojoules-kJ), sodium (mg), sugar (g) and saturated fat (g) per serving was calculated and described as medians (interquartile range); and minimum and maximum values. We calculated medians for these metrics because for some of the fast-food categories and combos, the nutrient contents were not normally distributed.

Percentage of adult daily recommendations/serving: The percentage contribution that each product and combo meal made to daily population Nutrient Reference Values (NRVs) [24] for energy, sodium, sugar and saturated fat was calculated. New Zealand and Australia share the same NRVs so we used the same energy benchmark as an Australian fast-food supply analysis of 8700 kJ [26]. We applied the recommended upper limit for sodium (2000 mg) [24] which is also the WHO upper limit [27], saturated and trans-fat ($\leq$10% of energy/day, 23 g for 8700 kJ diet) [24] and sugar (WHO recommendation for free sugars intake $\leq$10% of energy/day, 51 g for 8700 kJ diet) [25]. The free sugars content of fast foods is not available and intrinsic sugars are present in few fast foods in NZ, therefore total sugars were used as a proxy.

Proportion of products exceeding the UK sodium targets: UK sodium targets (2024 targets published in 2020) were used as there are no targets available for NZ. UK sodium targets are set per serving (or slice for pizza) for the takeaway sector [23] and per 100 g for other categories (used for 'cakes, muffins and pastry'), and were assessed for seven of 16 categories included in this study.

Proportion of sugar-sweetened beverages exceeding sugar thresholds for the UK soft drinks industry levy: The UK SDIL [16] is based on total sugar content of beverages per 100 mL. The UK soft drinks levy thresholds (>5 g and ≤8 g/100 mL and >8 g/100 mL) were used as NZ does not have sugar targets for beverages.

Analysis

All analyses were performed using SPSS software (version 25, IBM SPSS Statistics).

3. Results

3.1. Fast-Food Categories and Products

Table 1 presents the median (interquartile range) serving size, energy and nutrients/serving (sodium, total sugar, saturated fat), and percent contribution to daily recommendations by fast-food category. Table S1 (Supplementary Material) presents, for each fast-food category, the minimum and maximum values for energy and nutrient contents and for percent contributions of energy and nutrients to daily recommendations.

The categories with the highest median energy per serving and percentage contribution to recommended daily energy intake (8700 kJ) were Burgers (2585 kJ, 29.7%, respectively), followed by Fries (2010 kJ, 23.1%) and Asian meals (2015 kJ, 23.2.0%). Half of the Pastry, savory, Cakes, muffins and pastries and Milkshakes/smoothies, considered to be snack items, contributed, respectively, at least, 22.0%, 22.4% and 18.6%/serving to the daily recommended energy intake for a NZ adult (Table 1).

The categories with the highest median sodium per serving and percentage contribution to maximum recommended daily sodium intake (2000 mg) intake were Burgers (1090.6 mg, 54.5%, respectively), followed by Breakfast, savory (1075 mg, 53.8%,) and Sandwiches and wraps (900 mg, 45.0%).) The categories with the highest median total sugar per serving and percentage contribution to maximum recommended daily free sugar intake (51 g) were Milkshakes/smoothies (49.0 g, 96.0%, respectively) followed by sugar-sweetened soft drinks, (33.8 g, 66.3%) and Cakes/muffins/pastries (32.8 g, 64.3%). The highest median values of saturated fat per serving and percentage contribution to maximum daily recommended intake (23 g) were observed for Pastry, savory (13.0 g, 56.8%, respectively) followed by Breakfast, savory (9.5 g, 41.3%) and Burgers (9.2 g, 40.0%) (Table 1).

Analysis involving all products showed that most (n = 1562; 89.1%) contributed 30% or less/serving to the daily recommended energy intake for an average NZ adult. In total, 235 products (13.4%) contributed 50% or more/serving of the maximum daily recommended sodium intake. Twenty-three products (1.3%) exceeded the maximum recommended daily sodium intake (data not shown in table).

Figure 2 shows the proportions of products above the UK sodium target [24] among the fast-food categories where a target existed. Overall, almost half the products (46.5%) exceeded the target/serving. The categories with the largest percentage of products exceeding the UK sodium target were Fries (100% of products), Pizzas (57.1%) and Pastries, savory (52.6%). The categories with the lowest percentage exceeding the UK sodium target were Sandwiches and wraps (38.1%), Burgers (36.5%) and Cakes, muffins and sweet pastry products (35.8%).

Of the 73 sugar-sweetened soft drinks assessed, the majority (n = 67; 91.8%) exceeded the UK-SDIL thresholds [16], where 58 (74.5%) had a sugar content >8 g/100 mL and 9 (12.3%) had a sugar content >5 g & ≤8 g/100 mL (data not shown).

Table 1. NZ fast food supply 2020, by food category: Median (interquartile ranges) for: total energy, sodium, sugar and saturated fat content by serving size and percent energy, sodium, sugar and saturated fat contributions to recommended daily intakes.

Major and Minor Fast-Food Category	Serve Size (g/mL)		Energy and Nutrients									
				Energy—Kilojoules/Serving		Sodium—Milligrams/Serving		Total Sugar—Grams/Serving		Saturated Fat—Grams/Serving	
	N	Median (IQR)	N	Content	% of Daily Recommendation *	Content	% of Daily Recommendation *	Content	% of Daily Recommendation *	Content	% of Daily Recommendation *
				Median (IQR)	Median (IQR)	Median (IQR)	Median (IQR)	Median (IQR)	Median (IQR)	Median (IQR)	Median (IQR)
Asian (sushi, katsu chicken, noodles)	47	302.0 (267.0–322.0)	48	2015.0 (1794.6–2280.0)	23.2 (20.6–26.2)	673.0 (535.5–1028.8)	33.7 (26.8–51.4)	12.2 (11.2–13.2)	23.9 (22.0–25.9)	2.1 (1.2–3.6)	9.0 (5.0–15.8)
Beverages											
Milkshakes, smoothies	99	475.0 (402.0–480.0	130	1617.0 (1220.0–2519.5)	18.6 (14.0–29.0)	154.5 (100.0–252.6)	7.7 (5.0–12.6)	49.0 (39.5–70.6)	96.0 (77.5–138.5)	6.3 (2.7–10.7)	27.4 (11.6–48.6)
Soft Drinks, sugar sweetened	51	328.0 (250.0–375.0)	72	590.5 (457.0–774.0)	6.8 (5.3–8.9)	27.3 (12.1–55.3)	1.4 (0.6–2.8)	33.8 (27.0–45.2)	66.3 (52.9–88.5)	0.0 (0.0–0.0)	0.0 (0.0–0.0)
Soft Drinks, artificially sweetened	15	330.0 (330.0–355.0)	24	8.5 (5.0–35.8)	0.1 (0.6–0.4)	32.5 (5.0–53.8)	1.6 (0.3–2.7)	0.0 (0.0–0.0)	0.0 (0.0–0.0)	0.0 (0.0–0.1)	0.0 (0.0–0.3)
Breakfast, savory (egg dishes, bread dishes)	28	212.0 (167.3–245.8)	36	1920.0 (1552.5–2224.8)	22.1 (17.8–25.6)	1075.0 (758.0–1287.5)	53.8 (37.9–64.4)	3.2 (2.3–6.3)	6.3 (4.6–12.3)	9.5 (5.9–14.8)	41.3 (25.7–64.1)
Pastry, savory	87	200.0 (180.0–220.0)	88	1935.0 (1630–2225.0)	22.2 (18.7–25.6)	725.5 (604.3–910.8)	36.3 (30.2–45.5)	2.6 (1.3–4.0)	5.0 (2.6–7.8)	13.0 (9.3–16.2)	56.8 (40.2–70.3)
Cakes, muffins and pastry	296	150.0 (100.0–167.0)	315	1950.0 (1510.0–2454.0)	22.4 (17.4–28.2)	299.0 (223.0–366.0)	15.0 (11.2–18.3)	32.8 (24.7–42.2)	64.3 (48.4–82.7)	6.3 (4.2–10.1)	27.4 (18.3–43.9)
Desserts	60	105.5 (90.0–160.0)	75	1250.0 (833.0–1700.0)	14.4 (9.6–19.5)	133.0 (70.0–221.0)	6.7 (3.5–11.1)	26.5 (18.0–38.6)	52.0 (35.3–75.7)	7.5 (4.2–11.7)	32.6 (18.3–50.9)
Burgers	118	300.4 (197.5–373.0)	149	2585.0 (1999.5–3370.0)	29.7 (23.0–38.7)	1090.6 (746.5–1462.0)	54.5 (37.3–73.1)	6.7 (8.5–11.9)	16.7 (13.1–23.3)	9.2 (5.5–17.4)	40.0 (23.7–75.7)
Chicken	57	95.0 (52.0–200.0)	63	927.0 (421.0–1770.0)	10.7 (4.8–20.3)	594.0 (335.0–964.0)	29.7 (16.8–48.2)	0.5 (0.2–1.6)	1.0 (0.3–3.1)	2.0 (1.4–4.0)	8.7 (6.1–17.4)
Pizza	414	93.5 (74.0–126.0)	416	850.5 (694.5–1199.8)	9.8 (8.0–13.8)	433.5 (336.0–599.8)	21.7 (16.8–30.0)	2.6 (1.7–4.0)	5.1 (3.3–7.8)	3.3 (2.4–5.0)	14.5 (10.4–21.7)
Salads	54	266.3 (148.5–332.3)	59	916.0 (524.0–1230.0)	10.5 (6.0–14.1)	542.0 (350.0–758.0)	27.1 (17.5–37.9)	5.5 (3.1–7.4)	10.8 (6.0–14.5)	2.3 (0.9–4.2)	10.0 (3.9–18.3)
Sandwiches and wraps	101	192.5 (238.0–258.5)	113	1748.0 (1385.0–2165.0)	20.1 (15.9–24.9)	900.0 (660.5–1320.0)	45.0 (33.0–66.0)	5.4 (4.0–7.8)	10.6 (7.8–15.3)	5.8 (3.0–8.5)	25.2 (12.8–37.0)
Fries	22	246.0 (156.0–302.0)	25	2010.0 (1370.0–2790.0)	23.1 (15.7–32.1)	650.0 (284.0–1334.5)	32.5 (14.2–66.7)	0.7 (0.5–1.8)	1.4 (1.0–3.5)	2.6 (1.7–5.9)	11.3 (7.2–25.4)
Sides, other	45	120.0 (35.0–253.0)	52	867.5 (305.5–1677.5)	10.0 (3.5–19.3)	321.5 (69.3–682.5)	16.1 (3.5–34.1)	3.3 (0.5–8.0)	6.4 (1.0–15.6)	1.9 (0.7–5.0)	8.3 (3.2–21.7)
Dressings/condiments, savory	49	21.0 (16.0–30.0)	88	207.0 (135.3–371.0)	2.4 (1.6–4.3)	135.0 (88.8–194.5)	6.8 (4.4–9.7)	2.0 (1.0–5.9)	3.9 (2.0–11.6)	0.4 (0.0–1.0)	1.5 (0.0–4.3)

IQR: Interquartile ranges. * Percentage calculated having as reference the recommended adult average daily energy intake (8700 kilojoules/day), sodium intake (2000 mg/day), free sugars intake (maximum 51 g/day based on 8700 kJ), saturated fat intake (maximum of 23 g/day based on 8700 kJ). Missing (n) Serving size-g/mL: Asian (12); Milkshakes, smoothies (31); Soft drinks-sugar sweetened (22); Soft drinks-artificially sweetened (9); Breakfast-savory (8); Pastry-savory (1); Cakes, muffins and pastry (19); Desserts (15); Burgers (31); Chicken (12); Pizza (3); Salads (5); Sandwiches and wraps (12); Fries (3); Sides, other (7); Dressings/condiments-savory (39). Missing (n) Nutrient content and % of daily recommendation: Asian (11); Milkshakes, smoothies (0); Soft drinks-sugar sweetened (1); Soft drinks-artificially sweetened (0); Breakfast-savory (0); Pastry-savory (0); Cakes, muffins and pastry (0); Desserts (0); Burgers (0); Chicken (6); Pizza (1); Salads (0); Sandwiches and wraps (0); Fries (0); Sides, other (0); Dressings/condiments-savory (0).

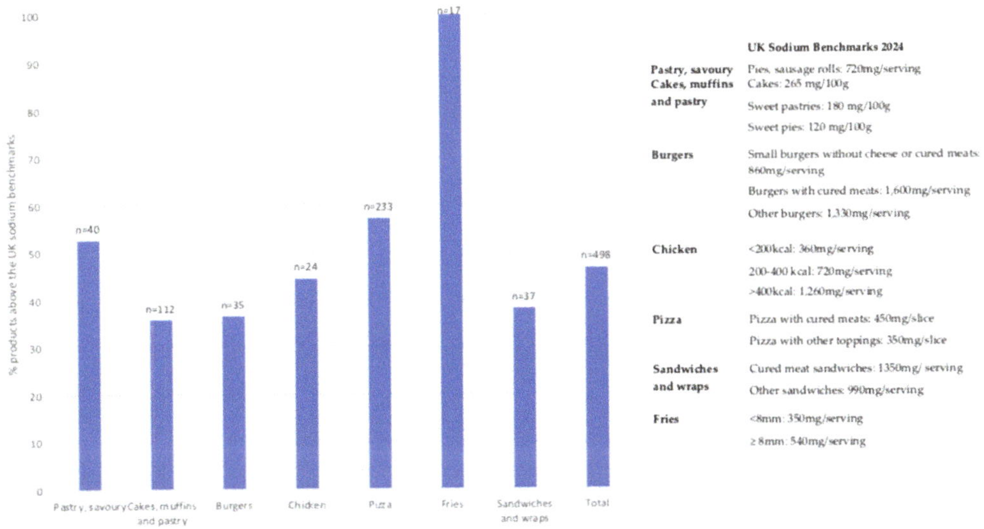

Figure 2. Proportions of products above the 2024 UK sodium targets for the respective food category.

3.2. Fast-Food Combo Meals

Table 2 presents the medians (interquartile range) for energy and nutrients per combo meal (serving for one person) as well as the median percent contributions of energy and nutrients (sodium, total sugar, saturated fat) to daily recommendations. Table S2 (Supplementary Material) presents, for each combo meal, the minimum and maximum values for energy and nutrient contents and for percent contributions of energy and nutrients to daily recommendations.

Overall, the median energy content of the combo meals/serving was 4381 kJ, meaning that 50% of the combo meals were contributing at least 50% of the average energy intake recommended for a NZ adult (8700 kJ). The combo meal categories with the highest median energy/serving and median energy contribution to daily recommended intake/serving were 'Burger(s), fries, dessert, SSB' (7531 kJ, 86.6%, respectively) and 'Burger(s), fries, dessert, ASB' (6463 kJ, 77.7%). The combo categories with the highest median energy/serving were those that contained more items and/or a dessert, and a sugary drink. The combo category with the lowest median energy/serving was 'Burger or chicken and fries' (2780 kJ, 32.0%). The combo categories with lower median energy were those constituted only by burger (or chicken) and fries and combos containing an ASB, sandwich or salad (Table 2).

Overall, the median sodium content of the meal combos was 1771.0 mg/serving, which corresponds to 88.6% of the maximum recommended daily sodium intake for adults (2000 mg) [24]. The meal combo categories with the highest median sodium/serving and highest median sodium contribution to daily maximum recommended intake/serving were 'Chicken, potato, fries, additional item, ASB (2852.7 mg, 142.6%, respectively), 'Chicken, potato, fries, additional item, SSB' (2830.2 mg, 141.5), 'Chicken, fries or potato, dessert, ASB' (2353.5 mg, 117.47%) and 'Chicken, fries or potato, dessert, SSB' (2339.5 mg, 117.0%). The meal combo category with the lowest median sodium/serving was 'Salad or wrap, smoothie' (398.0 mg, 19.9% of daily maximum recommended intake). The meal combos with lower median sodium/serving were those containing regular rather than large serving sizes, combos with fewer items such as 'burger, fries and drink' and 'chicken, fries and drink' and combos based on sandwiches, pizza, pies, salad, or wraps (Table 2). Overall, among all meal combos examined, 84.1% (n = 148) contributed 50% or more of the maximum daily recommended sodium intake (data not shown).

Table 2. NZ fast food supply 2020, by combo types: Median (interquartile ranges) for: energy, sodium sugar and saturated fat content by serving size and for percent energy, sodium, sugar and saturated fat contributions to recommended daily intakes.

Fast-Food Meal Combo Category	N	Energy—Kilojoules/Serving		Sodium—Milligrams/Serving		Total Sugar—Grams/Serving		Saturated Fat—Grams/Serving	
		Content	% of Daily Recommendation *	Content	% of Daily Recommendation *	Content	% of Daily Recommendation *	Content	% of Daily Recommendation *
		Median (IQR **)	Median (IQR **)	Median (IQR **)	Median (IQR **)	Median (IQR **)	Median (IQR **)	Median (IQR **)	Median (IQR **)
Burger(s), fries, drink-SSB	20	4042.0 (3574.5–4942.5)	46.5 (41.1–56.8)	1299.1 (1157.8–2000.5)	65.0 (57.9–100.0)	55.1 (40.2–62.7)	108.0 (78.8–123.0)	10.1 (7.1–14.8)	43.7 (31.0–64.2)
Burger(s), fries, drink-ASB	20	3380.0 (2805.1–4108.3)	38.9 (32.2–47.2)	1319.5 (1173.9–2023.0)	66.0 (58.7–101.2)	8.1 (6.9–11.5)	15.8 (13.6–22.5)	10.1 (7.1–14.8)	43.7 (31.0–64.2)
Burger(s), fries, dessert, drink-SSB	9	7531 (5985.8–8301.0)	86.6 (68.8–95.4)	2413.0 (1843.6–2550.0)	120.7 (92.2–127.5)	80.3 (71.4–89.5)	157.5 (139.9–175.5)	19.1 (17.9–30.2)	83.0 (77.6–131.3)
Burger(s), fries, dessert, drink-ASB	9	6763.0 (5433.3–7533.0)	77.7 (62.5–86.6)	2433.0 (1866.6–2570.0)	121.7 (93.3–128.5)	39.4 (33.4–43.9)	77.3 (65.5–86.1)	19.1 (17.9–30.2)	83.0 (77.6–131.3)
Chicken, fries, drink-SSB	3	4090.0	47.0	1492.0	74.6	49.2	94.5	8.3	47.0
Chicken, fries, drink-ASB	3	3286.8	37.8	1514.5	75.7	1.5	2.9	8.3	36.2
Chicken, fries or potato, dessert, drink-SSB	10	5957.0 (4588.1–7615.5)	68.5 (52.7–87.5)	2339.5 (1657.6–2948.0)	117.0 (82.8–147.0)	50.6 (42.7–63.5)	99.1 (83.8–124.4)	17.3 (10.0–21.3)	75.0 (43.4–92.5)
Chicken, fries or potato, dessert, drink-ASB	10	5404.0 (3839.4–6847.5)	62.1 (44.1–78.7)	2353.5 (1668.9–2968.0)	117.7 (83.4–148.4)	13.6 (7.3–17.9)	26.6 (14.2–35.0)	17.3 (10.0–21.3)	75.0 (43.4–92.5)
Sandwich, chips, drink-SSB	6	3789.9 (3438.8–4073.2)	43.6 (39.5–46.8)	1218.0 (800.3–1523.4)	60.9 (40.0–76.2)	53.3 (51.8–54.1)	104.4 (101.5–106.1)	4.1 (3.3–5.9)	17.6 (14.1–25.5)
Sandwich, chips, drink-ASB	6	2986.7 (2635.6–3270.0)	34.3 (30.3–37.6)	1240.5 (822.8–1545.9)	62.0 (41.1–77.3)	5.6 (4.1–6.4)	10.9 (8.0–12.6)	4.1 (3.3–5.9)	17.6 (14.1–25.5)
Pizza(s), Side(s)	13	3302.0 (2938.8–3892.0)	38.0 (33.8–44.7)	1033.0 (793.0–1270.0)	51.7 (39.7–63.5)	42.6 (10.1–46.9)	83.5 (19.7–91.9)	11.7 (9.9–13.4)	50.9 (43.0–58.3)
Pizza(s), Side(s), Drink-SSB	5	4517.0 (4352.0–4748.5)	51.9 (50.0–54.6)	891.0 (767.0–1200.0)	44.6 (38.4–60.0)	77.9 (74.8–82.6)	152.7 (146.6–161.9)	13.6 (11.1–15.9)	59.1 (48.3–69.1)
Pizza(s), Side(s), Drink-ASB	5	3927.6 (3762.6–4159.1)	45.1 (43.2–47.8)	894.3 (770.3–1203.3)	44.7 (38.5–60.2)	42.9 (39.8–47.6)	84.1 (77.9–93.2)	13.6 (11.1–15.9)	59.1 (48.3–69.1)
Pie, side (optional), drink-SSB	4	4062.0 (3727–4547.0)	46.7 (42.8–52.3)	1239.0 (952.5–1377.8)	62.0 (47.6–68.9)	49.6 (47.4–63.3)	97.3 (93.0–124.0)	13.9 (9.4–27.0)	60.4 (40.8–117.3)
Pie, side (optional), drink-ASB	4	3311.0 (2976.0–3796.0)	38.1 (34.2–43.6)	1260.0 (973.5–1398.8)	63.0 (48.7–69.9)	4.6 (2.4–18.3)	9.0 (4.8–35.8)	13.9 (9.4–27.0)	60.4 (40.8–117.3)
Breakfast, hot chocolate	3	4581.0	52.7	2163.0	108.2	29.1	57.1	17.3	75.2
Salad or wrap, smoothie	3	3051.0	35.1	398.0	19.9	50.6	99.2	6.1	26.5
Burger or chicken and fries	3	2780.0	32.0	1261.0	63.1	5.9	11.6	8.0	34.8
Chicken, potato, chips, additional item ***, drink-SSB	20	5816.1 (5433.5–6658.8)	66.9 (62.5–76.5)	2830.2 (2452.0–3173.7)	141.5 (122.6–158.7)	59.5 (58.0–65.7)	116.7 (113.8–128.8)	10.1 (7.7–12.6)	43.7 (33.5–54.8)
Chicken, potato, chips, additional item ***, drink-ASB	20	5012.0 (4630.3–5855.6)	57.6 (53.2–67.3)	2852.7 (2474.5–3196.2)	142.6 (123.7–159.8)	11.8 (10.3–18.0)	23.1 (20.2–35.2)	10.1 (7.7–12.6)	43.7 (33.5–54.8)
Total	176	4381.0 (3505.8–5660.9)	50.3 (40.3–65.1)	1771.0 (1172.5–2571.0)	88.6 (58.6–128.6)	41.3 (10.6–57.7)	81.0 (20.7–113.1)	10.6 (8.0–16.8)	46.1 (34.8–73.0)

IQR: Interquartile ranges. * Percentage calculated having as reference the recommended adult average daily energy intake (8700 kilojoules/day), sodium intake (2000 mg/day), free sugars intake (maximum 51 g/day based on 8700 kJ), saturated fat intake (maximum of 23 g/day based on 8700 kJ). ** Only median presented if n = 3. *** Burger, sandwich, bread roll or coleslaw. SSB: Sugar-sweetened beverage, ASB: Artificially-sweetened beverage.

Overall, the median total sugar content of the combo meals was 41.3 g/serving, corresponding to 81.0% of the maximum recommended intake of free sugar (51 g) [25]. Most (160/176) combos included a beverage option and those containing an SSB had the highest median total sugar per serving in relation to combos containing an ASB (Table 2). Overall, among all meal combos examined, one in three exceeded the WHO [25] maximum recommended intake of free sugars (data not shown).

Overall, the median saturated fat content of the combo meals was 10.6 g, corresponding to 46% of the maximum recommended intake (23 g) [24]. The combo meal categories with the highest median saturated fat/serving and median saturated contribution to daily maximum recommended intake/serving were 'Burger(s), fries, dessert, SSB/ASB' (both 19.1 g, 83.0%), 'Chicken, fries or potato, dessert, SSB/ASB' (both 17.3 g, 75.0%) and 'Breakfast, hot chocolate (17.3 g, 75.2%). The fast-food combo meal category with the lowest median saturated fat/ serving was 'Sandwich, chips, SSB/ASB' (both 4.1 g, 17.6%) (Table 2). Overall, among all meal combos examined, one in ten exceeded the maximum daily recommended intake [24] for saturated fat.

4. Discussion

4.1. Summary of Findings

For NZ fast-food products for which nutrition information could be sourced (30%), many product categories and meal combos were high in energy and sodium and within some categories many products were high in total sugar and saturated fat. For products with a relevant UK sodium target, almost half exceeded the target. Over 90% of sugar sweetened soft drinks available in NZ fast-food outlets would be liable for the UK soft industry drinks levy. Burgers in particular had a high energy, sodium and saturated fat content.

The meal combos usually replace one of three usual main meal occasions in a day. However, half of the combos examined provided at least 50% of the daily energy requirement and 89%, 81% and 46% of the maximum recommended intake for sodium, sugar, and saturated fat. There was a wide range of combo options, and some provided a choice of product options, particularly sides or drinks. Combos with fewer items or smaller burgers, no dessert and an ASB (rather than a sugary drink) provided less energy, sodium, and sugar. For example, the median sugar and energy of the 'chicken, fries and drink' combo category was 49.2 g and 4090 kJ when it contained a sugary drink but only 1.5 g and 3286 kJ when it contained an ASB. The options based on sandwiches/wraps also tended to have less energy and sodium although there were not many of these combos.

4.2. Comparisons of Findings to Previous Studies

These findings are consistent with assessments of the fast-food supply in other countries including Australia, the US and Canada [26,28,29]. The Australian report on The State of the Fast-Food Supply in 2019 [26] concluded that most products were unhealthy, sold in oversized portions, and high in salt, sugar and harmful fats. The authors also commented on the lack of nutrition information with over half of Australian chains not providing sufficient data. In the United States, combination meals at chain restaurants were high in energy, sodium, saturated fat and sugar and most default options in meal combos exceeded national guidelines for calories and sodium [28]. In Canada, meals from fast-food chain restaurants were high in saturated fat, sodium and sugar [29]. An analysis of combo meals offered by quick-service restaurants in Australia [30] also found many combos provided more than 30% of an adult's average daily energy intake. An earlier analysis conducted in NZ reported mean serving size, energy per serving and sodium per serving [13]. While we cannot directly compare with the 2016 NZ data, we found that Burgers and Asian meals were still in the top three categories for energy per serving, and burgers and sandwiches/wraps were still in the top three for sodium content per serving.

Of particular concern were the high sodium levels of many products, with many exceeding the UK benchmarks and almost half the combos exceeding the daily recommended maximum sodium intake. However, for every category where a benchmark existed, except for fries, there were also products that did not exceed that benchmark. This indicates that in most cases, it is possible to offer lower sodium options. There are no recent data to indicate the contribution of fast-food to New Zealanders' sodium intake. However, a survey conducted in 2012 indicates that New Zealanders consume considerably more sodium (3373 mg) [31] than the Suggested Dietary Target (SDT) for NZ adults of 2000 mg/day [24]. One NZ study estimated that the mean daily sodium intake from savory fast foods for regular fast-food consumers was 1229 mg/day [32]. Another NZ study estimated the percentage contribution of sodium from takeaway and restaurant foods at 887 mg/day, 26.3% of sodium sources in NZ diet [33] As fast food consumption is growing, this contribution is likely to be higher now.

While the fast-food industry has grown, it appears that it has done little to improve the overall healthiness of the fast-food supply despite the recommendations made by the Food Industry Taskforce convened by the Ministers of Health and Primary Industries in 2018 [18]. Eyles et al. 2018 [13] found moderate to large increases in product serving size, and energy and sodium per serving from 2012 to 2016. An Australian analysis [26] that looked at changes in categories, rather than individual products found there was little change in the healthiness of products between 2016 and 2019.

4.3. Strengths and Limitations

A strength of this study is the systematic data collection from a large number of NZ fast-food chains, covering at least 60% of the fast-food sales [12], however data collection did not include small chains and independent retailers. Combination meals have not been assessed in NZ before and are useful to analyze as they provide the context of a meal when benchmarking against daily recommendations. As combination meals involve several options, the analysis of combo meals was carefully conducted; with two options analyzed for most combos: the healthiest (less energy and sugar) and the least healthy. There are also some important limitations to consider. Most NZ fast-food chains did not provide nutrition information on their products, so it was impossible to undertake a comprehensive analysis of the nutritional state of the national fast-food supply. Half of the chains that provided some nutrition information were international chains and half were national chains. However, of the six chains that did not provide any nutrition information about their products, five were international chains. There is also a chance that the products from chains that provided some nutrition information may have had a better nutrition profile in relation to products from chains that did not provide any information. This means that this study may have underestimated portion size and overestimated the healthiness of the fast-food supply. This study's data were not sales weighted and therefore do not reflect the healthiness of items by frequency of consumption, though our analysis does include the most commonly consumed fast foods (bread-based dishes, fries, non-alcoholic beverages, poultry) in New Zealand [22].

4.4. Implications and Recommendations

This research highlights the need for policies, guidelines, and targets to improve the healthiness of fast food and provision of nutrition information as these do not currently exist in NZ. The Government needs to ensure the Food Industry Taskforce acts on the taskforce recommendations, and provide leadership by setting guidance for serving sizes, maximum targets for sodium content that are specific and measurable, and requiring fast-food outlets to provide nutrition information. A systematic review [14] found no standardized assessment methods or metrics to evaluate transnational chain restaurants' practices to improve the healthiness of menu items. Public health experts recommend a robust, independent regulatory system with targets set by government and regular monitoring [34].

The wide range of serving sizes within food categories in fast-food outlets makes it difficult to compare products, apply benchmarks, and for consumers to choose healthier options and appropriate serving sizes. For example, in this study the serving size ranged from 43–298 g for pizzas and from 79–513 g for fries. Advice should include maximum energy values for combos, given that some were very high in energy. A scoping review found some expert-recommended targets for restaurants to improve products, but no internationally accepted standard for serving sizes [14]. In the US, some organisations provide targets for serving sizes for healthy meals such as the Healthy Restaurant Meal Standards [35] and the Heart-Check certification [36]. The Australian Healthy Food Partnership had a portion size working group (now disbanded) [37] to develop recommended portion sizes, including for fast food, and published targets for some nutrients and limited fast-food categories [38]. The UK has calorie reduction guidance for the eating out of home sector [39].

The substantial amount of sodium in the fast-food supply and the increasing consumption of takeaways in NZ warrants reformulation of fast foods to lower sodium and monitoring of the sodium content of fast-food products [40]. Excess sodium intake is a major preventable risk factor for hypertension [41], a leading cause of heart disease and stroke in NZ [1]. Few countries have targets for out-of-home foods. The UK sodium reduction targets are government-led, though voluntary but are regularly monitored [23]. WHO have recently published global sodium benchmarks but only a small number are applicable to fast food [42]. In addition, warning labels should be placed on those products and combos that exceed sodium targets. In 2015 New York City passed a sodium warning label rule, requiring chain restaurants to add a salt shaker icon beside menu items or combos containing more than 2300 mg of sodium [43].

Consumers have a right to know what is in their food and menu nutrition labeling is a strategy to provide this and can also encourage reformulation. Research in the U.S. suggests that the 2010 national menu labeling law may have influenced chain restaurants to reduce the energy content of newly introduced items [44–46]. Menu labeling is under consideration by Food Standards Australia NZ, with a consultation conducted in 2021 on a range of options for labeling the energy content of foods on the menu, including voluntary and mandatory options [47]. Menu labeling, particularly for energy content, is mandatory for fast-food chains in some countries such as Australia (5 jurisdictions) [47], Canada [48] and the U.S. [49] and will be mandatory for large businesses in the UK from 2022 [50].

Other areas that could improve the healthiness of fast-food menu offers include reformulation to reduce saturated fat and sugar content across menu items, healthier items (such as ASBs) to be the default option in combos and deals, introduction of healthier menu items, and marketing and pricing strategies to encourage purchasing of healthier items.

5. Conclusions

Nutrition information was available for one-third of products of major fast-food chains in New Zealand, limiting the generalizability of findings for the whole NZ fast food supply. Among products with information available, the majority had a high median content of energy and sodium. Some fast-food product categories had a high median content of sugar and saturated fat. Many serving sizes were large and varied considerably within a category. The majority of fast-food combo meals/serving provided a considerable contribution towards the daily recommended energy intake and the maximum daily sodium and sugar intake recommendations. This is the first comprehensive study of fast-food combo meals in NZ. This research benchmarks the current healthiness of the fast-food supply providing evidence to encourage Government to: (i) develop policy to ensure that NZ fast-food chains make nutrition information on their products readily available and, (ii) implement government-led guidance on serving sizes for fast foods including combos and (iii) set targets for sodium and sugar content, including warning labels for products that exceed such targets.

Supplementary Materials: The following are available online at https://www.mdpi.com/article/10.3390/nu13114010/s1, Table S1: NZ fast food supply 2020, by food category: Minimum and maximum energy, sodium, total sugar and saturated fat content per serving and percentage contribution to recommended daily intakes of energy, sodium, sugar and saturated fat, Table S2: NZ fast food supply

2020, by meal combo: Minimum and maximum energy, sodium, total sugar and saturated fat content per serving and percentage contribution to recommended daily intakes of energy, sodium, sugar and saturated fat.

Author Contributions: Conceptualization, S.M., T.G.d.C., C.N.M., H.E. and L.Y.; methodology, T.G.d.C., S.M. and H.E.; formal analysis, T.G.d.C., S.M. and G.S.; writing—original draft preparation, S.M.; T.G.d.C. and G.S.; writing—review and editing, S.M., T.G.d.C., C.N.M., H.E., G.S. and L.Y. All authors have read and agreed to the published version of the manuscript.

Funding: This research was supported by the National Health and Medical Research Council (NHMRC) funded Centre of Research Excellence in Reducing Salt Intake using Food Policy Interventions (APP1117300). The opinions, analysis and conclusions in this paper are those of the authors and should not be attributed to the NHMRC. The Nutritrack data collections and database are funded by a Health Research Council of New Zealand program grant (18/672).

Institutional Review Board Statement: Not applicable.

Informed Consent Statement: Not applicable.

Data Availability Statement: Because of the commercial and legal restrictions to the use of copyrighted material, it is not possible to share data openly, but unredacted versions of the dataset are available with a licensed agreement that they will be restricted to non-commercial use. For access to Nutritrack, please contact the National Institute for Health Innovation at the University of Auckland at enquiries@nihi.auckland.ac.nz.

Acknowledgments: Thank you to Summer Wright for collecting the data from fast-food chains and Christina Cao for developing the meal combo database.

Conflicts of Interest: The authors declare no conflict of interest.

References

1. Ministry of Health. New Zealand Health Survey 2019/20. 2020. Available online: https://www.health.govt.nz/publication/annual-update-key-results-2019-20-new-zealand-health-survey (accessed on 8 July 2021).
2. GBD 2017 Diet Collaborators. Health effects of dietary risks in 195 countries, 1990-2017: A systematic analysis for the Global Burden of Disease. *Lancet* **2017**, *393*, 1958–1972. [CrossRef]
3. Swinburn, B.A.; Sacks, G.; Hall, K.D.; McPherson, K.; Finegood, D.T.; Moodie, M.L.; Gortmaker, S.L. The global obesity pandemic: Shaped by global drivers and local environments. *Lancet* **2011**, *378*, 804–814. [CrossRef]
4. Chandon, P.; Wansink, B. Does food marketing need to make us fat? A review and solutions. *Nutr. Rev.* **2012**, *70*, 571–593. [CrossRef]
5. Dunford, E.; Webster, J.; Barzi, F.; Neal, B. Nutrient content of products served by leading Australian fast food chains. *Appetite* **2010**, *55*, 484–489. [CrossRef]
6. Rosenheck, R. Fast food consumption and increased caloric intake: A systematic review of a trajectory towards weight gain and obesity risk. *Obes. Rev.* **2008**, *9*, 535–547. [CrossRef]
7. Nguyen, B.T.; Powell, L.M. The impact of restaurant consumption among US adults: Effects on energy and nutrient intakes. *Public Health Nutr.* **2014**, *30*, 2445–2452. [CrossRef]
8. De Vogli, R.; Kouvonen, A.; Gimeno, D. The influence of market deregulation on fast food consumption and body mass index: A cross-national time series analysis. *Bull. World Health Organ.* **2014**, *92*, 99–107A. [CrossRef]
9. Mazariegos, S.; Chacón, V.; Cole, A.; Barnoya, J. Nutritional quality and marketing strategies of fast food children's combo meals in Guatemala. *BMC Obes.* **2016**, *3*, 52. [CrossRef]
10. Statistics New Zealand. Food Price Index: June 2021. Wellington: Stats NZ; 2021. Available online: https://www.stats.govt.nz/information-releases/food-price-index-june-2021 (accessed on 21 July 2021).
11. Bayer, N.Z.; Nutrition Foundation, Auckland University of Technology. The Bayer Food Focus Project. 2019. Available online: https://nutritionfoundation.org.nz/sites/default/files/Bayer_Survey%20Report_v10.pdf (accessed on 17 August 2021).
12. Passport Global Market Information Database. Limited-Service restaurants in New Zealand 2021. 2021. Available online: http://www.portal.euromonitor.com.ezproxy.auckland.ac.nz/poral/analysis/tab (accessed on 20 June 2021).
13. Eyles, H.; Jiang, Y.; Blakely, T.; Neal, B.; Crowley, J.; Cleghorn, C.; Mhurchu, C.N. Five year trends in the serve size, energy, and sodium contents of New Zealand fast foods: 2012 to 2016. *Nutr. J.* **2018**, *17*, 65. [CrossRef]
14. Kraak, V.; Patiño, S.R.; Renukuntla, D.; Kim, E. Progress evaluation for transnational restaurant chains to reformulate products and standardize portions to meet healthy dietary guidelines and reduce obesity and non-communicable disease risks, 2000–2018: A scoping and systematic review to inform policy. *Int. J. Environ. Res. Public Health* **2019**, *16*, 15. [CrossRef]
15. UK Government. Sugar Reduction and Wider Reformulation. 2021. Available online: https://www.gov.uk/government/collections/sugar-reduction#sugar-reduction (accessed on 24 October 2021).

16. UK Government. Soft Drinks Industry Levy: Detailed Information. 2020. Available online: https://www.gov.uk/guidance/check-if-your-drink-is-liable-for-the-soft-drinks-industry-levy (accessed on 10 November 2020).
17. Food Standards Australia New Zealand. Nutrition Information User Guide to Standard 1.2.8 Nutrition Information Requirements. 2013. Available online: http://www.foodstandards.govt.nz/code/userguide/Documents/Userguide_PrescribedNutritionInformationNov13Dec2013.pdf (accessed on 10 November 2020).
18. Food Industry Taskforce. Food Industry Taskforce on Addressing Factors Contributing to Obesity Final Report to Ministers of Health and Food Safety. 2018. Available online: https://www.health.govt.nz/system/files/documents/pages/food-industry-taskforce-final-report.pdf (accessed on 24 August 2021).
19. Heart Foundation. Heart Foundation Food Reformulation Targets. 2019. Available online: https://assets.heartfoundation.org.nz/documents/food-industry/food-reformulation/heart-foundation-reformulation-targets-2020.pdf (accessed on 12 July 2021).
20. Nutritrack Database. Nutritrack. 2020. Available online: https://diet.auckland.ac.nz/sites/default/files/2019-08/The_Nutritrack_Database.pdf (accessed on 15 March 2020).
21. Fleischhacker, S.; Evenson, K.; Rodriguez, D.; Ammerman, A. A systematic review of fast food access studies. *Obes. Rev.* **2011**, *12*, e460–e471. [CrossRef]
22. Smith, C.; Gray, A.R.; Fleming, E.A.; Parnell, W.R. Characteristics of fast-food/takeaway-food and restaurant/café-food consumers among New Zealand adults. *Public Health Nutr.* **2014**, *17*, 2368–2377. [CrossRef]
23. Public Health England. Salt Reduction Targets for 2024 September 2020. 2020. Available online: https://www.gov.uk/government/publications/salt-reduction-targets-for-2024 (accessed on 1 September 2021).
24. National Health and Medical Research Council. Nutrient Reference Values. 2020. Available online: https://www.nrv.gov.au/ (accessed on 19 May 2021).
25. World Health Organization. Guideline: Sugars Intake for Adults and Children. Geneva. 2015. Available online: https://apps.who.int/iris/rest/bitstreams/668769/retrieve (accessed on 20 July 2021).
26. The George Institute for Global Health. FoodSwitch: State of the Fast Food Supply. 2020. Available online: https://cdn.georgeinstitute.org/sites/default/files/2020-05/state-of-food-supply-2020-report.pdf (accessed on 10 July 2021).
27. World Health Organization. Guideline: Sodium Intake for Adults and Children. Geneva. 2012. Available online: https://apps.who.int/iris/rest/bitstreams/110243/retrieve (accessed on 10 July 2021).
28. Vercammen, K.A.; Frelier, J.M.; Moran, A.J.; Dunn, C.G.; Musicus, A.A.; Wolfson, J.A.; Bleich, S.N. Calorie and nutrient profile of combination meals at U.S. fast food and fast casual restaurants. *Am. J. Prev. Med* **2019**, *57*, e77–e85. [CrossRef] [PubMed]
29. Murphy, S.A.; Weippert, M.V.; Dickinson, K.M.; Scourboutakos, M.J.; L'Abbé, M.R. Cross-Sectional analysis of calories and nutrients of concern in Canadian chain restaurant menu i.0tems in 2016. *Am. J. Prev. Med.* **2020**, *59*, e149–e159. [CrossRef]
30. Looi, E.S.; Backholer, K.; Cameron, A.J.; Grigsby-Duffy, L.; Orellana, L.; Sacks, G. Price promotions offered by quick service restaurants in Australia: Analysis from an obesity prevention perspective. *Public Health Nutr.* **2021**, 1–15. [CrossRef]
31. Skeaff, S.; McLean, R.; Mann, J.; Williams, S. The Impact of Mandatory Fortification of Bread with Iodine. Wellington. 2013. Available online: http://www.mpi.govt.nz/news-resources/publications.aspx (accessed on 1 July 2021).
32. Prentice, C.A.; Smith, C. McLean RM. Sodium in commonly consumed fast foods in New Zealand: A public health opportunity. *Public Health Nutr.* **2016**, *19*, 958–966. [CrossRef] [PubMed]
33. Eyles, H.; Shields, E.; Webster, J.; Ni Mhurchu, C. Achieving the WHO sodium target: Estimation of reductions required in the sodium content of packaged foods and other sources of dietary sodium. *Am. J. Clin. Nutr.* **2016**, *104*, 470–479. [CrossRef] [PubMed]
34. Seferidi, P.; Millett, C.; Laverty, A.A. Industry self-regulation fails to deliver healthier diets, again. *BMJ* **2021**, *372*, m4762. [CrossRef]
35. Cohen, D.; Bhatia, R.; Story, M.T.; Wootan, M.; Economos, C.D.; Van Horn, L.; Whitsel, L.P.; Roberts, S.; Powell, L.M.; ; et al. *Performance Standards for Restaurants. A New Approach to AddressOdoms-Young, A.ing the Obesity Epidemic*; RAND Corporation: Santa Monica, CA, USA, 2013.
36. American Heart Association. American Heart Association. Heart-Check Meal Certification Program Nutrition Requirements. 2021. Available online: https://www.heart.org/en/healthy-living/company-collaboration/heart-check-certification/heart-check-meal-certification-program-foodservice/heart-check-meal-certification-program-nutrition-requirements (accessed on 1 September 2021).
37. Healthy Food Partnership. Portion Size Working Group. 2019. Available online: https://webarchive.nla.gov.au/awa/20210604022950/http://medicareforall.health.gov.au/internet/main/publishing.nsf/Content/portion-control (accessed on 1 March 2021).
38. Healthy Food Partnership. Partnership Reformulation Program—Summary of Food Categories and Reformulation Targets. 2021. Available online: https://www.health.gov.au/resources/publications/partnership-reformulation-program-summary-of-food-categories-and-reformulation-targets (accessed on 1 September 2021).
39. Public Health England. Calorie Reduction. Technical Report: Guidelines for Industry, 2017 Baseline Calorie Levels and the Next Steps. 2020. Available online: https://assets.publishing.service.gov.uk/government/uploads/system/uploads/attachment_data/file/915367/Calorie_reduction_guidelines-Technical_report_070920-FINAL.pdf (accessed on 24 October 2021).
40. Maalouf, J.; Cogswell, M.E.; Gunn, J.P.; Curtis, C.J.; Rhodes, D.; Hoy, K.; Pehrsson, P.; Nickle, M.; Merritt, R. Monitoring the sodium content of restaurant foods: Public health challenges and opportunities. *Am. J. Public Health* **2013**, *103*, e21–e30. [CrossRef]

41. Aburto, N.J.; Ziolkovska, A.; Hooper, L.; Elliott, P.; Cappuccio, F.P.; Meerpohl, J.J. Effect of lower sodium intake on health: Systematic review and meta-analyses. *BMJ* **2013**, *346*, f1326. [CrossRef]
42. World Health Organization. WHO Global Sodium Benchmarks for Different Food Categories. Geneva. 2021. Available online: https://apps.who.int/iris/rest/bitstreams/1344511/retrieve (accessed on 12 August 2021).
43. New York City Health. Sodium Initiatives. 2021. Available online: https://www1.nyc.gov/site/doh/health/health-topics/national-salt-reduction-initiative.page (accessed on 1 July 2021).
44. Bleich, S.N.; Wolfson, J.A.; Jarlenski, M.P. Calorie changes in chain restaurant menu items: Implications for obesity and evaluations of menu labeling. *Am. J. Prev. Med.* **2015**, *48*, 70–75. [CrossRef] [PubMed]
45. Bleich, S.N.; Wolfson, J.A.; Jarlenski, M.P. Calorie changes in large chain restaurants. *Am. J. Prev. Med.* **2016**, *50*, e1–e8. [CrossRef] [PubMed]
46. Bleich, S.N.; Wolfson, J.A.; Jarlenski, M.P. Calorie changes in large chain restaurants from 2008 to 2015. *Prev. Med.* **2017**, *100*, 112–116. [CrossRef] [PubMed]
47. Food Regulation. Policy Guidance for Menu Labelling in Australia and New Zealand. 2021. Available online: https://foodregulation.gov.au/internet/fr/publishing.nsf/Content/Policy-Guidance-for-Menu-Labelling-in-Australia-and-New-Zealand (accessed on 1 September 2021).
48. Ontario Government. Guide to Menu-Labelling Requirements. 2019. Available online: https://www.ontario.ca/document/guide-menu-labelling-requirements (accessed on 1 September 2021).
49. US Food & Drug Administration. Labeling & Nutrition Guidance Documents & Regulatory Information. 2021. Available online: https://www.fda.gov/food/guidance-documents-regulatory-information-topic-food-and-dietary-supplements/labeling-nutrition-guidance-documents-regulatory-information (accessed on 1 September 2021).
50. UK Department of Health & Social Care. Mandating Calorie Labelling in the Out-of-Home Sector: Consultation Response on Policy Enforcement. 2021. Available online: https://www.gov.uk/government/consultations/calorie-labelling-for-food-and-drink-served-outside-of-the-home/outcome/mandating-calorie-labelling-in-the-out-of-home-sector-consultation-response-on-policy-enforcement#outcomes-and-next-steps (accessed on 22 July 2021).

nutrients

MDPI

Article

Development of an Unified Food Composition Database for the European Project "Stance4Health"

Daniel Hinojosa-Nogueira [1], Sergio Pérez-Burillo [1,2], Beatriz Navajas-Porras [1], Bartolomé Ortiz-Viso [3], Silvia Pastoriza de la Cueva [1], Fabio Lauria [4], Alexandra Fatouros [5], Kostas N. Priftis [6], Verónica González-Vigil [7] and José Ángel Rufián-Henares [1,8,*]

1 Centro de Investigación Biomédica, Departamento de Nutrición y Bromatología, Instituto de Nutrición y Tecnología de los Alimentos, Universidad de Granada, 18071 Granada, Spain; dhinojosa@ugr.es (D.H.-N.); spburillo@ugr.es (S.P.-B.); beatriznavajas@ugr.es (B.N.-P.); spdelacueva@ugr.es (S.P.d.l.C.)
2 Department of Biochemistry and Molecular Biology, Boonshoft School of Medicine, Wright State University, Dayton, OH 45435, USA
3 Departamento de Ciencias de la Computación e Inteligencia Artificial, Universidad de Granada, 18071 Granada, Spain; bortiz@ugr.es
4 Institute of Food Sciences, National Research Council, 83100 Avellino, Italy; fabio.lauria@isa.cnr.it
5 Department of Food Chemistry and Analysis, Institute of Food Technology and Food Chemistry, Technische Universität Berlin, 13355 Berlin, Germany; a.urbisch@tu-berlin.de
6 Allergology and Pulmonology Unit, 3rd Paediatric Department, National and Kapodistrian University of Athens, 12462 Athens, Greece; kpriftis@otenet.gr
7 Gestión de Salud y Nutrición S.L., 33003 Oviedo, Spain; vgonzalez@gsnsoft.es
8 Instituto de Investigación Biosanitaria ibs.GRANADA, Universidad de Granada, 18071 Granada, Spain
* Correspondence: jarufian@ugr.es

Citation: Hinojosa-Nogueira, D.; Pérez-Burillo, S.; Navajas-Porras, B.; Ortiz-Viso, B.; de la Cueva, S.P.; Lauria, F.; Fatouros, A.; Priftis, K.N.; González-Vigil, V.; Rufián-Henares, J.Á. Development of an Unified Food Composition Database for the European Project "Stance4Health". *Nutrients* **2021**, *13*, 4206. https://doi.org/10.3390/nu13124206

Academic Editors: Alessandra Durazzo, Rosa Casas and Massimo Lucarini

Received: 27 October 2021
Accepted: 22 November 2021
Published: 24 November 2021

Abstract: The European Commission funded project Stance4Health (S4H) aims to develop a complete personalised nutrition service. In order to succeed, sources of information on nutritional composition and other characteristics of foods need to be as comprehensive as possible. Food composition tables or databases (FCT/FCDB) are the most commonly used tools for this purpose. The aim of this study is to describe the harmonisation efforts carried out to obtain the Stance4Health FCDB. A total of 10 FCT/FCDB were selected from different countries and organizations. Data were classified using FoodEx2 and INFOODS tagnames to harmonise the information. Hazard analysis and critical control points analysis was applied as the quality control method. Data were processed by spreadsheets and MySQL. S4H's FCDB is composed of 880 elements, including nutrients and bioactive compounds. A total of 2648 unified foods were used to complete the missing values of the national FCDB used. Recipes and dishes were estimated following EuroFIR standards via linked tables. S4H's FCDB will be part of the smartphone app developed in the framework of the Stance4Health European project, which will be used in different personalized nutrition intervention studies. S4H FCDB has great perspectives, being one of the most complete in terms of number of harmonized foods, nutrients and bioactive compounds included.

Keywords: food composition database; food standardization; food data; nutrients; bioactive compounds; public health; personalized nutrition

1. Introduction

There is a close relationship between eating habits, nutrition and health [1]. Many efforts have been made to investigate the nutrient composition of foods consumed by the population [2]. Food composition data describe the content in terms of energy, macronutrients and micronutrients, as well as other compounds such as phytochemicals, antinutrients, bioactive compounds or toxic compounds in foods [3]. Generally, food composition data are published via food composition tables (FCT) and more recently as food composition databases (FCDB) [4–7].

The FCT/FCDB provide data of foods and beverages consumed by the largest portion of a population [8–10]. Currently, there are new agents to take into account, such as climate change [11–13] or the loss of biodiversity [12,14]. Add to this the constant change in consumer preferences [15,16], such as the increased consumption of processed products [17], novel foods [12,18], and an increase in global trade [5,19]. Due to these factors, FCDB are increasingly trying to collect a greater number of nutrients, bioactive compounds and foods.

FCDB data can come from: (i) original analytical data; (ii) published or imputed values of a specific or similar food; and (iii) calculated values or data provided by other FCDB [6,9,10]. On the other hand, food composition can be influenced by different factors [5,6,11,15,20–22] as depicted in Figure 1. All these factors can result in a somewhat different food composition between countries, and even between regions from the same country, thus requiring the development of more detailed and higher quality FCDB [14,15].

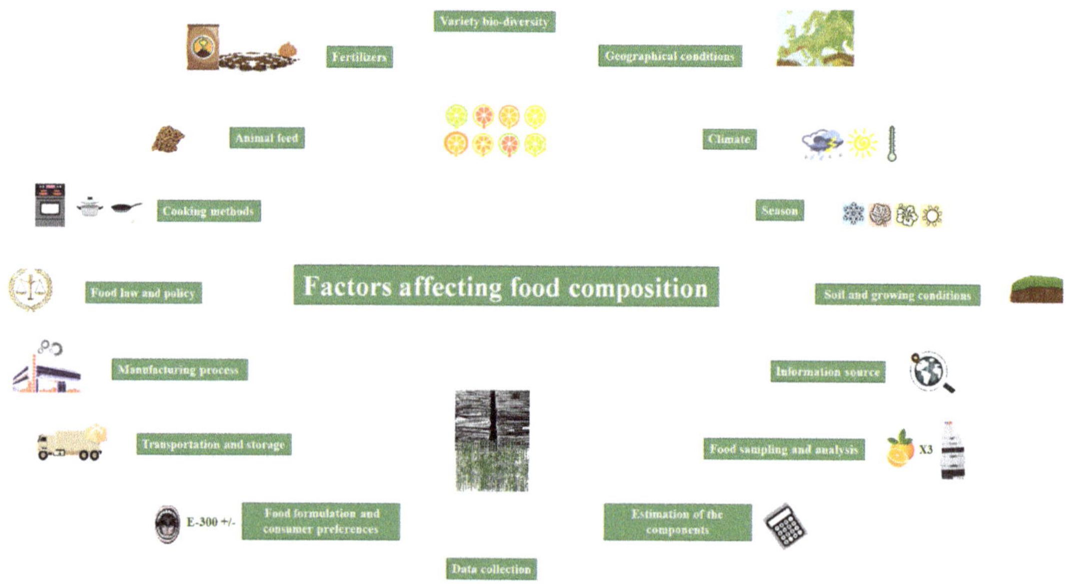

Figure 1. Main factors affecting the nutrient content and food composition. Numbers correspond to appropriate references.

FCT/FCDB are an essential tool in a wide range of areas. For example, in the field of public health [10], health programs and clinical practice [10,23], nutritional epidemiology [24], in research and food safety [18,20,25], in the food industry [20], and in agricultural programs and policies [7,23].

Currently, a growing number of countries are updating their FCT/FCDB, for example, McCance and Widdowson's food composition table [26], the Dutch Food Composition Database (NEVO) [27] and Frida Food Data [28] which include a wide range of foods and compounds, making them a reference at an international level [12]. However, several countries still lack their own data sets [10,15,20,22,29] so they often resort to foreign FCT/FCDB [9,24] such as the Food and Agriculture Organization (FAO) [30]; or the United States Department of Agriculture (USDA) FCDB [31], among others. Nevertheless, most food composition data are based on fresh foods, while information on processed foods, recipes or fortified foods is usually missing or not up to date [9,15,16,29]. Organizations such as the International Network of Food Data Systems (INFOODS) [23,32,33] are making great efforts to provide information about different FCT/FCDB, promoting the reliability and up-to-date nature of the data [34,35].

Therefore, by having a harmonized and standardized FCT/FCDB, comparisons between countries would be possible and nutritional data would be more accurate and comprehensive [10,15,34]. In order to standardize terms between data bases, different ontologies are used in nutritional research [9,36,37]. The most common is LanguaL™ [38,39]. LanguaL™ is based on the concept that any food (or food product) may be described systematically by a combination of characteristics [39]. There are other descriptors such as those developed by the European Food Safety Authority (EFSA), and FoodEx2 [40]. FoodEx2 is a standardized food classification that consists of individual food items aggregated into food groups and food categories in a hierarchical structure [15,37,40–42]. Recently, efforts have been made to map FoodEx2 facet descriptors with LanguaL codes [39,42].

In order to overcome this challenge, in the last decades great progress has been made to develop standards and guidelines focused on the harmonization and standardization of FCDB [10,15,23]. Among the most important is FAO/INFOODS that coordinates food composition activities at the international level [6,32]. FAO/INFOODS has developed different strategies in an attempt to harmonize data and make it comparable across countries [4,6,32,33,43–46].

Additionally, there have also been numerous EU-funded initiatives to standardize and harmonize food compositional data [15] such EUROFOODS, COST99 or NORFOODS [16]. More recently, they continued via the European Food Information Resource Network (EuroFIR), now known as EuroFIR AISBL [2,10,16,47]. The main objective of EuroFIR is to contribute to the harmonization of high-quality food composition data in Europe [47–49]. For this purpose, the EuroFIR project has developed different tools like its own LanguaL™ descriptors; EuroFIR Theasauri or FoodEXplorer. FoodEXplorer is a query tool that includes food composition data across more than 30 countries [20,47] and is updated regularly [2,47,50]. In Europe, these networks allowed the development of large multicenter nutritional studies. For example, the European Prospective Investigation into Cancer and Nutrition (EPIC) [51]. Notably, food composition analysis is very expensive and can be time consuming [22,46]. However, an increasing number of FCDB are introducing as many nutrients and bioactive compounds available as possible [30,52–54]. FooDB (https://www.foodb.ca, accessed on 27 October 2021) represents the most comprehensive effort to integrate food composition data [24] and a large amount of different compounds [55].

Stance4Health (Smart Technologies for personalized Nutrition and Consumer Engagement) (S4H) is a project funded by European Union's Horizon 2020 research and innovation program, aimed at evaluating the benefits of a novel smart personalized nutrition service in a large clinical study [56]. One of the main tasks of the project is to build a nutritional database (with as many foods and nutrients as possible) to complete the national FCDBs from the countries involved in the project. As the FCT/FCDB of the countries is completed, a more accurate approximation of the users' diets will be achieved.

The aim of the present study is to describe all the harmonization efforts and introduce this novel and unified Stance4Health's FCDB (S4H FCDB). This database will be part of the app developed in the framework of the European project, which will be used in different personalized nutritional intervention studies (Trial ID: ISRCTN63745549).

2. Materials and Methods

2.1. Working Group Organization and Training

A working team composed of two coordinators and a committee (including researchers, computer scientists and compilers, all of whom were dietitians and nutritionists) was established for the preparation of the S4H FCDB. Both the coordinators and the compilers completed the e-learning course offered free of charge by FAO/INFOODS [57]. The e-FoodComp course on food composition was designed by experts to be used by different professional users. The course consisted of 14 lessons structured in five units, for a total of approximately 10 h. The course offers a large number of examples and exercises suitable for on-the-job training. In addition, different guides and research were chosen to be used as a reference for the standardization and harmonization processes [3,25,32,33,41–51,57–68]. The coordinators established the general guidelines, and also helped choosing and obtaining the FCDBs used. In addition, they were subsequently responsible for checking and assessing the quality of the harmonized procedures and data. The remaining committee members performed the rest of the tasks.

2.2. Data Collection, Harmonization and Standardization Methods

A personalized nutrition intervention for different populations in Spain, Germany and Greece will be carried out within the S4H Project [56]. For this reason, the three national FCT/FCDB of the intervention countries were used as references [69–71]. These FCT/FCDB were completed with values of nutrients, bioactive compounds, such as polyphenols, and foods from different databases [14,26,27,30,31,72–75] (Table 1). All FCT/FCDB were either free of charge or permissions were granted when needed. The original FCDB data, such as original name or food identifier, were kept for the purpose of future checks or updates. In addition, quality and traceability of the documented data was guaranteed. However, the data needed to undergo some conversions before being added to our FCDB. All data were harmonized in order to obtain standardized foods and nutrients. Subsequently, all the information was entered into dynamic spread-sheets that related the data and characteristics to each other. As all foods were not in one single language, names and recipes were translated into English. All foods were uniquely identified using the standardized food classification and description system proposed by EFSA FoodEx2 [40,42]. The coding was carried out by qualified compilers and the last version of FoodEx2 system was used [40]. FoodEx2 allowed coding of all foods and beverages present in the FCDB into 20 main food categories, divided into subgroups up to a maximum of four levels [68]. Fortified foods, dietary supplements, food commercial brands, recipes or prepared dishes were discarded from the FCDB. Cooked foods were included, and the cooking method was extracted as an additional data element. Generic unbranded processed foods (such as canned foods, pickles, processed meats or pastries, among others) were also included.

Table 1. Compilation of FCDBs used in the construction of the S4H FCDB.

Name of the FCDB	Last Update	N° FoodsT [1]	N° Foods [2]	N° Items [3]	References
Tabla de Composición de Alimentos de Martin Peña actualizada de la version original por i-Diet (Spain)	2019	726	711	90	[69]
Composition tables of foods and Greek dishes (Greece)	2007	88	84	18	[71]
Bundeslebensmittelschlüssel (BLS) (Germany)	2014	936	715	146	[70]
Banca Dati di Composizione degli Alimenti per Studi Epidemiologici in Italia (Italy)	2015	978	976	97	[72]
Dutch Food Composition Database (NEVO) (Netherlands)	2019	2152	949	144	[27]
McCance and Widdowson's 'composition of foods (United Kingdom)	2019	2910	1208	280	[26]
Food and Nutrient Database for Dietary Studies (FNDDS) (United States)	2018	7083	609	69	[31]
FAO/INFOODS Analytical food composition database version 2.0 (AnFooD2. 0)	2017	2953	346	378	[30]
FAO/INFOODS food composition database for biodiversity (BioFoodComp4.0)	2017	7953	355	538	[14]
Phenol-Explorer 3.6 database on polyphenol content in food	2015	458	457	520	[73–75]

[1] Number of foods used out of total. [2] Number of foods included. [3] Number of items collected, including information on food nutrients and other compounds and data.

The complete dataset was examined and converted into standard units [3,43]. The tagnames for food components developed by INFOODS were used for this purpose [33,60]. In order to ensure harmonization, standard tagnames were designed for each compound. The original FCDB compounds that were in different units or did not correspond to those described in the INFOODS tagnames, were transformed and recalculated to match the one expressed in the standard tagname (i.e., change of units from grams to milligrams) [33,43].

Only in specific cases were tagnames not modified (as in the case of some polyphenols) where the coordinators decided that it was more functional to leave all compounds with the same units. Those compounds that did not have labels were assigned one that was proposed by compilers. The labels and units can be found in Supplementary Material S1 (Excel sheet). All compounds were expressed in amount per 100 g or 100 mL of food and edible portion values were extracted for further calculations as recommended [3,33]. All changes were made manually or semi-automatically in spreadsheets. All changes were monitored and subsequently validated as described in Section 2.5.

2.3. Mapping and Unification Process

Once the data were harmonized, a single FCDB was created. The data were differentiated by origin, but organized in a homogeneous structure. The mapping process involved matching foods based on the FoodEx2 identification code. The data were cleaned by eliminating 0 values and treated as missing to eliminate possible errors in the matching. Standard rounding values were taken [43]. Statistical parameters (mean, median, standard deviation) were calculated for each compound whenever a food had the same code. After all the data were evaluated, the coordinators decided to use the median as the final value. Unification was applied to foods with the same codes. The median was used in order to unify and complete the values of a food as long as the matchings were identical. The results were filtered using different filters as values to locate the values of the outer layers. Afterwards, the quality of the data was evaluated. All changes were made in spreadsheets, and Python 3.0 was used for unification and statistical calculations. The scripts used are shown in Supplementary Material S2. For the S4H FCDB, energy was recalculated using the Atwater factors [62]. Once the values were obtained, they could be inputted in the national FCDB for those foods that are not yet included, or for those nutrients or compounds that were missing.

2.4. Recipe Calculation and Additional Factors

Recipes or prepared dishes will be introduced as part of another database. Recipes will be linked to the S4H FCDB in order to obtain all the necessary information. For the calculations, the edible portion, cooking method and those factors that can generate changes in the nutrient content (such as retention factors (RF) and yield factors (YF)) will be taken into account. In addition, allergen data and preparation methods will be implemented.

For the harmonized calculation of recipes, a mixed model was used, since it is the most widely used and accepted [3,76]. This method was proposed as standard by EuroFIR, and consist of applying YF at the recipe level and the RF to each individual ingredient [48,77]. This procedure requires incorporating beforehand the standardized YF and RF based on the food group classification system [25,78]. YF and RF values were obtained from different sources in order to cover the largest number of foods and cooking methods [26,50,76–80]. For the RF of polyphenols, in addition to those given in Phenol-Explorer 3.6 [75], the values retrieved from the EPIC study [61] were also used. The calculation method involved the following steps: first, weights of the raw ingredients were collected. Second, nutrient and compound levels were corrected for edible portions, if applicable. Next, ingredients were modified to account for the effects of cooking by using yield factors to adjust the raw weights. In addition, retention factors were also applied for nutrient losses or gains during cooking. Finally, the ingredient values were summed to obtain recipe values. Final values were expressed per 100 g of recipe and per total recipe weight. The estimates were performed automatically and entered as recipes in the database.

2.5. Information Management and Data Quality

Tables and FCDB were implemented in MySQL open-source software. MySQL is a cross-platform relational database management system. A total of eight tables were implemented and interrelated. Tables were disaggregated to provide more versatility and security. All values were subjected to a variation range. Organizations such as INFOODS or

EUROFIR propose different methodologies to ensure and validate data quality [25,33,50]. However, in this case, the coordinators decided to follow a system of hazard analysis and critical control points (HACCP) [50]. For each data input, an original document and a working document identified with the same code were stored. For each step identified as HACCP, a series of validation tests were performed. These tests were based on different recommendations [3,25,33,50,57]. The validation procedure was followed by corrections, if necessary. The corrections of the conflicting foods were checked by data traceability extending to the original FCDB. The verifications performed are shown in Table 2. Those processes were applied at each stage of quality control, trying to minimize systematic and random errors. All tests were performed manually or semi-automatically by the coordinators, except for the recipes, which were automated.

Table 2. Steps identified as HACCP and validation testing.

Validation Testing	Step as HACCP
Verification of food name and description, possible misspellings or translation errors.	Harmonization
Classification and consistency verification of food name and grouping.	Harmonization
Verification of FoodEx2 coding and INFOODS compound tagging.	Harmonization
Sum of (water + protein + fat + total carbohydrates + alcohol + ash) is within the range: 95–105 g.	Unification
Implausible values, such total fat value, is = 0, Fatty acids = 0 and Cholesterol = 0, or fiber in fish.	Unification
Outlier values within each nutrient or compound.	Unification
Spreadsheet to MySQL data transfer checking.	Management
Model recipe testing.	Recipes Calculation

3. Results

Around 26,200 foods were collected from different FCDBs. Branded foods, recipes or ready-to-eat products, among others, were excluded and a total of 6410 foods were obtained. The Netherlands, the Italian and the United Kingdom's FCDB were the ones that contributed the largest number of foods in the unification process. A large number of foods were excluded from the FAO FCDB due to incomplete information. Subsequent to unification, filtering and quality validation, 2648 foods were obtained for the S4H FCDB and 47% of them had an equivalent food in another FCDB, so that achieved unified values. The foods were grouped by food groups and shown in Supplementary Material S1 (Excel sheet).

Regarding nutrients, bioactive compounds and other information, 880 items were collected. About 95% of the items corresponded to nutrients or other food compounds. Only 5% corresponded to other items such as the food group, its code or some additional factors. During harmonization and standardization, 78.7% of the tagnames were kept with the recommended INFOOD standards units [33,60], without taking into account the polyphenol tagnames. However, the majority of the polyphenols did not have standard tagnames and represented 55.7% of the total of items. Only 5.3% of other compounds did not have standard tagnames. The standard units of 8.4% of the total number of compounds was modified to more functional units.

Germany contributed the highest percentage (15%) of total nutrients, Spain 9% and Greece 2%. It should be noted that 65.5% of the nutrients included in the database were polyphenols from Phenol-Explorer 3.6. If we do not take this into account, the percentages are tripled, as shown in Figure 2. For example, Spain and Germany had around 88% of the 40 most used nutrients in epidemiology, while Greece had only 40%. After Phenol-Explorer, the FAO FCDB is the one with the highest percentage of compounds, around 28.2%. However, the English and Italian FCDBs were the ones with the highest percentage of nutrient values used in epidemiology, with more than 95%.

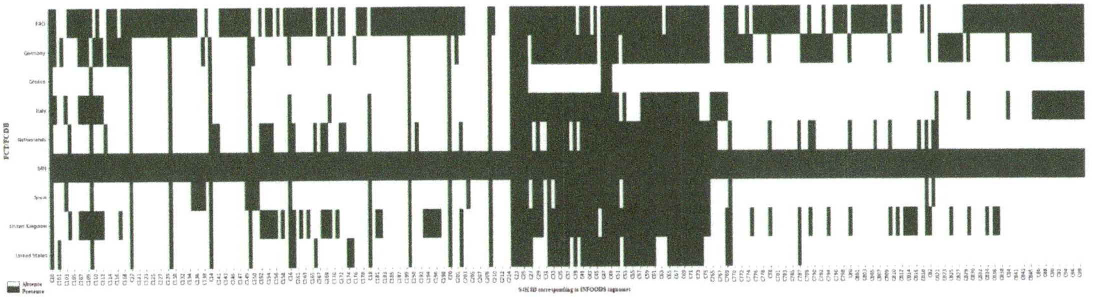

Figure 2. Absence or presence of different compounds and nutrients in the FCDB. The FAO/INFOODS tagnames are expressed with the S4H IDs, listed in the Supplementary Material Table S1. Not all tagnames are shown; the Phenol-Explorer Database is not included and the complete figure is depicted as Supplementary Material Table S3.

Figure 3 shows an example of the values of the unification process for the item A00MH Spinaches, raw. Raw spinach was selected because it was included in most FCT/FCDB. The value of total proteins is quite similar, which confirms a correct classification of the food. However, micronutrient values were more heterogeneous among the different FCT/FCDB. With the unification, the S4H FCDB obtained intermediate values considering the possible variability and also, in the case of Selenium, it retrieved values similar to those of the national FCT/FCDB.

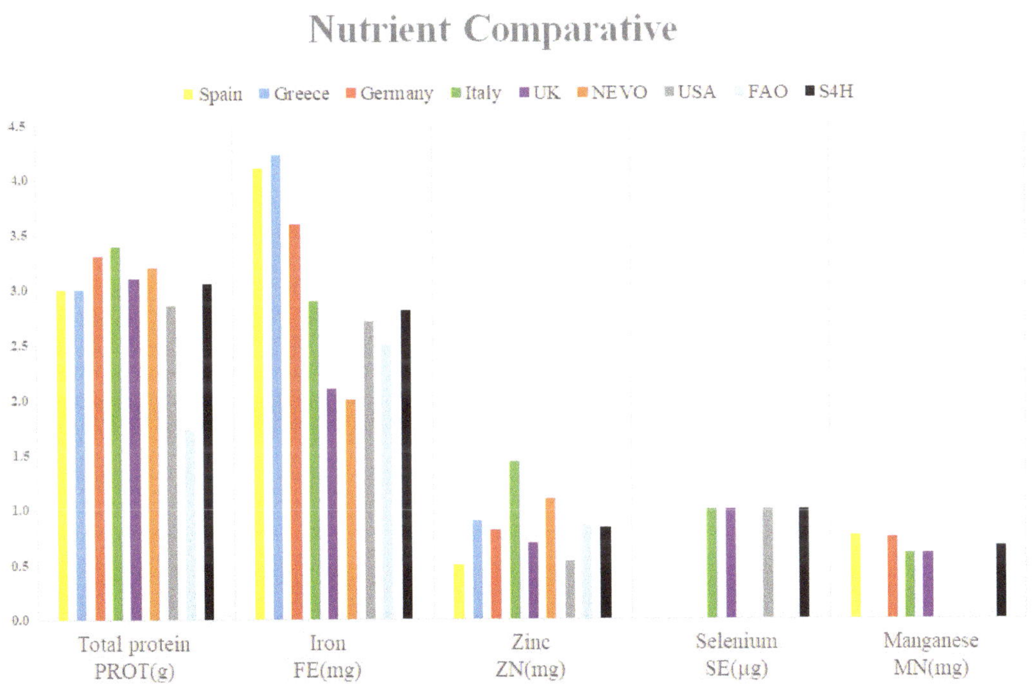

Figure 3. Protein, zinc, selenium and manganese content of the food categorized as spinach in different FCDBs and the unified values corresponding to the S4H FCDB.

Regarding recipes, tables and interrelations for energy, protein, carbohydrate, fat, sodium, calcium, riboflavin, Vitamin C, the flavonols group and (-)-epicatechin were checked for correctness. A set of recipes was selected from the database to perform manual and automatic calculations; the results were identical in 80% of the cases, and when they were not, mismatches came from the compilers' failure to choose performance or retention factors. This problem disappears when automated.

After data validation, no errors were detected in the transformation of units because there were no systemic deviations detected in any specific nutrient or compound. In 1.8% of the foods, some nutrients showed extreme standard deviations, most likely coming from the original FCDB. In addition, 7.5% of the foods had high deviations in some nutrients; all of these values, coming from the harmonization and coding phase, were reviewed and corrected. No differences were detected when using either mean or median values, except in some specific cases, such as unified foods with more than six FCDB. Nevertheless, the median value gave estimates closer to the overall computation of the data. In addition, 4.9% of foods had macronutrients that did not meet the established quality limits; the same happened with the sum of total fats, where 2.8% presented mismatches. Therefore, 17% of the food products had some type of error. Of this percentage, about 88% could be resolved by excluding 54 food items, resulting in a total of 2648 foods. The data were transferred properly and all MySQL interrelations were checked.

4. Discussion

The aim of this study was to develop a FCDB as complete as possible in terms of food, nutrients and other compounds. This is especially important because globally there is a large nutritional data gap [33]. This trend is changing, since according to Finglas et al. [49] many countries are making efforts to create or update their FCDB. Epidemiological studies where several countries are involved, such as the EPIC study, are becoming more and more common. According to Slimani et al. [51] during the EPIC study a total of 26 nutrients for more than 550 foods per country were selected; after appropriate standardization they were used for cross-country comparisons.

Since the S4H project involves several countries, we used the EPIC study as our reference [51]. All the databases were chosen by agreement between coordinators and researchers. The three countries involved in the intervention were used as the main sources for FCT/FCDB [69–71]. The national FCT/FCDB were selected assuming that the most reliable values were available at the local level. Some nutrients were missing from one or more FCDBs. Therefore, we decided to include three more to make ours more representative of European foods [10]. These widely recognized databases were from Italy [72], the Netherlands [26,27] and the United Kingdom, since these countries had more updated versions [81,82]. Finally, four more international FCT/FCDB were included to enrich nutritional composition: the USDA FCDB, since it is widely used [31], the INFOODS/FAO—FCT/FCDB [14,30] to increase the number of nutrients and to take into account the biodiversity of some foods, and, finally, Phenol explorer 3.6 was chosen [73–75] due to the great implication that polyphenols have on diet and health [61]; this allowed for the enhancement of national FCT/CBDT through the addition of more foods and the inclusion of more than 600 bioactive compounds.. We discarded 75% of the foods since quality issues were reported in the estimates when introducing new commercial foods [83], emerging dietary components [61], fortified foods or dietary supplements [84], since these are specific to each country. Recipes and prepared dishes were also not incorporated due to the great variability of preparations in each country [85]. Recipes will be linked from another interconnected database under construction. The national FCDB will input those foods or missing values from the S4H FCDB.

4.1. Standardization and Unification

The use of food composition data from different countries needs a high level of harmonization of both food values and the nutrients that are included [48]. Data processing requires precise nomenclature and standardized methods, such as the use of ontologies or tags that allow correct classification and description [86]. Nutrients from the TEDDY study were compared between four countries. According to Uusitalo et al. [67], harmonizing datasets before calculations generally made the results comparable, as systematic and random errors were minimized. This approach was previously used for ten European countries in the EPIC study, producing similar results [51,66].

Due to the large amount of available food items, the implementation of artificial intelligence and computational approaches is recommended [87]. Currently there are many automatic and semi-automatic tools that are extensively used to classify FCDBs [9,41,87,88]. A clear example is the ASA24 system that uses automated methods for several databases [16,87]. Another example is StandFood, a semi-automated system that obtained an overall result accuracy of 79% [41]. New techniques of natural language processing [88], machine learning, and statistical models, such as Monte Carlo simulations [12] or extraction of 'big data' [20], make the process faster than manual work [16]. However, due to the complex work, a manual *post hoc* review is always required [82,87]. After a first approach using different methodologies, manual and semi-automated harmonization and standardization work was decided to be performed in the S4H FDCB. Although human errors are still possible, this work guarantees a higher accuracy when comparing the same foods from different FCDB than automated predictions [41].

The first step was to achieve harmonization to classify foods. Durazzo et al. [37] classified foods based on different criteria. One of the classifications used is the FoodEx2 classification implemented by EFSA [40]. We selected these classification criteria due to its hierarchical nature and its widespread use. All foods were harmonized and linked between the different FCDBs. This classification provides the possibility to match foods, although full comparability is not guaranteed [2,59]. Secondly, since all nutrients and compounds have to be made comparable, they were defined in the same way, according to measurement units [51,67]. The tagnames proposed by INFOODS [43], indicating the name of the component, units and analytical method [60], have been implemented in different FCDBs around the world [45]. INFOODS tagnames allowed us to normalize variables from all databases to reference units (such as µg or mg) with faster results. Also, when unifying two nutrients, it allowed us to ensure that they were expressed in the same way and could be comparable. We modified 8.4% of the units of the tagnames to obtain a more functional FCDB. Most of the individual phenolic compounds did not have a tagname, and a new tagname was created to facilitate their integration into S4H FCDB. After standardizing both foods and nutrients, we had the opportunity to unify those foods that were categorized as identical. This would allow the inputting of those missing data and foods in the national FCDB.

Several studies claim that for research purposes in nutritional epidemiology, it is better to approximate nutrient values than to leave them as missing [51]. Not imputing data could lead to systematic underestimations of nutrient intake [18]. Although authors and institutions recognize this as a reliable method [33], others are critical, arguing that food composition changes considerably from one country to another [2]. S4H FCDB inputted the values of a weighted estimate of several FCDBs, making the values of high quality and taking into account the biodiversity of foods, thereby improving the estimations [14]. The inputting of missing values are frequent mechanisms that are performed when using FCDBs with recognized data quality [88]; typically, the data come from FCDBs from the United States, Europe, or other countries in the same region [5,12]. An example is the FCDBs from countries in sub-Saharan Africa, which import up to 88% of data about animal-source food [22]. Another example is the Middle East FCDBs, which inputted food from the United Kingdom FCDB [81]. The S4H FCDB uses an ad hoc approach to standardize the FCDB, as was done in the EPIC study [66]. This approach will make it possible to add foods or replace the value of a missing compound from other FCDBs with comparable estimated and weighted values [51,66].

During the first unification tests between foods, large standard deviations were identified in some macronutrient or micronutrient, largely coming from beverages and spices. The reason was that most of the 0 values for a compound or nutrient were not of the 'logical zero' type. Authors such as Pérez Grana or Westenbrink et al. [1,25] recommend that missing values should never be replaced by 0 and even modify the 'logical zero' values so as to avoid affecting the estimations. Before unifying the values, all the 0 values were removed. Then, by unifying the values, most of the data were homogenized, thus improving the results. The loss of the 'logical zero' values would not affect the calculations since they should remain at 0 and can be incorporated later.

On the other hand, although the mean and median were calculated, median values were chosen as the reference value after unification. Although some authors choose the mean [9,25], the median value is, in some cases, a better measure of central tendency [88], especially for extreme values from national FCDB. This ensures homogenization of the data and prevents wrong estimations. The unification allowed the inclusion of many foods and compounds. Figure 3 shows how the unification guarantees the homogenization of values. Once the values were unified and cleaned, as recommended by FAO or EuroFIR [2,33], estimated energy was recalculated using the Atwater coefficients [62].

Organizations such as FAO work with spreadsheets due to their simplicity, wide availability and familiarity to users [44,89]. Our work started out using spreadsheets, although the amount of data quickly became rather difficult to handle [89]. Therefore, the software MySQL was used, which allowed us to send and retrieve data through its interrelated tables [45,90]. This ensured traceability and quality controls, and also facilitated the relationship of S4H FCDB with the recipe tables for subsequent calculations.

4.2. *Data Quality and Recipe Calculation*

High quality data are essential for nutritional studies [48]. The use of the HACCP system [50] allowed us to quickly and sensibly evaluate data quality at different stages. In addition, the FAO guidelines served as a reference in the detection of critical points at any stage of the process [33]. Initial training was essential to successfully complete all the tasks, while guaranteeing the highest possible accuracy and quality.

For S4H FCDB, name verification and food description, as well as translations, were corrected thanks to the collaboration of researchers whose first language was mostly the language of the FCDB.

An FCDB should be frequently updated. For example, in the TEDDY study, the FCDB was updated at least once a year [67]. The incorporation of the original food IDs to guarantee the traceability of the food was a critical control point. Original food IDs allowed us to identify and correct errors and even to retrieve or update the information. Failures in the classification and verification of food grouping and compound labeling were detected due to outliers or manual coding by using standard deviations. Three different checking approaches were used: (i) Checking that the sum of macronutrients was within the range or the presence of implausible values detected semi-automatically in the spreadsheets; (ii) Checking for data transfer from spreadsheet to MySQL by direct verifications between versions and table relationships; and (iii) Checking the model recipe by manual verifications by compilers and automatic verifications by interconnecting the different databases. These verifications made it possible to ensure the comparability and reliability of the data.

Performing chemical analyses for all recipes and complex food matrices is not achievable. Calculations are performed indirectly using each ingredient's nutritional information [10,91]. In order to properly calculate a recipe, different parameters must be taken into account, such RF or YF. One of them is that values should not be missing, since these may lead to a biased underestimation of nutrient intake [43]. During unification and the inputting of values, this problem was solved to a large extent. The EuroFIR recipe calculation procedure was selected as a reference because it is one of the most commonly used [76]. There are several studies that use an app or software to estimate or perform interventions in nutrition and health [61,92,93]. Accordingly, the S4H FCDB will be interlinked with a recipe database. It will therefore make possible the automatic calculation of recipe values, taking into account all necessary parameters, such as edible portion, retention factors and yield factors or even allergens. Thus, the recipes will be as adequate and representative as possible to cover the needs of the population.

4.3. Strengths and Limitations of the S4H FCDB

With the continuous expansion of food trade worldwide [10], climate change or innovation in agriculture [13], international FCDBs are essential. For this reason, S4H FCDB wants to be a reference in the creation of a unified FCDB. Much effort has been made to overcome the common drawbacks that are generally associated with the FCDB's construction. The variability in food composition (when using different FCDBs) is one of the most detected limitations [7,20]. S4H FCDB attempts to address this limitation by using the median value as the reference estimation. Additionally, there is no guarantee that national FCDB data are free of errors [2]. However, all national FCDBs are used in their own country. The unification gave us a global view of possible wrong values, allowing them to be corrected. Another limitation was represented by missing foods and nutrients from the national FCDB [47]. The S4H FCDB inputs those missing foods and compounds giving coverage and completing those values in the national FCDB. Discrepancies may exist between the tagnames proposed by FAO/INFOODS and their units [15]. However, the decisions to change units were consensual and made to improve their functionality. Moreover, inputted values from other datasets, especially dishes and recipes, did not guarantee directly related values [10,65]; for this reason, recipes and ready-to-eat products were removed. Recipes will be calculated thanks to the interconnection between the S4H FCDB and a recipe database.

The work was complex, and although the compilers were experts in nutrition, mistakes may have been made when choosing codes for harmonization [15,23]. However, the use of guidelines and data validation throughout the whole process allowed for the verification and correction of possible mistakes. The preparation of this material required a long time, and perhaps with automated methods and a subsequent exhaustive check, similar results could have been obtained [88]. There may have been failures during the translation of some foods [47], especially regional foods, although if no reliable translation was found, foods were discarded. Even so, our results are encouraging. Misspellings and translation mistakes were detected while manually identifying and classifying. Thus, one of the limitations may have actually been a strength.

In most nutritional epidemiological studies, results are similarly interpreted regardless of how they make estimations or which FCDB is used. This generates an unrealistic relationship of nutrient intakes and their impact on health [94]. An increasingly large number of epidemiological studies attempt to make their data comparable [51,67,95,96]. One of the strengths of the S4H FCDB is that with unified values, data from different countries could be compared, as it would take biodiversity and different parameters affecting the same kind of food into account. Another option is to use national FCDB data and only fill in the missing nutrients and compounds to avoid underestimations [6,18]. Organizations such as EUROFIR have the potential to create a standardized FCDB which should be free to use [48]. EFSA already has a tool as a first step towards the unification of nutrients [97]. The S4H FCDB is one of the most comprehensive FCDB regarding the number of foods and nutrients, being able to collect more than 800 compounds from each foodstuff. Thus, to date it is only surpassed by the https://foodb.ca (accessed on 27 October 2021) project supported by the Canadian Institutes of Health Research and by The Metabolomics Innovation Centre. This Database includes not only nutritional information, but also a large amount of bioactive compounds [24,55,98]. However, it must be noted that the S4H FCDB uses different FCT/FCDB, giving much more homogeneous and comparable nutritional values.

4.4. S4H FCDB's Future Perspective

The S4H FCDB consists of interlinked tables that make a complete nutritional information system. S4H FCDB not only allows accurate calculations, but also provides the user with information on different aspects integrated in the personalized nutrition system. The purpose of the S4H FCDB was for it to be used in epidemiological studies, in particular precision nutritional studies. This S4H FCDB will be connected to an app that will be used during the nutritional intervention of the project. The study aims to generate personalized nutritional recommendations to different populations, more specifically adults and children [56]. The app derived from the S4H Project will be an automated diet evaluator and generator used from smartphones. A set of more than 10,000 recipes from all countries is expected to be available. All recipes will be implemented in a mobile app for future nutritional intervention. An example is depicted in Figure 4. Other similar apps have a smaller number of foods and were developed from a smaller number of food data sources [82].

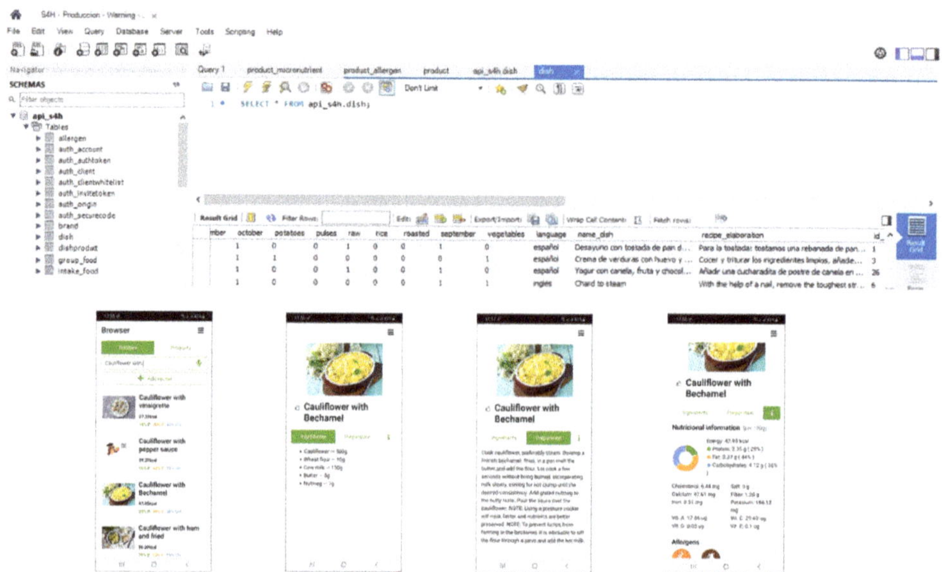

Figure 4. Shows the relationship between MySQL tables and how the recipe is generated in the APP.

The S4H FCDB will be able to connect to other tools. One of the milestones of personalized nutrition is to understand the health level of the gut microbiota of a given patient. The S4H FCDB generated data will also be completed with the use of AGREDA [99], an extended reconstruction of diet metabolism by the human gut microbiota. The S4H FCDB will also introduce commercial products, incorporating allergens and different scores as used in Open Food Facts [16]. These products and fast foods from the different countries of the project will make S4H FCDB more comprehensive and representative [38]. The data will also be updated periodically to avoid obsolescence [5], which will be possible thanks to traceability.

In the future it is expected that the S4H FCDB will be extended by implementing toxic substances, such as food processing contaminants, as few FCDBs contain these components [15,88,100–102]. Due to the importance of climate change in nutrition, sustainability parameters and different markers of climate change would be an added value to be included [13]. Finally, in order to identify food-disease associations [55], food biomarkers could be introduced by linking them to FOBI (Food-Biomarker Ontology) [36], or extending the compounds related to https://foodb.ca (accessed on 27 October 2021) or other big data sources [24,98].

5. Conclusions

S4H FCDB was built through a huge scientific work to collect and harmonize all the nutritional data. S4H FCDB is one of the most comprehensive FCDB with more than ten FCDBs used, which is one of its main unique characteristics. This food database is comparable to that used in other relevant studies, such as EPIC. A large number of harmonized foods (over 2000) and more than 800 nutrients and bioactive compounds (such as polyphenols) have been included, the inclusion of such a large number of bioactive compounds being another unique strength of the paper. S4H FCDB attempts to mitigate the usual limitations, such as variability in food composition, errors, and missing values in the national FCT/FCDB databases. Trained personnel following the guidelines of official agencies were able to homogenize the information. This made it possible to unify foods, their nutrients and bioactive compounds among the FCT/FCDBs using the median value as the reference value. The values obtained were less extreme and made it possible to complete the national FCT/FCDB. The S4H FCDB has many perspectives, not only the implementation in nutritional studies through an application. But it is also capable of being part of other tools and has the versatility to be continuously enhanced with much more information. Thus, S4H FCDB becomes a solid and indispensable tool to approach the age of personalized nutrition.

Supplementary Materials: The following are available online at https://www.mdpi.com/article/10.3390/nu13124206/s1, Table S1: INFOODS Tagnames used in S4H FCDB, Code S2: Code used in the S4H FCDB unification stage. Figure S3: Figure with the absence or presence of different compounds and nutrients in the FCDB. All compounds are included and expressed with FAO/INFOODS names.

Author Contributions: Conceptualization, V.G.-V. and J.Á.R.-H.; Methodology, D.H.-N., B.N.-P. and S.P.d.l.C.; Software, B.O.-V. and V.G.-V.; Validation, D.H.-N. and S.P.-B.; Formal analysis, D.H.-N., S.P.-B., B.N.-P. and B.O.-V.; Investigation, A.F. and S.P.d.l.C.; Resources, S.P.-B., F.L. and V.G.-V.; Data curation, D.H.-N. and B.O.-V.; Writing—Original draft preparation, D.H.-N. and S.P.-B.; Writing—Review and editing, A.F., F.L., K.N.P., S.P.-B. and J.Á.R.-H.; Supervision, V.G.-V. and J.Á.R.-H.; Funding acquisition, J.Á.R.-H. All authors have read and agreed to the published version of the manuscript.

Funding: This research was funded by the European Research Commission (Research Executive Agency) under the research project Stance4Health (Grant Contract No. 816303) to José Á. Rufián-Henares. This work was supported by the "Plan propio de Investigación y Transferencia" of the University of Granada under the programs "Intensificación de la Investigación, modalidad B", granted to José Á. Rufián-Henares.

Institutional Review Board Statement: Not applicable.

Informed Consent Statement: Not applicable.

Data Availability Statement: The S4H FCDB is not freely available at the present moment, since it will be used along the research project. Those researchers aiming at using the S4H FCDB should request a license to José A. Rufián-Henares, who will transfer the request to the relevant S4H committee.

Acknowledgments: This work will be a part of the doctoral thesis of Daniel Hinojosa-Nogueira conducted within the context of the 'Program of Nutrition and Food Sciences' at the University of Granada.

Conflicts of Interest: The authors declare no conflict of interest.

References

1. Pérez Grana, R. Exactitud de las tablas de composición de alimentos en la determinación de nutrientes. *Sanid. Mil.* **2013**, *69*, 102–111. [CrossRef]
2. Westenbrink, S.; Kadvan, A.; Roe, M.; Koroušić Seljak, B.; Mantur-Vierendeel, A.; Finglas, P. 12th IFDC 2017 special issue—Evaluation of harmonized eurofir documentation for macronutrient values in 26 European food composition databases. *J. Food Compos. Anal.* **2019**, *80*, 40–50. [CrossRef]
3. Greenfield, H.; Southgate, D.A. *Food Composition Data: Production, Management, and Use*; Food and Agriculture Organization: Roma, Italy, 2003; ISBN 92-5-104949-1.
4. Charrondiere, U.; Stadlmayr, B.; Haytowitz, D.; Oseredczuk, M.; Ireland, J.; Wolmarans, P.; Rittenschober, D.; Selley, B.; Puwastien, P.; Reykdal, Ó. *FAO/INFOODS Guidelines for Checking Food Composition Data Prior to the Publication of a User Table/Database Version 1.0*; FAO: Roma, Italy, 2012.
5. Wafa, S.; Colaiezzi, B.; Some, J.; Bell, W.; Prata, C.; Perlick, H.; Rogers, B.; Coates, J. INDDEX24: An innovative global dietary assessment platform to scale up the availability, access, and use of dietary data. *Curr. Dev. Nutr.* **2020**, *4*, 1180. [CrossRef]
6. Ene-Obong, H.; Schönfeldt, H.C.; Campaore, E.; Kimani, A.; Mwaisaka, R.; Vincent, A.; El Ati, J.; Kouebou, P.; Presser, K.; Finglas, P.; et al. Importance and use of reliable food composition data generation by nutrition/dietetic professionals towards solving africa's nutrition problem: Constraints and the role of FAO/INFOODS/AFROFOODS and other stakeholders in future initiatives. *Proc. Nutr. Soc.* **2019**, *78*, 496–505. [CrossRef]
7. Lupiañez-Barbero, A.; González Blanco, C.; de Leiva Hidalgo, A. Tablas y bases de datos de composición de alimentos españolas: Necesidad de un referente para los profesionales de la salud. *Endocrinol. Diabetes Nutr.* **2018**, *65*, 361–373. [CrossRef]
8. Emmett, P.M.; Jones, L.R.; Northstone, K.; Pounis, G.; Taylor, C.M. Collection and management of dietary data. In *Analysis in Nutrition Research*; Elsevier: Amsterdam, The Netherlands, 2019; pp. 43–73. ISBN 978-0-12-814556-2.
9. Ispirova, G.; Eftimov, T.; Koroušić Seljak, B.; Korošec, P. Mapping food composition data from various data sources to a domain-specific ontology. In Proceedings of the 9th International Joint Conference on Knowledge Discovery, Knowledge Engineering and Knowledge Management, Funchal, Portugal, 1–3 November 2017; SCITEPRESS—Science and Technology Publications: Funchal, Portugal, 2017; pp. 203–210.
10. Marconi, S.; Durazzo, A.; Camilli, E.; Lisciani, S.; Gabrielli, P.; Aguzzi, A.; Gambelli, L.; Lucarini, M.; Marletta, L. Food composition databases: Considerations about complex food matrices. *Foods* **2018**, *7*, 2. [CrossRef] [PubMed]
11. Ortiz, J.; Astudillo, G.; Castro, M.; Castro, C.; Astudillo, S.; Donoso, S. 12th IFDC 2017 special issue—Seasonal variations in nutrient composition of plant-based foods produced at the southern highlands of ecuador. *J. Food Compos. Anal.* **2019**, *83*, 103284. [CrossRef]
12. Smith, M.R.; Micha, R.; Golden, C.D.; Mozaffarian, D.; Myers, S.S. Global expanded nutrient supply (genus) model: A new method for estimating the global dietary supply of nutrients. *PLoS ONE* **2016**, *11*, e0146976. [CrossRef] [PubMed]
13. Drabo, A. Climate change mitigation and agricultural development models: Primary commodity exports or local consumption production? *Ecol. Econ.* **2017**, *137*, 110–125. [CrossRef]
14. Charrondière, U.R.; Stadlmayr, B.; Rittenschober, D.; Mouille, B.; Nilsson, E.; Medhammar, E.; Olango, T.; Eisenwagen, S.; Persijn, D.; Ebanks, K. FAO/INFOODS food composition database for biodiversity. *Food Chem.* **2013**, *140*, 408–412. [CrossRef] [PubMed]
15. Kapsokefalou, M.; Roe, M.; Turrini, A.; Costa, H.S.; Martinez-Victoria, E.; Marletta, L.; Berry, R.; Finglas, P. Food composition at present: New challenges. *Nutrients* **2019**, *11*, 1714. [CrossRef]
16. Zeb, A.; Soininen, J.-P.; Sozer, N. Data harmonisation as a key to enable digitalisation of the food sector: A review. *Food Bioprod. Process.* **2021**, *127*, 360–370. [CrossRef]
17. Lauria, F.; Russo, M.D.; Formisano, A.; De Henauw, S.; Hebestreit, A.; Hunsberger, M.; Kroghe, V.; Intemann, T.; Lissnerd, L.; Molnar, D.; et al. Ultra-processed foods consumption and diet quality of European children, adolescents and adults: Results from the I.Family study. *Nutr. Metab. Cardiovasc. Dis.* **2021**, *31*, 3031–3043. [CrossRef] [PubMed]
18. Ocké, M.C.; Westenbrink, S.; van Rossum, C.T.M.; Temme, E.H.M.; van der Vossen-Wijmenga, W.; Verkaik-Kloosterman, J. The essential role of food composition databases for public health nutrition—Experiences from the Netherlands. *J. Food Compos. Anal.* **2021**, *101*, 103967. [CrossRef]
19. Egan, M.B.; Fragodt, A.; Raats, M.M.; Hodgkins, C.; Lumbers, M. The importance of harmonizing food composition data across Europe. *Eur. J. Clin. Nutr.* **2007**, *61*, 813–821. [CrossRef] [PubMed]
20. Traka, M.H.; Plumb, J.; Berry, R.; Pinchen, H.; Finglas, P.M. Maintaining and updating food composition datasets for multiple users and novel technologies: Current challenges from a UK perspective. *Nutr. Bull.* **2020**, *45*, 230–240. [CrossRef]
21. Delgado-Andrade, C.; Rufián-Henares, J.A.; Morales, F.J. Lysine availability is diminished in commercial fibre-enriched breakfast cereals. *Food Chem.* **2007**, *100*, 725–731. [CrossRef]
22. Grande, F.; Vincent, A. The importance of food composition data for estimating micronutrient intake: What do we know now and into the future? In *Global Landscape of Nutrition Challenges in Infants and Children*; Michaelsen, K.F., Neufeld, L.M., Prentice, A.M., Eds.; Nestlé Nutrition Institute Workshop Series; S. Karger AG: Basel, Switzerland, 2020; Volume 93, pp. 39–50. ISBN 978-3-318-06648-7.
23. Micha, R.; Coates, J.; Leclercq, C.; Charrondiere, U.R.; Mozaffarian, D. Global dietary surveillance: Data gaps and challenges. *Food Nutr. Bull.* **2018**, *39*, 175–205. [CrossRef]

24. Barabási, A.-L.; Menichetti, G.; Loscalzo, J. The unmapped chemical complexity of our diet. *Nat. Food* **2020**, *1*, 33–37. [CrossRef]
25. Westenbrink, S.; Presser, K.; Roe, M.; Ireland, J.; Finglas, P. Documentation of aggregated/compiled values in food composition databases; EuroFIR default to improve harmonization. *J. Food Compos. Anal.* **2021**, *101*, 103968. [CrossRef]
26. Roe, M.; Pinchen, H.; Church, S.; Finglas, P. McC ance and widdowson's the composition of foods seventh summary edition and updated composition of foods integrated dataset. *Nutr. Bull.* **2015**, *40*, 36–39. [CrossRef]
27. Westenbrink, S.; Jansen-van der Vliet, M.; van Rossum, C. Updated folate data in the dutch food composition database and implications for intake estimates. *Food Nutr. Res.* **2012**, *56*, 5449. [CrossRef]
28. Nielsen, C.W. *Fødevaredatabanken August 2019*; DTU: Lyngby, Denmark, 2019.
29. Delgado, A.; Issaoui, M.; Vieira, M.C.; Saraiva de Carvalho, I.; Fardet, A. Food composition databases: Does it matter to human health? *Nutrients* **2021**, *13*, 2816. [CrossRef] [PubMed]
30. Charrondière, U.; Rittenschober, D.; Grande, F.; Nowak, V.; Stadlmayr, B. *FAO/INFOODS Analytical Food Composition Database Version 2.0*; FAO: Roma, Italy, 2017.
31. Montville, J.B.; Ahuja, J.K.; Martin, C.L.; Heendeniya, K.Y.; Omolewa-Tomobi, G.; Steinfeldt, L.C.; Anand, J.; Adler, M.E.; LaComb, R.P.; Moshfegh, A. USDA food and nutrient database for dietary studies (FNDDS), 5.0. *Procedia Food Sci.* **2013**, *2*, 99–112. [CrossRef]
32. Murphy, S.P.; Charrondiere, U.R.; Burlingame, B. Thirty years of progress in harmonizing and compiling food data as a result of the establishment of INFOODS. *Food Chem.* **2016**, *193*, 2–5. [CrossRef] [PubMed]
33. Charrondiere, U.R.; Rittenschober, D.; Nowak, V.; Stadlmayr, B.; Wijesinha-Bettoni, R.; Haytowitz, D. Improving food composition data quality: Three new FAO/INFOODS guidelines on conversions, data evaluation and food matching. *Food Chem.* **2016**, *193*, 75–81. [CrossRef] [PubMed]
34. Apekey, T.A.; Copeman, J.; Kime, N.H.; Tashani, O.A.; Kittaneh, M.; Walsh, D.; Maynard, M.J. Nutrient composition of popularly consumed african and caribbean foods in the UK. *Foods* **2019**, *8*, 500. [CrossRef]
35. Hulshof, P.; Doets, E.; Seyha, S.; Bunthang, T.; Vonglokham, M.; Kounnavong, S.; Famida, U.; Muslimatun, S.; Santika, O.; Prihatini, S.; et al. Food composition tables in southeast asia: The contribution of the SMILING project. *Matern. Child Health J.* **2019**, *23*, 46–54. [CrossRef]
36. Castellano-Escuder, P.; González-Domínguez, R.; Wishart, D.S.; Andrés-Lacueva, C.; Sánchez-Pla, A. FOBI: An ontology to represent food intake data and associate it with metabolomic data. *Database* **2020**, *2020*, baaa033. [CrossRef]
37. Durazzo, A.; Camilli, E.; D'Addezio, L.; Sette, S.; Marconi, S.; Piccinelli, R.; Le Donne, C.; Turrini, A.; Marletta, L. Italian composite dishes: Description and classification by LanguaL^{TM} and FoodEx2. *Eur. Food Res. Technol.* **2020**, *246*, 287–295. [CrossRef]
38. Katidi, A.; Vlassopoulos, A.; Kapsokefalou, M. Development of the hellenic food thesaurus (HelTH), a branded food composition database: Aims, design and preliminary findings. *Food Chem.* **2021**, *347*, 129010. [CrossRef] [PubMed]
39. Møller, A.; Ireland, J. LanguaL^{TM} 2017—The LanguaL^{TM} thesaurus technical report. *Dan. Food Inform.* **2018**. [CrossRef]
40. European Food Safety Authority (EFSA). *The Food Classification and Description System FoodEx 2 (Revision 2)*; EFSA: Palma, Italy, 2015; Volume 12. [CrossRef]
41. Eftimov, T.; Korošec, P.; Koroušić Seljak, B. StandFood: Standardization of foods using a semi-automatic system for classifying and describing foods according to FoodEx2. *Nutrients* **2017**, *9*, 542. [CrossRef] [PubMed]
42. European Food Safety Authority (EFSA); Ioannidou, S. *EFSA FoodEx2 Interpreting and Checking Tool User Guide*; EFSA: Palma, Italy, 2019; Volume 16. [CrossRef]
43. Charrondière, U.R.; Rittenschober, D.; Nowak, V.; Wijesinha-Bettoni, R.; Stadlmayr, B.; Haytowitz, D.; Persijn, D. *FAO/INFOODS Guidelines for Converting Units, Denominators and Expressions, Version 1.0*; FAO: Roma, Italy, 2012.
44. Charrondiere, U.R.; Burlingame, B. Report on the FAO/INFOODS compilation tool: A simple system to manage food composition data. *J. Food Compos. Anal.* **2011**, *24*, 711–715. [CrossRef]
45. Schlotke, F.; Becker, W.; Ireland, J.; Møller, A.; Ovaskainen, M.-L.; Monspart, J.; Unwin, I. EUROFOODS Recommendations for Food Composition Database Management and Data Interchange. *J. Food Compos. Anal.* **2000**, *13*, 709–744. [CrossRef]
46. Merchant, A.T.; Dehghan, M. Food Composition Database Development for between Country Comparisons. *Nutr. J.* **2006**, *5*, 2. [CrossRef] [PubMed]
47. Finglas, P.M.; Berry, R.; Astley, S. Assessing and Improving the Quality of Food Composition Databases for Nutrition and Health Applications in Europe: The Contribution of EuroFIR. *Adv. Nutr.* **2014**, *5*, 608S–614S. [CrossRef]
48. Westenbrink, S.; Roe, M.; Oseredczuk, M.; Castanheira, I.; Finglas, P. EuroFIR Quality Approach for Managing Food Composition Data; Where Are We in 2014? *Food Chem.* **2016**, *193*, 69–74. [CrossRef] [PubMed]
49. Finglas, P.; Roe, M.; Pinchen, H.; Astley, S. The Contribution of Food Composition Resources to Nutrition Science Methodology. *Nutr. Bull.* **2017**, *42*, 198–206. [CrossRef]
50. Westenbrink, S.; Oseredczuk, M.; Castanheira, I.; Roe, M. Food Composition Databases: The EuroFIR Approach to Develop Tools to Assure the Quality of the Data Compilation Process. *Food Chem.* **2009**, *113*, 759–767. [CrossRef]
51. Slimani, N.; Deharveng, G.; Unwin, I.; Southgate, D.A.T.; Vignat, J.; Skeie, G.; Salvini, S.; Parpinel, M.; Møller, A.; Ireland, J.; et al. The EPIC Nutrient Database Project (ENDB): A First Attempt to Standardize Nutrient Databases across the 10 European Countries Participating in the EPIC Study. *Eur. J. Clin. Nutr.* **2007**, *61*, 1037–1056. [CrossRef] [PubMed]
52. Longvah, T.; Anantan, I.; Bhaskarachary, K.; Venkaiah, K.; Longvah, T. *Indian Food Composition Tables*; National Institute of Nutrition, Indian Council of Medical Research: Hyderabad, Indian, 2017; ISBN 93-5267-677-7.

53. Park, S.-H.; Kim, S.-N.; Lee, S.H.; Choe, J.-S.; Choi, Y. Development of 9th Revision Korean Food Composition Table and Its Major Changes. *Korean J. Community Nutr.* **2018**, *23*, 352–365. [CrossRef]
54. Coelho, K.C.; Giuntini, B.G.; Grande, F.; Dias, J.S.; Purgatto, E.; Franco, B.D.G.M.; Lajolo, F.M.; Menezes, E.W. Brazilian Food Composition Table (TBCA): Development and functionalities of the online version. *J. Food Comp. Anal.* **2019**, *84*, 103287. [CrossRef]
55. Naveja, J.J.; Rico-Hidalgo, M.P.; Medina-Franco, J.L. Analysis of a Large Food Chemical Database: Chemical Space, Diversity, and Complexity. *F1000Research* **2018**, *7*, 993. [CrossRef] [PubMed]
56. Pérez-Burillo, S.; Hinojosa-Nogueira, D.; Pastoriza, S.; Rufián-Henares, J.A. Nutrición personalizada inteligente. *Alimentaria* **2019**, *500*, 25–29.
57. Charrondiere, U.R.; Rittenschober, D.; Nowak, V.; Nicodemi, C.; Bruggeling, P.; Petracchi, C. FAO/INFOODS e-Learning Course on Food Composition Data. *Food Chem.* **2016**, *193*, 6–11. [CrossRef]
58. Castanheira, I.; André, C.; Oseredczuk, M.; Ireland, J.; Owen, L.; Robb, P.; Earnshaw, A.; Calhau, M.A. Improving Data Quality in Food Composition Databanks: A EuroFIR Contribution. *Accredit. Qual. Assur.* **2007**, *12*, 117–125. [CrossRef]
59. Dahdouh, S.; Grande, F.; Espinosa, S.N.; Vincent, A.; Gibson, R.; Bailey, K.; King, J.; Rittenschober, D.; Charrondière, U.R. Development of the FAO/INFOODS/IZINCG Global Food Composition Database for Phytate. *J. Food Compos. Anal.* **2019**, *78*, 42–48. [CrossRef]
60. Klensin, J.C.; Feskanich, D.; Lin, V.; Truswell, A.S.; Southgate, D.A. *Identification of Food Components for INFOODS Data Interchange*; United Nations University: Tokyo, Japan, 1989; ISBN 92-808-0734-X.
61. Knaze, V.; Rothwell, J.A.; Zamora-Ros, R.; Moskal, A.; Kyrø, C.; Jakszyn, P.; Skeie, G.; Weiderpass, E.; Santucci de Magistris, M.; Agnoli, C.; et al. A New Food-Composition Database for 437 Polyphenols in 19,899 Raw and Prepared Foods Used to Estimate Polyphenol Intakes in Adults from 10 European Countries. *Am. J. Clin. Nutr.* **2018**, *108*, 517–524. [CrossRef] [PubMed]
62. Maclean, W.; Harnly, J.; Chen, J.; Chevassus-Agnes, S.; Gilani, G.; Livesey, G.; Warwick, P. Food energy—Methods of analysis and conversion factors. In *Food and Agriculture Organization of the United Nations Technical Workshop Report*; Agricultural Research Service, U.S. Department of Agriculture: Beltsville, MD, USA, 2003; Volume 77, pp. 8–9.
63. Martinez-Victoria, E.; de Victoria, I.M.; Martinez-Burgos, M.A. Intake of Energy and Nutrients; Harmonization of Food Composition Databases. *Nutr. Hosp.* **2015**, *31*, 168–176.
64. Rand, W.M.; Pennington, J.A.; Murphy, S.P.; Klensin, J.C. *Compiling Data for Food Composition Data Bases*; United Nations University Press: Tokyo, Japan, 1991; ISBN 92-808-0772-2.
65. Roe, M.A.; Bell, S.; Oseredczuk, M.; Christensen, T.; Westenbrink, S.; Pakkala, H.; Presser, K.; Finglas, P.M. *Updated Food Composition Database for Nutrient Intake*; EFSA: Palma, Italy, 2013; Volume 10. [CrossRef]
66. Slimani, N.; Ruth Charrondière, U.; van Staveren, W.; Riboli, E. Standardization of Food Composition Databases for the European Prospective Investigation into Cancer and Nutrition (EPIC): General Theoretical Concept. *J. Food Compos. Anal.* **2000**, *13*, 567–584. [CrossRef]
67. Uusitalo, U.; Kronberg-Kippilä, C.; Aronsson, C.A.; Schakel, S.; Schoen, S.; Mattisson, I.; Reinivuo, H.; Silvis, K.; Sichert-Hellert, W.; Stevens, M.; et al. Food Composition Database Harmonization for Between-Country Comparisons of Nutrient Data in the TEDDY Study. *J. Food Compos. Anal.* **2011**, *24*, 494–505. [CrossRef]
68. Aouachria, N.B.; Bakker, M.; König, J.; Leblanc, J.-C.; Lindtner, O.; Tlustos, C.; Arcella, D.; Tard, A.; Tasiopoulou, S. European Food Safety Authority Use of the EFSA Comprehensive European Food Consumption Database in Exposure Assessment. *EFSA J.* **2011**, *9*, 2097. [CrossRef]
69. Gestión de Salud y Nutrición, S.L. I-Diet Food Composition Database, Updated from Original Version of G. Martín Peña FCD. Asturias, Spain, 2019.
70. Hartmann, B.M.; Bell, S.; Vásquez-Caicedo, A.L.; Götz, A.; Erhardt, J.; Brombach, C. *Der Bundeslebensmittelschlüssel*; German Nutrient Database; Federal Research Centre for Nutrition and Food (BfEL): Karlsruhe, Germany, 2005.
71. Trichopoulou, A.; Georga, K. *Composition Tables of Foods and Greek Dishes*, 3rd ed.; Parisianou Publications: Athens, Greece, 2004.
72. Gnagnarella, P.; Salvini, S.; Parpinel, M. Food Composition Database for Epidemiological Studies in Italy. Available online: http://www.bda-ieo.it/ (accessed on 31 January 2020).
73. Neveu, V.; Perez-Jimenez, J.; Vos, F.; Crespy, V.; du Chaffaut, L.; Mennen, L.; Knox, C.; Eisner, R.; Cruz, J.; Wishart, D.; et al. Phenol-Explorer: An Online Comprehensive Database on Polyphenol Contents in Foods. *Database* **2010**, *2010*, bap024. [CrossRef]
74. Rothwell, J.A.; Urpi-Sarda, M.; Boto-Ordonez, M.; Knox, C.; Llorach, R.; Eisner, R.; Cruz, J.; Neveu, V.; Wishart, D.; Manach, C.; et al. Phenol-Explorer 2.0: A Major Update of the Phenol-Explorer Database Integrating Data on Polyphenol Metabolism and Pharmacokinetics in Humans and Experimental Animals. *Database* **2012**, *2012*, bas031. [CrossRef]
75. Rothwell, J.A.; Perez-Jimenez, J.; Neveu, V.; Medina-Remon, A.; M'Hiri, N.; Garcia-Lobato, P.; Manach, C.; Knox, C.; Eisner, R.; Wishart, D.S.; et al. Phenol-Explorer 3.0: A Major Update of the Phenol-Explorer Database to Incorporate Data on the Effects of Food Processing on Polyphenol Content. *Database* **2013**, *2013*, bat070. [CrossRef] [PubMed]
76. Reinivuo, H.; Bell, S.; Ovaskainen, M.-L. Harmonisation of Recipe Calculation Procedures in European Food Composition Databases. *J. Food Compos. Anal.* **2009**, *22*, 410–413. [CrossRef]
77. Vásquez-Caicedo, A.; Bell, S.; Hartmann, B. *Report on Collection of Rules on Use of Recipe Calculation Procedures Including the Use of Yield and Retention Factors for Imputing Nutrient Values for Composite Foods*; European Food Information Resource Network: Brussels, Belgium, 2008.

78. Bell, S.; Becker, W.; Vásquez-Caicedo, A.; Hartmann, B.; Møller, A.; Butriss, J. *Report on Nutrient Losses and Gains Factors Used in European Food Composition Databases*; European Food Information Resource Network: Brussels, Belgium, 2006.
79. Bognár, A. *Tables on Weight Yield of Food and Retention Factors of Food Constituents for the Calculation of Nutrient Composition of Cooked Foods (Dishes)*; BFE: Karlsruhe, Germany, 2002; ISBN 0933-5463.
80. Charrondiere, U.R.; Sivieri, A.; Burlingame, B. Differences in nutrient values of recipes due to different calculation methods and sets of nutrient retention factors. In *Quality Food Composition Data: Key for Health and Trade, Proceedings of the 8th International Food Data Conference, Bangkok, Thailand, 1–3 October 2009*; Judprasong, K., Puwastien, P., Jittinandana, S., Eds.; Bangkok, Thailand, 2009.
81. Bawajeeh, A.; Kalendar, S.; Scarpa, G.; Hancock, N.; Beer, S.; Gibson, L.; Williams, G.; Dashti, B.; Albar, S.; Ensaff, H.; et al. Development of an Arabic Food Composition Database for Use in an Arabic Online Dietary Assessment Tool (Myfood24). *J. Food Compos. Anal.* **2021**, *102*, 104047. [CrossRef]
82. Carter, M.; Hancock, N.; Albar, S.; Brown, H.; Greenwood, D.; Hardie, L.; Frost, G.; Wark, P.; Cade, J. Development of a New Branded UK Food Composition Database for an Online Dietary Assessment Tool. *Nutrients* **2016**, *8*, 480. [CrossRef]
83. Nowak, V.; Persijn, D.; Rittenschober, D.; Charrondiere, U.R. Review of Food Composition Data for Edible Insects. *Food Chem.* **2016**, *193*, 39–46. [CrossRef]
84. Yoo, Y.M.; Atkin, R.; Pachón, H. Development of a Food Composition Table to Analyze Senegalese Food Expenditure Data. *Afr. J. Food Agric. Nutr. Dev.* **2019**, *19*, 1–21.
85. Durazzo, A.; Camilli, E.; Marconi, S.; Lisciani, S.; Gabrielli, P.; Gambelli, L.; Aguzzi, A.; Lucarini, M.; Kiefer, J.; Marletta, L. Nutritional Composition and Dietary Intake of Composite Dishes Traditionally Consumed in Italy. *J. Food Compos. Anal.* **2019**, *77*, 115–124. [CrossRef]
86. Ireland, J.D.; Møller, A. Review of International Food Classification and Description. *J. Food Compos. Anal.* **2000**, *13*, 529–538. [CrossRef]
87. Chin, E.L.; Simmons, G.; Bouzid, Y.Y.; Kan, A.; Burnett, D.J.; Tagkopoulos, I.; Lemay, D.G. Nutrient Estimation from 24-Hour Food Recalls Using Machine Learning and Database Mapping: A Case Study with Lactose. *Nutrients* **2019**, *11*, 3045. [CrossRef]
88. Ispirova, G.; Eftimov, T.; Korošec, P.; Koroušić Seljak, B. MIGHT: Statistical Methodology for Missing-Data Imputation in Food Composition Databases. *Appl. Sci.* **2019**, *9*, 4111. [CrossRef]
89. Presser, K.; Weber, D.; Norrie, M. FoodCASE: A System to Manage Food Composition, Consumption and TDS Data. *Food Chem.* **2018**, *238*, 166–172. [CrossRef] [PubMed]
90. Møller, A.; Christensen, T. *Danish Food Information: EuroFIR Web Services-EuroFIR Food Data Transport Package, Version 1.3*; DTU: Lyngby, Denmark, 2008.
91. Durazzo, A.; Camilli, E.; D'Addezio, L.; Le Donne, C.; Ferrari, M.; Marconi, S.; Marletta, L.; Mistura, L.; Piccinelli, R.; Scalvedi, M.L.; et al. Food Groups and Individual Foods: Nutritional Attributes and Dietary Importance. In *Reference Module in Food Science*; Elsevier: Amsterdam, The Netherlands, 2018; p. B9780081005965214000. ISBN 978-0-08-100596-5.
92. Jimoh, F.; Lund, E.K.; Harvey, L.J.; Frost, C.; Lay, W.J.; Roe, M.A.; Berry, R.; Finglas, P.M. Comparing Diet and Exercise Monitoring Using Smartphone App and Paper Diary: A Two-Phase Intervention Study. *JMIR mHealth uHealth* **2018**, *6*, e17. [CrossRef] [PubMed]
93. Kwon, D.Y. Personalized Diet Oriented by Artificial Intelligence and Ethnic Foods. *J. Ethn. Foods* **2020**, *7*, 10. [CrossRef]
94. Cespedes, E.M.; Hu, F.B. Dietary Patterns: From Nutritional Epidemiologic Analysis to National Guidelines. *Am. J. Clin. Nutr.* **2015**, *101*, 899–900. [CrossRef] [PubMed]
95. Ocké, M.; de Boer, E.; Brants, H.; van der Laan, J.; Niekerk, M.; van Rossum, C.; Temme, L.; Freisling, H.; Nicolas, G.; Casagrande, C.; et al. *PANCAKE—Pilot Study for the Assessment of Nutrient Intake and Food Consumption Among Kids in Europe*; EFSA: Palma, Italy, 2012; Volume 9. [CrossRef]
96. Wark, P.A.; Hardie, L.J.; Frost, G.S.; Alwan, N.A.; Carter, M.; Elliott, P.; Ford, H.E.; Hancock, N.; Morris, M.A.; Mulla, U.Z.; et al. Validity of an Online 24-h Recall Tool (Myfood24) for Dietary Assessment in Population Studies: Comparison with Biomarkers and Standard Interviews. *BMC Med.* **2018**, *16*, 136. [CrossRef] [PubMed]
97. European Food Safety Authority (EFSA). *Dietary Reference Values for Nutrients Summary Report*; EFSA: Palma, Italy, 2017.
98. Harrington, R.A.; Adhikari, V.; Rayner, M.; Scarborough, P. Nutrient Composition Databases in the Age of Big Data: FoodDB, a Comprehensive, Real-Time Database Infrastructure. *BMJ Open* **2019**, *9*, e026652. [CrossRef] [PubMed]
99. Blasco, T.; Pérez-Burillo, S.; Balzerani, F.; Hinojosa-Nogueira, D.; Lerma-Aguilera, A.; Pastoriza, S.; Cendoya, X.; Rubio, Á.; Gosalbes, M.J.; Jiménez-Hernández, N.; et al. An Extended Reconstruction of Human Gut Microbiota Metabolism of Dietary Compounds. *Nat. Commun.* **2021**, *12*, 4728. [CrossRef] [PubMed]
100. Kovarich, S.; Ceriani, L.; Ciacci, A.; Baldin, R.; Perez Miguel, M.; Gibin, D.; Carnesecchi, E.; Roncaglioni, A.; Mostrag, A.; Tarkhov, A.; et al. *OpenFoodTox: EFSA's Chemical Hazards Database*; EFSA: Palma, Italy, 2020.
101. Rufian-Henares, J.A.; De la Cueva, S.P. Assessment of Hydroxymethylfurfural Intake in the Spanish Diet. *Food Addit. Contam.* **2008**, *25*, 1306–1312. [CrossRef]
102. Rufian-Henares, J.A.; Garcia-Villanova, B.; Guerra-Hernández, E. Occurrence of furosine and hydroxymethylfurfural as markers of thermal damage in dehydrated vegetables. *Eur. Food Res. Technol.* **2008**, *228*, 249–256. [CrossRef]

nutrients MDPI

Article

Potential Effects of Sucralose and Saccharin on Gut Microbiota: A Review

Susana del Pozo [1,*], Sonia Gómez-Martínez [2], Ligia E. Díaz [2], Esther Nova [2], Rafael Urrialde [3,4] and Ascensión Marcos [2]

1 Department of Nutrition and Food Science, Faculty of Pharmacy, Complutense University of Madrid, 28040 Madrid, Spain
2 Department of Metabolism and Nutrition, Institute of Food Science, Technology and Nutrition (ICTAN), Spanish National Research Council (CSIC), 28040 Madrid, Spain; sgomez@ictan.csic.es (S.G.-M.); ldiaz@ictan.csic.es (L.E.D.); enova@ictan.csic.es (E.N.); amarcos@ictan.csic.es (A.M.)
3 Department of Genetics, Physiology and Microbiology, School of Biological Sciences, Complutense University of Madrid, 28040 Madrid, Spain; rurriald@ucm.es
4 Area of Nutrition and Bromatology, Department of Pharmaceutical and Health Sciences, San Pablo CEU University, 28925 Madrid, Spain
* Correspondence: spozocal@ucm.es

Abstract: Artificial sweeteners are additives widely used in our diet. Although there is no consensus, current evidence indicates that sucralose and saccharin could influence the gut microbiota. The aim of this study was to analyze the existing scientific evidence on the effects of saccharin and sucralose consumption on gut microbiota in humans. Different databases were used with the following search terms: sweeteners, non-caloric-sweeteners, sucralose, splenda, saccharin, sugartwin, sweet'n low, microbiota, gut microbiota, humans, animal model, mice, rats, and/or in vitro studies. In vitro and animal model studies indicate a dose-dependent relationship between the intake of both sweeteners and gut microbiota affecting both diversity and composition. In humans, long-term study suggests the existence of a positive correlation between sweetener consumption and some bacterial groups; however, most short-term interventions with saccharin and sucralose, in amounts below the ADI, found no significant effect on those groups, but there seems to be a different basal microbiota-dependent response of metabolic markers. Although studies in vitro and in animal models seem to relate saccharin and sucralose consumption to changes in the gut microbiota, more long-term studies are needed in humans considering the basal microbiota of participants and their dietary and lifestyle habits in all population groups. Toxicological and basal gut microbiota effects must be included as relevant factors to evaluate food safety and nutritional consequences of non-calorie sweeteners. In humans, doses, duration of interventions, and number of subjects included in the studies are key factors to interpret the results.

Keywords: saccharin; sucralose; gut microbiota; acceptable daily intake; short-term studies; long-term studies; short-chain fatty acids

Citation: del Pozo, S.; Gómez-Martínez, S.; Díaz, L.E.; Nova, E.; Urrialde, R.; Marcos, A. Potential Effects of Sucralose and Saccharin on Gut Microbiota: A Review. *Nutrients* **2022**, *14*, 1682. https://doi.org/10.3390/nu14081682

Academic Editor: Sheila Fleischhacker

Received: 21 March 2022
Accepted: 14 April 2022
Published: 18 April 2022

1. Introduction

Humans are drawn to sweetness, but the WHO directives state that free sugars should not represent more than 10% of the daily caloric contribution and propose a reduction to 5% [1]. Sweeteners are substances used to impart a sweet taste to foods either in food manufacturing or as tabletop sweeteners, substituting for sugars. Nowadays, they are much more abundant than they used to be in some types of popular foods consumed by adults and children, because of their lower calorie content [2,3]. They are used in very small amounts and either do not provide any calories or provide just a few. Indeed, they replace added sugars in a wide variety of foodstuffs [4]. For example, in the Spanish market the distribution of food and beverage subgroups (%) containing one or more low- and no-calorie sweeteners comprises bakery and pastry (16%); yogurt and fermented milks (10%);

chewing gums, candies, and sweets (10%); food supplements and substitutes (9%); diet soft drinks (7%); sugar soft drinks (7%); sausages and other meat products (6%); and others [5].

Intensive sweeteners have a negligible caloric contribution and high sweetening capacity, higher than sucrose, thus only being necessary in very low doses to obtain intense sweetness because of their high affinity for the tongue papillas. Sweeteners, like all other food additives, are subjected to strict safety control. There are currently 19 compounds authorized for use in food products by the European regulations, 7 of them being classified as polyols (low-calorie sweeteners) and the remaining 12 as non-calorie sweeteners, of which the most notable ones are acesulfame K (E950), aspartame (E951), cyclamates (E952), saccharin (E954), sucralose (E955), neotame (E961), and steviol glycosides (E960) [6]. These compounds have very different chemical structures, although all of them have in common the ability to potently activate some of the multiple potential ligand-binding sites of the sweet-taste receptors in human subjects [7]. In fact, with health concerns regarding currently available sweeteners, there is renewed interest in identifying a safe and palatable sweetener [8]. In addition, sweeteners, like any other element in the diet, can influence the gut microbiota [9].

The human body is inhabited by trillions of symbiotic microorganisms, most of which are found within the gastrointestinal tract, mainly in the large intestine, and they are collectively called the microbiota [10,11]. The gut microbiota are composed of several species of microorganisms, including more importantly bacteria, archaea, yeasts, and viruses, each individual being provided with a unique gut microbiota profile [12]. Eubiosis, the term used for a "healthy microbiota" can be considered the balance of the intestinal microbial ecosystem, with a preponderance of potentially beneficial bacteria species [13]. In opposition, an altered balance is termed dysbiosis. The optimal healthy gut microbiota composition is different for each individual [12]. Human gut microbiota depend on several factors, such as the type of birth (vaginal/caesarean), breast-feeding or bottle-feeding, type of dietary intake, especially during the first two years of life, as well as the environmental living conditions. This is called the basal commensal microbiota. However, microbiota continue to evolve and adapt throughout the whole life of each individual, taking into account certain factors, such as diet, eating behavior, physical activity, sedentary habits, weight and stress management, as well as sleep quality and quantity [14]. The Microbiome Project revealed that there are 600,000 microbial genes in the human gastrointestinal tract. Ninety-nine percent of these are of bacterial origin; the rest are from Archaea and a very small proportion are of viral origin. The core bacterial microbial genes mainly belong to the Firmicutes and Bacteroidetes phyla, followed by Actinobacteria, Proteobacteria, Fusobacteria, and Verrucomicrobia to lesser extents [15]. Typically, restricted anaerobes (such as *Bacteroides*, *Clostridium*, *Eubacterium*, *Ruminococcus*, *Peptococcus*, *Fusobacterium*, and *Bifidobacterium*) prevail over facultative anaerobic genera (such as *Lactobacillus*, *Escherichia*, *Enterobacter*, *Enterococcus*, *Proteus*, and *Klebsiella*), with *Cyanobacteria*, *Fusobacteria*, and *Spirochaeataceae* being less predominant [16].

The composition and activity of the gut microbiota during life is changing and shaped by several factors; most notably, diet and dietary factors are major determinants of gut microbiota composition and activity [14]. The gut microbiota of an individual can reflect his/her diet at any time. A recent study links the state of the gut microbiota and the Mediterranean diet, which was recognized in 2016 as an Intangible Cultural Heritage of Humanity and is associated with the prevention of cardiovascular and metabolic diseases. The study concluded that several beneficial bacteria (*Bifidobacterium animalis*, *Oscillibacter valericigenes*, and *Roseburia faecis*) are more abundant in individuals with greater adherence to the Mediterranean diet [17]. However, the current Western dietary pattern, rich in saturated fats and sugar, is related to an altered composition of the microbiota (often qualifying as less diverse), which seems to be involved in the development of inflammatory metabolic diseases such as obesity or diabetes [18]. Gut microbiota changes correlate with health status [19]. The activity of the gut microbiota in humans includes degradation of undigested proteins and carbohydrates (sugars, oligosaccharides, peptides, amino acids), amino acid and monosaccharide fermentation, hydrogen disposal, bile-acid transformation, and vitamin synthesis [9,20]. Any change in the profile of sugars/sweeteners we consume redefines the nutrient environments in our gut. How indigenous and exogenous microbes use these environments can result in benign, detrimental, or beneficial effects on the host [16].

Until a few years ago, non-caloric sweeteners were considered metabolically inert and without apparent physiological effects; however, some of them undergo multiple changes in the intestine, interacting with the gut microbiota and thus modifying their metabolites in different regions of the intestine [17]. Some studies have reported that sweeteners may have the ability to modify the gut microbiota [7,11,18–21]. Some of the previously published review works on sweeteners and gut microbiota indicate that, considering experimental studies and clinical trials in human, among the non-nutritive sweeteners, only saccharin and sucralose change gut microbiota populations [2,10,22], so in this review we will focus on these two sweeteners.

Saccharin (E 954) brand names include Sweet and Low®, Sweet Twin®, Sweet'N Low®, and Necta Sweet® [23]. In 1878, saccharin was the first intense sweetener discovered, being potassium, sodium, and calcium salts the most used. Taking sucrose as a reference, its sweetening power is 300–500 [24] and it does not provide any calories. A range of foods and beverages are sweetened by saccharin [2].

The acceptable daily intake (ADI) for saccharin and its sodium, potassium, and calcium salts, that is, the amount of food additive expressed on a body weight basis, established by the Joint FAO/WHO Expert Committee on Food Additives (JECFA) and the Scientific Committee on Food (SCF), is 5 milligrams per kilogram of body weight per day (mg/kg/d) [25] while other agencies are more restrictive, such as ANMAT, which indicates 2.5 mg/kg/d [26]. This is the amount that can be consumed daily throughout life without appreciable health risks (Table 1) [27].

Table 1. Acceptable daily intake (ADI) (mg/kg/bw).

	JECFA ADI [25,28]	EFSA ADI [29,30]	Health Canada	[31] ANMAT [26]
Saccharin	5	5	5	2.5
Sucralose	15	15	9	15

The study of its effect on the gut microbiota began at the end of the last century [11,23,32,33]. Saccharin is mostly absorbed in the stomach, with approximately 85% to 95% of ingested saccharin absorbed and eliminated in the urine, and the remainder excreted in the feces [22,24]. Only 15% of the consumed saccharin makes contact with the colonic microbiota, which suggests that only when consumed in high doses could it alter the intestinal microbiota composition [22].

Sucralose (E 955), FSA-Q-2011-00724, was discovered in 1976. Sucralose is sold under the brand name Splenda® [23]. Sucralose is a substituted disaccharide, a non-nutritive sweetener that is synthesized by the selective chlorination of sucrose in three of the primary hydroxyl groups [34]. The chemical name for sucralose is 1,6-dichloro-1,6-dideoxy-b-D-fructofuranosyl 4-chloro-4-deoxy-a-D-galactopyranoside [24]. Taking sucrose as a reference, its sweetening power is 600 [24]. Its ADI is 15 mg/kg/d of body weight by the JECFA (Joint Expert Committee on Food Additives) [28], EFSA (European Food Safety Agency) [29], and ANMAT (National Administration of Drugs, Foods and Medical Devices) [26] (Table 1).

Sucralose is poorly absorbed, undergoes little metabolism, and enters unchanged into the lower gastrointestinal tract, being excreted primarily unchanged in the feces in all species, including humans, and more than 85% of the consumed sucralose reaches the colon [23]. Therefore, sucralose could possibly either alter or change the gut microbiota composition, although it is scarcely metabolized by intestinal bacteria [24].

When evaluating the effects of saccharin and sucralose on the gut microbiota, several aspects must be considered, including the dose used in the studies and the average daily amount consumed by the population and the ADI of these sweeteners. In particular, the

ADI is used in many studies on gut microbiota and sweeteners as a reference dose. As an example of average consumption by a population, we can take the data on sweetener consumption by the Spanish population. In 2020, 0.11 kg/per capita was consumed, which was 26.2% more than in 2019 [35]. This amount represents 0.3 g/p/d of different sweeteners (Table 2). The ADIs for saccharin and sucralose, according to the JECFA, are 5 mg/kg/day and 15 mg/kg/day, respectively [25,28], which means that a 70 kg subject could consume a maximum of 350 mg of saccharin and 1050 mg sucralose. Based on this, the average consumption of the Spanish population would not exceed the ADI for either of the two sweeteners, but it should be considered that these are average data and there may be people with higher consumptions that are exceeding the ADI. Thus, evaluating how those doses may impact the microbiota composition is not without relevance.

Table 2. ADI. Mean consumption of sweeteners in the Spanish population.

	Saccharin	Sucralose
ADI mg/kg body wt (JECFA)	5 mg/kg	15 mg/kg
ADI subject 70 kg	350 mg	1050 mg
Average consumption of the Spanish population	300 mg/day	

In view of this knowledge on non-caloric sweeteners, the aim of this article was updating the existing evidence on the effect of consuming different amounts of saccharin and sucralose in short- and long-term studies on the composition of the gut microbiota.

2. Materials and Methods

A descriptive review was conducted to investigate whether there are potential effects of saccharin and sucralose consumption on gut microbiota composition.

The PubMed, Scopus, Google Scholar, ScienceDirect, and Scielo databases were used for the search. The terms entered in this search were as follows: sweeteners, non-calorie sweeteners, sucralose, splenda, saccharin, sugar-win, sweet'n low, microbiota, gut microbiota, human, animal model, mice, rat, and in vitro studies.

Using the term "sweeteners", for the last 5 years, 1573 clinical trials, meta-analyses, and randomized controlled trials, together with 2984 reviews and systematic reviews, were found. When narrowing the search also including the term "microbiota", we found 41 clinical trials, meta-analyses, and randomized controlled trials, plus 144 reviews and systematic reviews.

The following exclusion criteria were used: studies that focused on microbiota other than the gut microbiota, studies that did not include the effect of saccharin and sucralose on the gut microbiota, studies that included supplements and/or prebiotics and/or probiotics that affect the gut microbiota, and studies carried out in populations with diseases.

All these studies were divided into in vitro and in vivo studies, differentiating in the latter between studies in animal models and in humans. Finally, for the present review, 6 in vitro studies were evaluated, plus 14 in vivo studies in animal models and 4 in vivo studies in humans (Figure 1). Of the studies included in this publication, 10 were not present in previous reviews, 2 were studies in humans, 6 were studies in animal models, and 2 were in vitro studies.

The following formula was used to estimate the concentrations of saccharin and sucralose used in the animal studies with respect to the ADI in humans when the work did not indicate this, when it was possible with the published data.

ADI (EFSA/JECFA) (mg/kg/day) × Average animal weight (kg)/Average daily liquid intake (mL) (modified from Suez et al.) [34].

The amount of water consumed by the experimental animals was estimated according to the data indicated by Bachmanov et al. [36,37] and the animal care and use committee of the Johns Hopkins University [37].

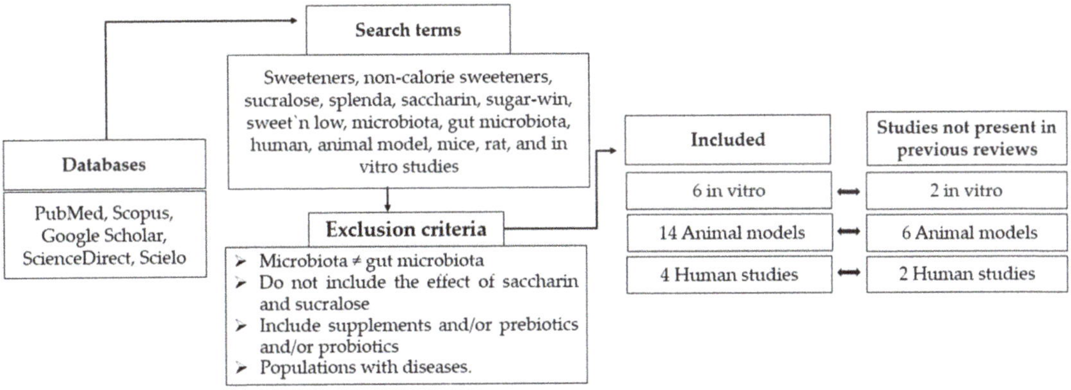

Figure 1. Flow chart regarding selection method.

2.1. Effects of Sweeteners on the Gut Microbiota: In Vitro Trials

In vitro models can be used to study the potential effects of sweeteners, specifically saccharin and sucralose, in humans. Data obtained from in vitro studies can serve as hypothesis generators and as indicators of possible interactions between these sweeteners and the gut microbiota.

In vitro studies focus on the changes in the main microbial groups and selected species together with their metabolites, analyzing the diversity, richness, and abundance in the community over time. The in vitro studies included in this review (Table 3) have commonly addressed the interactions between bacteria, intestinal epithelium, and simulated transit.

Table 3. Summary of the analyzed in vitro studies.

Reference	Sweeteners/Doses/Duration	Methods	Bacteria	Results/Conclusions Saccharine and/or Sucralose
Harpaz et al., 2018 [38]	Aspartame, sucralose, saccharine, neotame, advantame, and acesulfame potassium-k (ace-k). ADI (FDA)	Bioluminescent	*E. coli* strains (TV1061, DPD2544 and DPD2794)	Toxic effects
Wang et al., 2018 [39]	Sucralose, saccharin, acesulfame potassium, and rebaudioside Liquid assay: equal molarity of sodium chloride/5 h Agar: 1.25% (*w*/*v*) sucralose and 2.5% (*w*/*v*) sucralose/24 h	Liquid culture assay. LB agar plate assay	*E. coli* HB101 and *E. coli* K-12	Bacteriostatic effects
Markus V, et al., 2021 [40]	Aspartame, sucralose, saccharin Bioluminescence assay, growth assay: 10 μL non-calorie sweeteners or sports supplements. Swarming motility assay: aspartame (1.36 mM), sucralose (25.2 mM), or saccharine (2.72 mM) QS competition assay using Chromobacterium Violaceum CV026/20 h	Biosensor assays, biophysical protein characterization methods, microscale thermophoresis, swarming motility assays, growth assays, and molecular docking	*E coli* K802NR and *P. aeruginosa* lasRI *P. aeruginosa* PAO1 *C. violaceum* (CV026)	Inhibition of quorum sensing

Table 3. *Cont.*

Reference	Sweeteners/Doses/Duration	Methods	Bacteria	Results/Conclusions Saccharine and/or Sucralose
Gerasimidis C et al., 2020 [41]	Aspartame-based sweetener, sucralose, stevia 50% ADI (male, w: 75 kg)	Gas chromatography	Total bacteria (feces from healthy individuals) and 5 bacterial groups (Bacteroides/ Prevotella, *Bifidobacterium*, *B. coccoides*, *C. leptum* and *E. coli*)	Sucralose: shifted microbiome community structure ↔ bacterial populations ↑ Escherichia/Shigella
Shil A and Chichger, H, 2021 [42]	Saccharin, sucralose, and aspartameGrowth curve: 0.1 to 1000 μM/4 d Biofilm formation assay: 100 μM/48 h Haemolysis assay, adhesion assay, and invasion assay: 100 M/24 h Cytotoxicity assay: 100 M/48 h	Models of microbiota and the intestinal epithelium (Caco-2 cells)	*E. coli* NCTC10418 and *E. faecalis* ATCC19433 *S. aureus*	Saccharin bacteriostatic effects Saccharin, sucralose: ↑ biofilm formation ↑ ability of bacteria to adhere to, invade, and kill gut epithelial cells (exception saccharin on *E. coli*) Negative effect on intestinal epithelial cell apoptosis and permeability
Vamanu E et al., 2019 [43]	Sodium cyclamate, sucralose, sodium saccharin, steviol, white sugar 40 mg active substance (more than 90% purity)	Static GIS1 simulator (three segments of the human colon)	Total microbial (feces from healthy individuals)	Saccharin: ↓ number of microorganisms; ↓ SCFAs Both: ↓ phylum Firmicutes; ↓ fermentative processes; ↑ colonic pH; ↑ 10% ammonia synthesized; ↓ SCFAs

ADI: acceptable daily intake; SCFA: short-chain fatty acid. ↔: unmodified; ↑: increase; ↓: decrease.

In 2018, Harpaz et al., evaluated the relative toxicity for the bacteria of artificial sweeteners, approved by the FDA and in a range of concentrations based on acceptable daily intake (ADI). Genetically modified bacteria (*E. coli*) showing luminescence after exposure to certain stresses were used. Both the induced luminescent signals and bacterial growth were measured. The dose-dependent toxicity effect on *E. coli* in vitro was demonstrated [38]. In addition, Wang et al., (2018) evaluated the bacteriostatic effect of sucralose and saccharin on the growth of *E. coli* in liquid and solid media, finding that the ability to selectively inhibit the growth of enteric bacterial species may be due to inhibition of metabolic enzymes or alterations in nutrient transport [39,44,45].

According to Markus et al., using concentration ranges of non-calorie sweeteners, with comparable concentrations within FDA-approved acceptable daily intake (ADI), aspartame, sucralose, and saccharin are not bactericidal but may affect the bacterial communication system via a molecular system termed quorum sensing (QS)-inhibition and by extension may also affect the host metabolism. According to these authors, this outcome may be due to the significant inhibitory actions of these sweeteners on the Gram-negative bacteria N-acyl homoserine lactone-based (AHL) communication system. However, there is a need to continue to elucidate the mechanisms of action involved in the effects of these sweeteners and other related products on gut microbiota [40].

Gerasimidis et al., in 2020 investigated the effect of artificial sweeteners on the gut microbiome and fiber fermentation capacity. To conduct their study, they fermented fecal samples from 13 healthy volunteers in cultures with sweeteners (aspartame, sucralose, stevia-based sweetener). They measured short-chain fatty acid (SCFA) production by gas chromatography and characterized the composition of the microbiome with 16S rRNA sequencing and quantitative polymerase chain reaction (qPCR). Among their results they found that compared to the control, sucralose ($p = 0.025$) significantly increased valeric acid production and induced significant changes in microbiome community structure (β-diversity); using the Bray–Curtis dissimilarity index, it also increased the relative abundance of *Escherichia/Shigella* species as well as *Bilophila* [41].

However, Shil et al., conducted a study using gut microbiota and epithelial models on the role of commonly consumed sweeteners in the pathogenicity of gut bacteria. The effect of non-calorie sweeteners on *E. coli* and *E. faecalis* growth in planktonic culture was measured in vitro after exposure for 4 days to varying concentrations of non-calorie sweeteners (saccharin, sucralose, and aspartame). All these sweeteners increased the ability of model gut bacteria to adhere to and invade intestinal epithelial cells except for saccharin, which had no significant effect on *E. coli* invasion. Furthermore, a negative effect of these artificial sweeteners has been shown on intestinal epithelial cell apoptosis and permeability, thus further increasing the opportunity for bacteria to traverse the gut epithelium and cause septicemia [42].

Some authors (Vamanu et al., 2019), with the aim of establishing the effect of sweeteners on the microbiota pattern of healthy individuals, used a static in vitro system to simulate the transit through the three segments of the human colon. Under these conditions, both the fermentative response and microbial diversity were found to be altered after treatment with in vitro sweeteners, specifically sucralose and saccharin (equivalent to 9 g of sugar), also showing that non-nutritional sweeteners can induce toxicity, expressed by the establishment of dysbiosis [43].

All the reviewed in vitro studies allow us to hypothesize that in one way or another the consumption of artificial sweeteners can affect the bacteria present in the gut microbiota. We must be careful when interpreting the results and consider different aspects, such as the fact that the in vitro conditions may not correspond to the in vivo conditions of the organism. In addition, the different methodologies used in these studies may make it difficult to interpret the results.

2.2. Effects of Sweeteners on the Gut Microbiota in Animal Models

A summary of the "animal" studies analyzed is given in Table 4. Mainly murine species have been studied and the work focuses primarily on the number of total anaerobic and aerobic bacteria, bacterial diversity, the Bacteroidetes/Firmicutes ratio, fecal transplantation, and the effects of maternal intake of sweeteners on offspring in adulthood. In most studies, sweeteners were administered to the animals as part of the drinking water at different concentrations using the ADI for saccharin and sucralose as a reference (Table 4).

One of the first studies on saccharin and the intestinal microbiota was conducted in 1980 by Anderson and Kirkland in rats. They compared the total anaerobic and aerobic microbial populations of the cecum and the proportion of both in male rats fed 0 or 7.5% saccharin sodium, in Purina laboratory chow, for 10 days. After this period, the authors observed that the highest doses of saccharin in cecal content showed an increase in anaerobes and maintenance of aerobes, implying a downward shift in the anaerobic/aerobic ratio [33]. However, Serrano et al., showed that short-term saccharin supplementation with an equivalent dose to the highest acceptable level (JECFA) is insufficient to alter gut microbiota in apparently healthy mice [46].

Conversely, Falcon et al., found that chronic feeding of a commercial non-nutritive sweetened yogurt (0.3% sodium saccharin and sodium cyclamate, Zero-Cal, SP, Brazil) did not induce differences in the bacterial diversity of adult male Wistar rats, compared to animals fed a standard low-fat yogurt supplemented with 20% sucrose [47].

Table 4. Evidence from animal model studies relative to sucralose and saccharin effects on the gut microbiota.

Reference	Sweeteners/Doses/Duration	Animal Model	Results
Anderson & Kirkland, 1980 [33]	Treatment: 7.5% sodium saccharin in the Purina laboratory chow Control: Cellulose 7.5% in the Purina laboratory chow Duration: 10 d	Weaning male Charles River rats (Weight 55 ± 3 g) ($n = 7$)	↑ The numbers of aerobic microbes ↓ Anaerobic/aerobic ratio
Serrano et al., 2021 [46]	Treatment: saccharin average daily dose equal to 4 times (250 mg/kg) the human ADI (JECFA) Control: water Duration: 10 wk	8-wk-old mice	↔Alpha and beta diversity and relative microbial abundances
Falcon et al., 2020 [47]	Control: Sucrose-sweetened yogurt (suc): low-fat yogurt supplemented with 20% sucrose, final solution concentration 11.4% sucrose Treatment: NNS-supplemented yogurt: (0.3% sodium saccharin and sodium cyclamate). Final solution concentration 0.17% NNS Duration: 17 wk	Adult male Wistar rats (weight: 210 ± 6 g) SUC ($n = 9$ per group) NNS ($n = 10$ per group)	↔Species richness ↔ Shannon or Simpson diversity indices
Abou-Donia et al., 2008 [32]	Treatment: Splenda (Sucralose) oral gavage: 1.1; 3.3; 5.5 and 11 mg/kg/d sucralose concentrations. Control: water Duration: 12 wk	Male Sprague-Dawley rats (weight: 200–240 g) ($n = 10$ per group)	↓ Number of total anaerobes and other anaerobic bacteria (*Bifidobacteria, Lactobacilli, Bacteroides, and Clostridium*).
Uebanso et al., 2017 [48]	Treatment: LS (sucralose solution of 1.5 mg/kg bw/d). HS (sucralose solution of 15 mg/kg bw/d), which is equal to the maximum ADI. Control: distilled water Duration: 8 wk	Male and female C57Bl/6 J mice (4 wk old) ($n = 8$)	LS vs. HS ↔The relative amounts of fecal total bacteria LS vs. HS ↔ Firmicutes and Bacteroidetes phylum bacteria ↓ relative *Clostridium* cluster XIVa, dose-dependent
Sánchez-Tapia et al., 2020 [49]	Treatment: Sucralose: drinking water 1.5% sucralose Control: water Duration: 4 mo	Male Wistar rats (5 wk old) ($n = 6$ per group)	↓α-diversity ↑ *B. fragilis* abundance
Wang et al., 2018 [39]	Treatment: Sucralose: drinking water sucralose (2.5%, *w/v*) Duration: 8 wk	C57BL/6 mice (5 wk old)	↔ α-diversity, Actinobacteria, and Proteobacteria ↑ Abundance of Firmicutes
Zhang et al., 2021 [50]	Treatment: daily gavage of Sucralose ∼ 0.43 mg, sucralose ∼0.62 mg. Control: daily gavage of 2 mL normal saline Duration: 4 wk	Obese Sprague Dawley rats (4 wk old) (8 weeks after high fat diet (HFD)) ($n = 6$ per group)	0.43 mg sucralose: ↑ relative abundance of Firmicutes and ↓ Bacteroidetes 0.62 mg sucralose: ↓ relative abundance of Firmicutes ↑ Bacteroidetes The ratio of Firmicutes to Bacteroidetes in 0.43 mg sucralose was higher than that in 0.62 mg

Table 4. *Cont.*

Reference	Sweeteners/Doses/Duration	Animal Model	Results
Li et al., 2021 [51]	Treatment: Saccharin sodium in drinking water: 1.5 mM Control: water Duration: 4 wk	Female Harley-white guinea pigs (Cavia porcellus) (4 wk old) (weight: 240.7 ± 7.7 g) (*n*= 6 per group)	↑ Firmicutes and Lactobacillasceae-Lactobacillus abundance ↓ Relative abundance of *Erysipelotrichaceae*, *Eubacteriaceae*, and *Ileibacterium*
Bian et al., 2017 [52]	Treatment: Sucralose tap water (0.1 mg/mL). ADI (FDA) Control: tap water Duration: 6 mo	C57BL/6 male mice (~8 wk old) (*n* = 10 per group)	↑Numerous bacterial toxin genes (toxic shock syndrome toxin-1 and shiga toxin subunits) 14 genera exhibited different patterns over time in sucralose, different after 3 and/or 6 mo of treatment
Bian et al., 2017 [53]	Treatment: Saccharin, drinking water (0.3 mg/mL). ≈ ADI (FDA) Control: tap water Duration: 6 mo	C57BL/6 J male mice (Weight, ~23 g, ~8 wk old) (*n* = 10 per group)	Alterations of the gut metabolome with 1743 significant changes in molecular features 3 mo: ↑*Sporosarcina*, *Jeotgalicoccus*, *Akkermansia*, *Oscillospira*, and *Corynebacterium* ↓*Anaerostipes* and *Ruminococcus* 6 mo: ↑*Corynebacterium*, *Roseburia*, and *Turicibacter* ↓*Ruminococcus*, *Adlercreutzia*, and *Dorea*
Suez et al., 2014 [34]	Treatment: Commercial NAS in drinking water 10% solution: (5% saccharin, 95% glucose), (5% Sucralose), (4% Aspartame). Pure saccharin (0.1 mg ml^{-1}) in drinking Control: water or water with 10% glucose or 10% sucrose Duration: 11 wk NAS and 5 wk pure saccharin	Lean C57Bl/6 mice (10 wk old) with NAS treatment (*n* = 5 per group) C57Bl/6 mice fed on HFD with saccharin treatment (10 wk old) (*n* = 8 per group)	Saccharin: dysbiosis reflected by more than 40 operational taxonomic units (OTUs) abundances changed ↑ *Bacteroides* genus and Clostridiales order Dysbiosis in mice that consumed pure saccharin and HFD
Dai et al., 2020 [54]	MS treatment: gestation and lactation, sucralose 0.1 mg/mL (FDA ADI) Offspring treatment: weaned pups fed a control diet until 8 wk of age and treated with HDF for 4 wk Control: distilled water in MS maternal control and offspring fed with a control diet Duration: maternal treatment, 6 wk	C57BL/6 pregnant mice 3 wk old, weaned pups	MS: at phylum level ↑ the relative abundance of Verrucomicrobia and Proteobacteria and ↓Bacteroidetes At genus level ↑ abundance of *Akkermansia*, *Blautia*, *Corynebacterium*, *Robinsoniella*, and ↓ *Alistipes*, *Barnesiella*, *Paraprevotella*, *Saccharibacteria* genera incertae sedis, and *Streptococcus* MS alters the gut microbiota in the offspring, ↓alpha diversity of 3-wk-old pups

ADI: Acceptable daily intake; MS: maternal sucralose; d: day; wk: weeks, mo: months; HFD: high-fat diet; LS: low saccharin; HS: high saccharin. ↔: unmodified; ↑: increase; ↓: decrease.

In addition, the study by Abou-Donie et al., (2008) found adverse effects of sucralose on the gut microbiota. Splenda was administered to male Sprague-Dawley rats by oral gavage at 100, 300, 500, or 1000 mg/kg for 12 week, to evaluate the concentration of sucralose administered to these experimental animals. In the current review, an estimation was carried out taking into account the sucralose consumption of an adult rat drinking between 30 and 50 mL of the substance prepared in the study by Abou-Donia et al., according to the concentrations shown above and compared with the ADI (EFSA, JECFA), observing that all the values used exceeded admissible limits for humans. These data show that the consumption of sucralose produces an imbalance in the gut microbiota, specifically in the total numbers of anaerobic and aerobic bacteria that are reduced, with a significant decrease in beneficial anaerobic bacteria such as *Bifidobacteria*, *Lactobacilli*, and *Bacteroides*. In this study, equivalent levels of sucralose (Splenda®) in a single drink sweetened with sucralose per day were used [32]. Likewise, another study by Uebanson et al., using different doses of sucralose, found alterations in the microbiota, specifically suggesting that sucralose intake affected in a dose-dependent manner the relative amount of *Clostridium* cluster XIVa [48].

Sánchez-Tapia et al., studied whether the type of sweetener and the presence of a high-fat diet differentially could regulate the gut microbiota. Sucralose was dissolved in water to a concentration of 1.5%. Sucralose increased the Firmicutes abundance showing a decreasing trend in Bacteroidetes, with lower alpha diversity [49]. In this respect, Wang et al., in 2018 performed an 8 week sucralose treatment in mice; they found no changes in alpha diversity, Actinobacteria, and Proteobacteria, but they did find an increase in the abundance of the Firmicutes group [39].

Recently, Zhang et al., in their study with different low doses of sucralose in obese rats, found that ~0.43 mg (0.11 mg/kg translated to human) sucralose increased the relative abundance of Firmicutes but decreased the relative abundance of Bacteroidetes, and that ~0.62 mg sucralose (0.16 mg/kg translated to human) decreased the relative abundance of Firmicutes but increased that of Bacteroidetes. Therefore, the dose of sucralose consumed influenced the Bacteroidetes/Firmicutes ratio. There were no changes in alpha diversity. The authors concluded that the two lower doses of sucralose used in the study might alter the compositions of fecal microbiota [50]. However, in this study, the authors did not use a normal weight control animal model to evaluate the extent to which the establishment of obesity in these rats could modify the results.

Li et al., in 2021, evaluated the bacterial composition at different taxonomic levels in guinea pigs that for 28 days had received saccharin in their drinking water (5 mM). The abundance of Firmicutes tended to decrease in the saccharin-consuming group compared to the control group, while the abundance of Bacteroidetes increased. Therefore, the Bacteroidetes/Firmicutes ratio was affected. In addition, at the family level, the relative abundances of Muribaculaceae and Lactobacillaceae increased in the saccharin group and at the genus level, the relative abundance of *Lactobacillus* increased, while at the family level, the relative abundance of Erysipelotrichaceae and Eubacteriaceae decreased as well as *Ileibacterium* at the genus level [51].

Bian et al., conducted studies in male C57BL/6 J mice with sucralose and saccharin at concentrations equivalent to the ADI for humans (FDA). In 2017, concentrations of sucralose of 0.1 mg/mL [52] and concentrations of saccharin of 0.3 mg/mL administered to male mice [53], in a long-term study for 6 months, were found to induce gut microbiome perturbations, exemplified by the alteration of inflammation-related bacterial pathways and metabolites [52,53].

In 2014, Suez et al., had already demonstrated that the administration of saccharin, sucralose, and aspartame to mice can modulate gut microbiota composition and function, which leads to a higher risk of glucose intolerance, and this is associated with an increase in *Bacteroides* spp. and Clostridiales when performing fecal transplants in germ-free mice from the animals treated with commercial sweeteners. The sweeteners were dissolved in mouse drinking water to obtain a 10% solution: Sucrazit (5% saccharin, 95% glucose),

Sucralite (5% Sucralose), Sweet'n Low Gold (4% Aspartame). As controls, 10% glucose and 10% sucrose solutions were used [34].

In relation to the possible effect of sweeteners on the offspring, Dai et al., in 2020 investigated the effects of maternal sucralose (MS) intake on the offspring susceptibility to suffer from hepatic steatosis in adulthood. C57BL/6 pregnant mice were randomized into an MS group (MS during gestation and lactation) and a maternal control (MC) group (MC diet). MS group mice were given sucralose solution of 0.1 mg/mL, approximately 5–15 mg/kg BW/day, and equal to the upper limit of the FDA ADI. After weaning, all offspring were fed a control diet until 8 weeks of age, and then treated with a high-fat diet (HFD) for 4 weeks. The maternal intake of sucralose was found to inhibit intestinal development, induce intestinal dysbiosis, and decrease the production of butyrate-producing bacteria and butyrate in offspring through downregulation of G-protein-coupled receptor 43 (GPR43), and to exacerbate HFD-induced hepatic steatosis in adulthood. Likewise, at the phylum level, an increase in the relative abundance of Verrucomicrobia and Proteobacteria and a reduction in Bacteroidetes was observed in animals with MS. However, at the genus level, MS increased the abundance of *Akkermansia*, *Blautia*, *Corynebacterium*, and *Robinsoniella*, while, *Alistipes*, *Barnesiella*, *Paraprevotella*, *Saccharibacteria*_genera_inc_ertaesedis, and *Streptococcus* were reduced, with a decrease in alpha diversity [54].

However, we would like to emphasize that after reviewing the studies included in this review, not only the dilution of the sweetener in the drinking water should be considered, but also the adjustment to the amount of water ingested by the animals, because the consumption can vary among different species and strains. For example, the average dose/day of liquid drunk by one mouse can range from 3.9 ± 0.2 mL/mouse to 8.2 ± 0.3 mL [36]. There are also physiological and metabolic differences between rodents and humans [55], and, depending on the type of study and the duration of treatment, inferring the results of investigations using rodent models to those in humans may lead to misleading scientific interpretations. In addition, the metabolism of the sweeteners reviewed in this study can be different between animals and humans, and also among different types of animal species. In fact, in relation to sucralose, there is variability within the types of animals used. However, regarding sucralose (organochlorine), when administered orally, similar results have been found among all species evaluated, showing very low absorption levels and light metabolism. For saccharin, being a water-soluble acid with a pKa of 1.8, absorption is increased in those animal species with lower stomach pH, such as rabbits and humans, compared to those with higher stomach pH, including rats [24]. Thus, studies in animal models are a proxy to studying the potential human effects but human evidence should be gathered at the widest possible extent that the ethics premises in biomedicine and clinical trials may allow.

The animal studies reviewed, except that by Serrano et al. [46], show that saccharin and sucralose produce time- and dose-dependent changes in the gut microbiota. Some studies highlight the modification of the amount of anaerobic and aerobic microbiota, while others emphasize the effect of sucralose on the Bacteroidetes/Firmicutes ratio and others are focused on how maternal consumption can affect the offspring.

However, the mechanisms that mediate the physiological effects of low- or non-calorie sweeteners remain unclear and are most likely diverse. According to the literature, sucralose and saccharin, since they are not absorbed, can influence the maintenance of the pH of the bolus in its trajectory through the intestine, which implies a change in the microenvironmental conditions. Thus, this outcome could be a factor influencing the selective proliferation of certain bacterial groups. In addition, the presence in greater or lesser quantity of cells expressing the T1R2/T1R3 taste heterodimer would be related to the inflammatory effect and possible adaptations of the microbiota [45].

2.3. Effects of Sweeteners on the Gut Microbiota in Human Trials

Non-caloric sweeteners (sucralose and saccharin), as food additives, have been evaluated and approved for use in humans by the European Food Safety Authority and subsequently authorized by the European Commission, the Parliament, and the Council of the European Union. Currently, their consumption, as we have already mentioned, is very widespread in the population, especially in hypocaloric foods and diets as an adjuvant for weight loss or in diabetic patients. The fact that their industrial use in a great variety of products has increased favors the non-adverted consumption.

The human studies reviewed, described in Table 5, studied microbial diversity and metabolites, specifically changes in SCFAs, the main metabolites produced by the microbiota in the large intestine [56]. The SCFAs are bacterial metabolites produced during the colonic fermentation of undigested carbohydrates, such as dietary fiber and prebiotics, and can mediate the interaction between the diet, the microbiota, and the host [57]. SCFA levels are influenced by the proportion of intestinal bacteria, whose alteration (dysbiosis) can lead to an unbalanced composition of the gut SCFAs and therefore it has been concluded that supplementation with pure saccharin did not alter microbial diversity or composition [58].

Table 5. Summary of the analyzed in vivo studies. Humans.

Reference	Sweeteners/Doses/Duration	Design	Results/Conclusions Saccharin and Sucralose
Serrano et al., 2021 [46]	Saccharin 400 mg/d/2 wk	Randomized, double-blind, placebo-controlled interventional study	↔gut microbiota
Ahmad et al., 2020 [59]	Sucralose and aspartame 20% ADI sucralose (~0.136 g sucralose)/14 d	Randomized, double-blind crossover (12 wk) and controlled clinical trial.	↔ gut microbiota ↔ SCFAs
Thomson et al., 2019 [60]	Sucralose 780 mg/d/7 d	Randomized, double-blind study	↔ gut microbiota
Suez et al., 2014 [34]	Saccharin FDA maximal ADI/7 d	Intervention study	Response according to basal microbiota

ADI: Acceptable daily intake; SCFA: short-chain fatty acid; d: day; wk: weeks. ↔: unmodified

The following are the results of human studies, with a sweetener concentration not exceeding the ADI and short-term intake. Among the intervention studies carried out with saccharin, Serrano et al., performed a double-blind, placebo-controlled, parallel-arm study to explore the effects of pure saccharin compound on gut microbiota and glucose tolerance in healthy men and women (46 subjects completed the study; IMC $\leq$ 25). Participants were randomized into four treatment groups (placebo, saccharin, lactisole, or saccharin with lactisole) and consumed capsules containing pulp filler/placebo (1000 mg/d) sodium saccharin (400 mg/d), lactisole (670 mg/d), or sodium saccharin (400 mg/d) + lactisole (670 mg/d) twice daily for 2 weeks. The authors concluded that in these conditions, microbial diversity or composition at any taxonomic level were not changed by pure saccharin supplementation in humans. According to these results, short-term saccharin consumption at maximum acceptable levels (JECFA) is not sufficient to alter the gut microbiota or induce glucose intolerance in supposedly healthy humans [46]. However, Suez et al., did find some modifications in the gut microbiota in 4 of 7 healthy volunteers (5 men and 2 women, aged 28–36 years) from an ongoing clinical nutritional study who were selected as non-habitual sweetener consumers. A saccharin intervention was conducted for one week in which they consumed, on days 2 to 7, the FDA maximum acceptable daily intake (ADI) of commercial saccharin, in three daily doses (equivalent to 120 mg). Changes in the microbiota of only 4 participants, who had developed significantly worse glycemic responses in the study, were observed, and they suggest that humans exhibit a personalized response to non-caloric artificial sweeteners, possibly derived from differences in their basal microbiota [34].

In relation to sucralose, Thomson et al., (2019) conducted a randomized, double-blind study in 34 healthy men (18–50 years) with BMI 20–30 kg/m^2. Sixteen subjects were administered for one week a dose of 780 mg of sucralose per day that was divided into three-260 mg intakes; the control group received a placebo (n = 17). In this study, at the phylum level, the gut microbiome was not modified in healthy individuals [60].

Similar results were obtained in a randomized, double-blind, crossover, controlled clinical trial involving the follow-up of 17 healthy participants. They performed a crossover design for 12 weeks (two 14 day treatment periods separated by a 4 week washout period). In weeks 5 and 6, the volunteers consumed aspartame (n = 9) or sucralose (n = 8). Prior to the washout period, in which no artificial sweeteners were consumed in weeks 11 and 12, all participants consumed the sweetener that they had not previously consumed. The participants were administered 14% (0.425 g) of the ADI for aspartame and 20% of the ADI for sucralose (0.136 g) (approximately 10.5 packets of sucralose with beverages). To define the ADI, they used Health Canada data (sucralose as 9 mg/kg body weight and 40 mg/kg/bw for aspartame). The relative abundance of the five most abundant genus-level taxa within the four most dominant phyla (Actinobacteria, Bacteroidetes, Firmicutes, and Verrucomicrobia) before and after treatment were analyzed at the following days: 1, 28, 42, and 84. Alpha diversity estimation was performed with the Shannon index on the raw operational taxonomic unit. No changes were found for aspartame and sucralose in the gut microbiota composition or SCFAs after 14 days of a daily intake in healthy participants [59].

In relation to long-term studies with saccharin and sucralose in humans, there are not any studies to our knowledge. In the study conducted by Suez et al., in 2014 on the relation between artificial sweetener consumption and gut microbiota, the effect of long-term consumption of non-caloric artificial sweeteners was evaluated. To this end, a validated food frequency questionnaire comprising data collected from 381 non-diabetic individuals from an ongoing clinical nutritional study was used. The results show that artificial sweetener consumption increases the risk of glucose intolerance, these adverse metabolic effects being mediated by modulation of the composition, metabolic function, and the basal microbiota. In this regard, Aldrete-Velasco et al., pointed out in a review that under this design, eliminating completely the confounding variables was not possible, so changes in the microbiota and their metabolic characteristics could also be different due to other factors beyond the consumption of non-caloric sweeteners [61].

Considering the results mentioned above and according to other authors, by using high doses of saccharin and sucralose both in in vitro studies and in animal models, gut microbiota can be modified, whereas in human studies performed using amounts below the ADI and in short-term studies, no effects on gut microbiota are found [2,10,16,47–49]. Contrary to this outcome, Schiffman et al., in 2019 stated in an editorial regarding in vivo animal models, involving data on low- and non-caloric sweeteners and gut microbiota, that sucralose can unequivocally and irrefutably alter the gut microbiome at those levels approved by regulatory agencies, associated with human use. These authors also highlight that it is not appropriate to draw generalized conclusions about effects on the gut microbiota [62].

According to several studies, the explanation for these results may be due to the different doses used in in vitro and in animal model studies versus in human studies, where the doses are lower than the ADI [16,48]. In addition, in human clinical studies, the sample sizes are small, as well as the duration of the interventions. In addition, there is a relevant point to bear in mind like the failure in considering the knowledge regarding the basal gut microbiota of volunteers.

3. Conclusions

In conclusion, it is necessary to broaden the concept of food safety for sucralose and saccharin by re-evaluating toxicity referring to the effect on the gut microbiota and the possible consequences on health maintenance and disease amelioration in humans. Indeed, the mechanisms by which low-calorie and non-calorie sweeteners may alter the gut microbiota remain unclear, and it is not possible to conclude at present whether their effect is direct on the microbiota or mediated by the metabolic situation of the host, for which there are still no conclusive studies. In fact, the scientific literature in both health and disease sometimes refers to beneficial strains and other studies focus on pathogenic strains, which may be due to the lack of clarity regarding what defines dysbiosis or eubiosis. In order to obtain sufficient evidence in these types of studies, clinical trials should be conducted bearing in mind an adequate number of subjects, as well as considering their baseline gut microbiota, dietary habits, and lifestyles. Although the preferred population is healthy adults due to its easy accessibility, more studies must be conducted taking vulnerable population groups into account, such as children, the elderly, pregnant women, lactating women, or subjects with intestinal pathologies, obesity, diabetes, cardiovascular diseases, etc. and chronic and/or excessive consumers of low- and non-calorie sweeteners.

Author Contributions: Conceptualization, S.d.P., S.G.-M. and L.E.D.; data curation, S.d.P., S.G.-M. and L.E.D.; methodology, S.d.P., S.G.-M. and L.E.D.; writing—original draft, S.d.P., S.G.-M. and L.E.D.; writing—review and editing, S.d.P., S.G.-M., L.E.D., E.N., R.U. and A.M. All authors will be informed about each step of manuscript processing including submission, revision, etc. via emails from our system or assigned Assistant Editor. All authors have read and agreed to the published version of the manuscript.

Funding: This research received no external funding.

Conflicts of Interest: The authors declare no conflict of interest.

References

1. WHO. *Ingesta de Azúcares Para Adultos y Niños*; Organización Mundial de la Salud: Geneva, Switzerland, 2015.
2. Ruiz-Ojeda, F.J.; Plaza-Díaz, J.; Sáez-Lara, M.J.; Gil, A. Effects of Sweeteners on the Gut Microbiota: A Review of Experimental Studies and Clinical Trials. *Adv. Nutr.* **2019**, *10*, S31–S48. [CrossRef]
3. Sylvetsky, A.C.; Jin, Y.; Clark, E.J.; Welsh, J.A.; Rother, K.I.; Talegawkar, S.A. Consumption of Low-Calorie Sweeteners among Children and Adults in the United States. *J. Acad. Nutr. Diet* **2017**, *117*, 441–448.e2. [CrossRef] [PubMed]
4. Dunford, E.K.; Taillie, L.S.; Miles, D.R.; Eyles, H.; Tolentino-Mayo, L.; Ng, S.W. Non-Nutritive Sweeteners in the Packaged Food Supply—An Assessment across 4 Countries. *Nutrients* **2018**, *10*, 257. [CrossRef]
5. Samaniego-Vaesken, M.D.L.; González-Fernández, B.; Partearroyo, T.; Urrialde, R.; Varela-Moreiras, G. Updated Database and Trends of Declared Low- and No-Calorie Sweeteners from Foods and Beverages Marketed in Spain. *Front. Nutr.* **2021**, *8*, 670422. [CrossRef] [PubMed]
6. EUR-Lex-32011R1129-EN-EUR-Lex. Available online: https://eur-lex.europa.eu/legal-content/ES/ALL/?uri=CELEX:32011R1129 (accessed on 13 January 2022).
7. Pepino, M.Y. Metabolic effects of non-nutritive sweeteners. *Physiol. Behav.* **2015**, *152*, 450–455. [CrossRef] [PubMed]
8. Mooradian, A.D. In search for an alternative to sugar to reduce obesity. *Int. J. Vitam. Nutr. Res.* **2019**, *89*, 113–117. [CrossRef]
9. Barber, T.; Valsamakis, G.; Mastorakos, G.; Hanson, P.; Kyrou, I.; Randeva, H.; Weickert, M. Dietary Influences on the Microbiota–Gut–Brain Axis. *Int. J. Mol. Sci.* **2021**, *22*, 3502. [CrossRef]
10. Lobach, A.R.; Roberts, A.; Rowland, I.R. Assessing the in vivo data on low/no-calorie sweeteners and the gut microbiota. *Food Chem. Toxicol.* **2018**, *124*, 385–399. [CrossRef]
11. Daly, K.; Darby, A.C.; Shirazi-Beechey, S.P. Low calorie sweeteners and gut microbiota. *Physiol. Behav.* **2016**, *164*, 494–500. [CrossRef]
12. Rinninella, E.; Cintoni, M.; Raoul, P.; Gasbarrini, A.; Mele, M.C. Food Additives, Gut Microbiota, and Irritable Bowel Syndrome: A Hidden Track. *Int. J. Environ. Res. Public Health* **2020**, *17*, 8816. [CrossRef]
13. Iebba, V.; Totino, V.; Gagliardi, A.; Santangelo, F.; Cacciotti, F.; Trancassini, M.; Mancini, C.; Cicerone, C.; Corazziari, E.; Pantanella, F.; et al. Eubiosis and Dysbiosis: The Two Sides of the Microbiota SuMMAry. *New Microbiol.* **2016**, *39*, 1–12. [PubMed]
14. Redondo-Useros, N.; Nova, E.; González-Zancada, N.; Díaz, L.E.; Gómez-Martínez, S.; Marcos, A. Microbiota and Lifestyle: A Special Focus on Diet. *Nutrients* **2020**, *12*, 1776. [CrossRef] [PubMed]
15. Qin, X. Letter: Non-Nutritive Sweeteners May Have a Bell-Shaped Dose Response for Dysbiosis-Related Adverse Effects That Contributed to Discrepancies between Prospective Cohort and Randomized Controlled Studies. *J. Obes. Metab. Syndr.* **2020**, *29*, 237–239. [CrossRef] [PubMed]
16. Malesza, I.J.; Malesza, M.; Walkowiak, J.; Mussin, N.; Walkowiak, D.; Aringazina, R.; Bartkowiak-Wieczorek, J.; Mądry, E. High-Fat, Western-Style Diet, Systemic Inflammation, and Gut Microbiota: A Narrative Review. *Cells* **2021**, *10*, 3164. [CrossRef] [PubMed]
17. Rosés, C.; Cuevas-Sierra, A.; Quintana, S.; Riezu-Boj, J.; Martínez, J.; Milagro, F.; Barceló, A. Gut Microbiota Bacterial Species Associated with Mediterranean Diet-Related Food Groups in a Northern Spanish Population. *Nutrients* **2021**, *13*, 636. [CrossRef]
18. Boulangé, C.L.; Neves, A.L.; Chilloux, J.; Nicholson, J.K.; Dumas, M.-E. Impact of the gut microbiota on inflammation, obesity, and metabolic disease. *Genome Med.* **2016**, *8*, 42. [CrossRef]
19. Merra, G.; Noce, A.; Marrone, G.; Cintoni, M.; Tarsitano, M.G.; Capacci, A.; De Lorenzo, A. Influence of Mediterranean Diet on Human Gut Microbiota. *Nutrients* **2020**, *13*, 7. [CrossRef]

20. Rowland, I.; Gibson, G.; Heinken, A.; Scott, K.; Swann, J.; Thiele, I.; Tuohy, K. Gut microbiota functions: Metabolism of nutrients and other food components. *Eur. J. Nutr.* **2018**, *57*, 1–24. [CrossRef]
21. Carocho, M.; Morales, P.; Ferreira, I.C.F.R. Sweeteners as food additives in the XXI century: A review of what is known, and what is to come. *Food Chem. Toxicol.* **2017**, *107*, 302–317. [CrossRef]
22. Plaza-Diaz, J.; Pastor-Villaescusa, B.; Rueda, A.; Abadia-Molina, F.; Ruiz-Ojeda, F.J. Plausible Biological Interactions of Low- and Non-Calorie Sweeteners with the Intestinal Microbiota: An Update of Recent Studies. *Nutrients* **2020**, *12*, 1153. [CrossRef]
23. FDA. Additional Information about High-Intensity Sweeteners Permitted for Use in Food in the United States. Available online: https://www.fda.gov/food/food-additives-petitions/additional-information-about-high-intensity-sweeteners-permitted-use-food-united-states (accessed on 19 November 2021).
24. Magnuson, B.A.; Carakostas, M.C.; Moore, N.H.; Poulos, S.; Renwick, A.G. Biological fate of low-calorie sweeteners. *Nutr. Rev.* **2016**, *74*, 670–689. [CrossRef] [PubMed]
25. WHO/JECFA. Saccharin. Available online: https://apps.who.int/food-additives-contaminants-jecfa-database/Home/Chemical/3164 (accessed on 7 April 2022).
26. Administración Nacional de Medicamentos, Alimentos y Tecnología Médica. (A.N.M.A.T.). Código Alimentario Argentino. Alimentos régimen o dietéticos (2021). Available online: http://www.anmat.gov.ar/resultados.asp?cx=018082787451070703178%3Arx-vbt5pdfu&cof=FORID%3A10&ie=UTF-8&q=CAPITULO_XVII.pdf (accessed on 16 March 2022).
27. EUFIC. What Is an Acceptable Daily Intake (ADI)? Available online: https://www.eufic.org/en/understanding-science/article/qas-on-acceptable-daily-intakes-adis (accessed on 14 January 2022).
28. WHO/JECFA. Sucralose. Available online: https://apps.who.int/food-additives-contaminants-jecfa-database/Home/Chemical/2340 (accessed on 7 April 2022).
29. EFSA. Safety of the proposed extension of use of sucralose (E 955) in foods for special medical purposes in young children. *EFSA J.* **2016**, *14*, e04361. [CrossRef]
30. Rychen, G.; Aquilina, G.; Azimonti, G.; Bampidis, V.; de Bastos, M.L.; Bories, G.; Cocconcelli, P.S.; Flachowsky, G.; Gropp, J.; Kolar, B.; et al. Safety and Efficacy of Sodium Saccharin When Used as a Feed Flavour for Piglets, Pigs for Fattening, Calves for Rearing and Calves for Fattening. *EFSA J.* **2018**, *16*, e05208. [CrossRef]
31. Diabetes Canada. Sweeteners that increase blood sugar levels. 2018. Available online: https://guidelines.diabetes.ca/docs/patient-resources/sugars-and-sweeteners.pdf (accessed on 28 January 2022).
32. Abou-Donia, M.B.; El-Masry, E.M.; Abdel-Rahman, A.A.; McLendon, R.E.; Schiffman, S.S. Splenda Alters Gut Microflora and Increases Intestinal P-Glycoprotein and Cytochrome P-450 in Male Rats. *J. Toxicol. Environ. Health Part A* **2008**, *71*, 1415–1429. [CrossRef]
33. Anderson, R.; Kirkland, J. The effect of sodium saccharin in the diet on caecal microflora. *Food Cosmet. Toxicol.* **1980**, *18*, 353–355. [CrossRef]
34. Suez, J.; Korem, T.; Zeevi, D.; Zilberman-Schapira, G.; Thaiss, C.A.; Maza, O.; Israeli, D.; Zmora, N.; Gilad, S.; Weinberger, A.; et al. Artificial sweeteners induce glucose intolerance by altering the gut microbiota. *Nature* **2014**, *514*, 181–186. [CrossRef]
35. Ministerio de Agricultura, Pesca y Alimentación. Informe del Consumo Alimentario en España. 2020. Available online: https://www.mapa.gob.es/es/alimentacion/temas/consumo-tendencias/informe-anual-consumo-2020-v2-nov2021-baja-res_tcm30-562704.pdf (accessed on 15 February 2022).
36. Bachmanov, A.A.; Reed, D.R.; Beauchamp, G.K.; Tordoff, M.G. Food Intake, Water Intake, and Drinking Spout Side Preference of 28 Mouse Strains. *Behav. Genet.* **2002**, *32*, 435. [CrossRef]
37. John Hopkin University Species Specific Information: Rat. Available online: https://web.jhu.edu/animalcare/procedures/rat.html#general (accessed on 1 March 2022).
38. Harpaz, D.; Yeo, L.P.; Cecchini, F.; Koon, T.H.P.; Kushmaro, A.; Tok, A.I.Y.; Marks, R.S.; Eltzov, E. Measuring Artificial Sweeteners Toxicity Using a Bioluminescent Bacterial Panel. *Molecules* **2018**, *23*, 2454. [CrossRef]
39. Wang, Q.-P.; Browman, D.; Herzog, H.; Neely, G.G. Non-nutritive sweeteners possess a bacteriostatic effect and alter gut microbiota in mice. *PLoS ONE* **2018**, *13*, e0199080. [CrossRef]
40. Markus, V.; Share, O.; Shagan, M.; Halpern, B.; Bar, T.; Kramarsky-Winter, E.; Teralı, K.; Özer, N.; Marks, R.S.; Kushmaro, A.; et al. Inhibitory Effects of Artificial Sweeteners on Bacterial Quorum Sensing. *Int. J. Mol. Sci.* **2021**, *22*, 9863. [CrossRef]
41. Gerasimidis, K.; Bryden, K.; Chen, X.; Papachristou, E.; Verney, A.; Roig, M.; Hansen, R.; Nichols, B.; Papadopoulou, R.; Parrett, A. The impact of food additives, artificial sweeteners and domestic hygiene products on the human gut microbiome and its fibre fermentation capacity. *Eur. J. Nutr.* **2019**, *59*, 3213–3230. [CrossRef] [PubMed]
42. Shil, A.; Chichger, H. Artificial Sweeteners Negatively Regulate Pathogenic Characteristics of Two Model Gut Bacteria, *E. coli* and *E. faecalis*. *Int. J. Mol. Sci.* **2021**, *22*, 5228. [CrossRef] [PubMed]
43. Vamanu, E.; Pelinescu, D.; Gatea, F.; Sârbu, I. Altered in Vitro Metabolomic Response of the Human Microbiota to Sweeteners. *Genes* **2019**, *10*, 535. [CrossRef] [PubMed]
44. Omran, A.; Baker, R.; Coughlin, C. Differential Bacteriostatic Effects of Sucralose on Various Species of Environmental Bacteria. *Int. Sch. Res. Not.* **2013**, *2013*, 415070. [CrossRef]
45. Sünderhauf, A.; Pagel, R.; Künstner, A.; Wagner, A.E.; Rupp, J.; Ibrahim, S.M.; Derer, S.; Sina, C. Saccharin Supplementation Inhibits Bacterial Growth and Reduces Experimental Colitis in Mice. *Nutrients* **2020**, *12*, 1122. [CrossRef]
46. Serrano, J.; Smith, K.R.; Crouch, A.L.; Sharma, V.; Yi, F.; Vargova, V.; LaMoia, T.E.; Dupont, L.M.; Serna, V.; Tang, F.; et al. High-dose saccharin supplementation does not induce gut microbiota changes or glucose intolerance in healthy humans and mice. *Microbiome* **2021**, *9*, 11. [CrossRef]

47. Falcon, T.; Foletto, K.C.; Siebert, M.; Pinto, D.E.; Andrades, M.; Bertoluci, M.C. Metabarcoding reveals that a non-nutritive sweetener and sucrose yield similar gut microbiota patterns in Wistar rats. *Genet. Mol. Biol.* **2020**, *43*, e20190028. [CrossRef]
48. Uebanso, T.; Ohnishi, A.; Kitayama, R.; Yoshimoto, A.; Nakahashi, M.; Shimohata, T.; Mawatari, K.; Takahashi, A. Effects of Low-Dose Non-Caloric Sweetener Consumption on Gut Microbiota in Mice. *Nutrients* **2017**, *9*, 560. [CrossRef]
49. Sánchez-Tapia, M.; Miller, A.W.; Granados-Portillo, O.; Tovar, A.R.; Torres, N. The development of metabolic endotoxemia is dependent on the type of sweetener and the presence of saturated fat in the diet. *Gut Microbes* **2020**, *12*, 1840766. [CrossRef]
50. Zhang, M.; Chen, J.; Yang, M.; Qian, C.; Liu, Y.; Qi, Y.; Feng, R.; Yang, M.; Liu, W.; Ma, J. Low Doses of Sucralose Alter Fecal Microbiota in High-Fat Diet-Induced Obese Rats. *Front. Nutr.* **2021**, *8*, 787055. [CrossRef]
51. Li, J.; Zhu, S.; Lv, Z.; Dai, H.; Wang, Z.; Wei, Q.; Hamdard, E.; Mustafa, S.; Shi, F.; Fu, Y. Drinking Water with Saccharin Sodium Alters the Microbiota-Gut-Hypothalamus Axis in Guinea Pig. *Animals* **2021**, *11*, 1875. [CrossRef] [PubMed]
52. Bian, X.; Chi, L.; Gao, B.; Tu, P.; Ru, H.; Lu, K. Gut Microbiome Response to Sucralose and Its Potential Role in Inducing Liver Inflammation in Mice. *Front. Physiol.* **2017**, *8*, 487. [CrossRef] [PubMed]
53. Bian, X.; Tu, P.; Chi, L.; Gao, B.; Ru, H.; Lu, K. Saccharin induced liver inflammation in mice by altering the gut microbiota and its metabolic functions. *Food Chem. Toxicol.* **2017**, *107*, 530–539. [CrossRef] [PubMed]
54. Dai, X.; Guo, Z.; Chen, D.; Li, L.; Song, X.; Liu, T.; Jin, G.; Li, Y.; Liu, Y.; Ajiguli, A.; et al. Maternal sucralose intake alters gut microbiota of offspring and exacerbates hepatic steatosis in adulthood. *Gut Microbes* **2020**, *11*, 1043–1063. [CrossRef] [PubMed]
55. Robinson, N.B.; Krieger, K.; Khan, F.M.; Huffman, W.; Chang, M.; Naik, A.; Yongle, R.; Hameed, I.; Krieger, K.; Girardi, L.N.; et al. The current state of animal models in research: A review. *Int. J. Surg.* **2019**, *72*, 9–13. [CrossRef]
56. Silva, Y.P.; Bernardi, A.; Frozza, R.L. The Role of Short-Chain Fatty Acids from Gut Microbiota in Gut-Brain Communication. *Front. Endocrinol.* **2020**, *11*, 25. [CrossRef]
57. Boets, E.; Gomand, S.V.; Deroover, L.; Preston, T.; Vermeulen, K.; De Preter, V.; Hamer, H.M.; Van den Mooter, G.; De Vuyst, L.; Courtin, C.M.; et al. Systemic availability and metabolism of colonic-derived short-chain fatty acids in healthy subjects: A stable isotope study. *J. Physiol.* **2017**, *595*, 541–555. [CrossRef]
58. Ratajczak, W.; Rył, A.; Mizerski, A.; Walczakiewicz, K.; Sipak, O.; Laszczyńska, M. Immunomodulatory potential of gut microbiome-derived short-chain fatty acids (SCFAs). *Acta Biochim. Pol.* **2019**, *66*, 1–12. [CrossRef]
59. Ahmad, S.; Friel, J.; Mackay, D. The Effects of Non-Nutritive Artificial Sweeteners, Aspartame and Sucralose, on the Gut Microbiome in Healthy Adults: Secondary Outcomes of a Randomized Double-Blinded Crossover Clinical Trial. *Nutrients* **2020**, *12*, 3408. [CrossRef]
60. Thomson, P.; Santibañez, R.; Aguirre, C.; Galgani, J.E.; Garrido, D. Short-term impact of sucralose consumption on the metabolic response and gut microbiome of healthy adults. *Br. J. Nutr.* **2019**, *122*, 856–862. [CrossRef]
61. Aldrete-Velasco, J.; López-García, R.; Zúñiga-Guajardo, S.; Riobó-Serván, P.; Serra-Majem, L.; Suverza-Fernández, A.; Esquivel-Flores, M.; MOlina-Segui, F.; Pedroza-Islas, R.; Rascón-Hernández, M.; et al. Análisis de La Evidencia Disponible Para El Consumo de Edulcorantes No Calóricos. Documento de Expertos. *Med. Interna México* **2017**, *33*, 61–83.
62. Schiffman, S.S.; Nagle, H.T. Revisited: Assessing the in vivo data on low/no-calorie sweeteners and the gut microbiota. *Food Chem. Toxicol.* **2019**, *132*, 110692. [CrossRef] [PubMed]

nutrients

MDPI

Review

Management Strategies and Nursing Activities for Nutritional Care in Hospitalized Patients with Cognitive Decline: A Scoping Review

Gloria Liquori [1], Aurora De Leo [1,2], Daniele De Nuzzo [3], Victoria D'Inzeo [3], Rosario Marco Arancio [3], Emanuele Di Simone [2], Sara Dionisi [1], Noemi Giannetta [4], Francesco Ricciardi [3], Fabio Fabbian [5], Giovanni Battista Orsi [6], Marco Di Muzio [3] and Christian Napoli [7,*]

1 Department of Biomedicine and Prevention, Tor Vergata University of Rome, 00133 Rome, Italy
2 Nursing, Technical, Rehabilitation, Assistance and Research Direction—IRCCS Istituti Fisioterapici Ospitalieri—IFO, 00144 Rome, Italy
3 Department of Clinical and Molecular Medicine, Sapienza University of Rome, 00185 Rome, Italy
4 School of Nursing, UniCamillus—Saint Camillus International University of Health and Medical Sciences, 00131 Rome, Italy
5 Department of Medical Sciences, University of Ferrara, 44121 Ferrara, Italy
6 Department of Public Health and Infectious Diseases, Sapienza University of Rome, 00185 Roma, Italy
7 Department of Surgical and Medical Sciences and Translational Medicine, Sapienza University of Rome, 00185 Rome, Italy
* Correspondence: christian.napoli@uniroma1.it; Tel.: +39-06-3377-5529

Abstract: Cognitive impairment and dementia can negatively impact the nutritional capacities of older people. Malnutrition is common in hospitalized frail elderly people with cognitive impairment and negatively affects prognosis. Malnutrition worsens the quality of life and increases morbidity and mortality. This scoping review aimed to identify factors affecting the risk of malnutrition and preventive strategies in hospitalized patients with cognitive impairment, focusing on nursing interventions. The authors researched population, context, and concept in international databases of nursing interest. Full texts that met the inclusion criteria were selected and reviewed. The extracted data were subject to thematic analysis. A five-stage approach, already reported in the scientific literature, was utilized in the following scoping review. Of 638 articles yielded, 9 were included. Two focus areas were identified as follows: (1) prevalence and risk factors of malnutrition in older patients with cognitive decline; (2) nursing strategies used to enhance clinical outcomes. Nursing health interventions aim to recognize and reduce malnutrition risk, positively impacting this phenomenon. A multidisciplinary team is essential to meet the nutritional needs of these patients.

Keywords: cognitive impairment; cognitive decline; nutritional care; hospitalized patient; elderly patient

Citation: Liquori, G.; De Leo, A.; De Nuzzo, D.; D'Inzeo, V.; Arancio, R.M.; Di Simone, E.; Dionisi, S.; Giannetta, N.; Ricciardi, F.; Fabbian, F.; et al. Management Strategies and Nursing Activities for Nutritional Care in Hospitalized Patients with Cognitive Decline: A Scoping Review. *Nutrients* **2022**, *14*, 4036. https://doi.org/10.3390/nu14194036

Academic Editors: Alessandra Durazzo and Massimo Lucarini

Received: 9 August 2022
Accepted: 26 September 2022
Published: 28 September 2022

1. Introduction

Due to the global increase in aging, the elderly population is constantly growing [1]. During the period 2015–2030, the elderly population is expected to grow from 901 million to 1.4 billion people (by 56%), and the 2015 population is expected to double by 2050 [2].

This increase in the elderly population is already placing substantial extra strain on healthcare and support services [3], increasing the costs due to health-related complications [4].

Among the main concerns linked to age, malnutrition is a common health problem in people older than 65 [5]; in fact, nutritional fragility is a frequent condition in vulnerable elderly people and is related to an increased incidence of mortality in this population [6]. Moreover, it is often associated with a reduced adaptive response to physiological and pathological conditions [7]; for instance, the elderly population experiences a physiological loss of taste, which impacts the frailty condition. Assessment of nutritional status through anthropometric measurements in these patients is essential to ensure healthy aging and adequate food intake [8].

Malnutrition has a negative impact on both people living independently and, overall, on those admitted to healthcare facilities, affecting up to 60% of hospitalized older adults [9]. In fact, according to estimates, up to 50% of patients are undernourished when they are admitted, increasing their malnutrition while being treated in a hospital [10]. This issue is linked to lengthier hospital stays, increased morbidity (pressure ulcers, infections, and falls), and mortality, especially in patients affected by chronic diseases [11,12].

In elderly people with cognitive impairment, the phenomenon is also more serious, since malnutrition irreversibly worsens other health conditions [13]. At the same time, mental status significantly affects nutritional status; people with lower cognitive levels tend to face a higher risk of malnutrition, especially during hospitalization [14,15]. The relationships between nutritional status, cognitive decline, and performance are complex and reciprocal: the presence or the risk of malnutrition may influence cognitive performance, and the presence of cognitive decline may affect the activities of daily living (ADL), also affecting food intake [16].

Unfortunately, malnutrition is also a frequently underdiagnosed entity, capable of subtly impacting patient outcomes, length of stay, hospital costs, and readmissions [9]. Recent studies suggest the crucial role of nurses in preventing, assessing, and treating malnutrition in this fragile population [2,7]. One of the possible key points could be implementing all known strategies to avoid worsening nutritional status, improving health status, and reducing mortality risk [17].

Therefore, this scoping review was targeted at evaluating the relationship between nursing activities and the identification, prevention, and management of malnutrition among hospitalized elderly individuals with cognitive impairment. In particular, our aims were collecting best practice and scientific evidence with regard to: (i) risk factors for developing low food intake in hospitalized older patients with cognitive impairment; (ii) prevalence of malnutrition in older patients with cognitive impairment; (iii) identify the nursing strategies to enhance clinical outcomes and care of patients with cognitive impairment in the hospital environments.

2. Materials and Methods

2.1. Literature Search

The five-step approach presented by Arksey and O'Malley [18], and advanced by Levac and collaborators [19] was utilized in the following scoping review. The foreseen steps were: 1. determining the study problem, 2. outlining relevant investigations, 3. studies selection, 4. data charting, 5. collating, summarization, and presenting the findings.

The choice of a scoping review was based on the need to identify the nature and extent of the research evidence in accordance with Grant et al. [20].

The study was conducted according to the Preferred Reporting Item for Systematic Review and Meta-analysis for Scoping Review (PRISMA-ScR) [21] (Supplementary Table S1).

2.2. Step 1: Determining the Study Problem

The objectives of the review were to provide answers to the following questions:

1. What is the prevalence of malnutrition in older patients with cognitive impairment?
2. What are the risk factors for developing low food intake in hospitalized older patients with cognitive impairment?
3. Which nursing strategies are used to enhance clinical outcomes and patients with cognitive impairment care in the hospital environments?

The PCCT (population, concept, context, and type of study) methodology was utilized to identify search questions according to Peters et al. [22]. Specifically, the population included hospitalized elderly patients (aged > 65 years) with cognitive impairment and malnutrition. The concept was prevalence of phenomenon, nursing health interventions aimed at recognizing, reducing the risk of malnutrition, and positively affecting this phenomenon. The context was health care services admitting older people with cognitive impairment. With regards to type of study, all observational, experimental, and quasi-experimental studies with available full text in English, Spanish, and Italian were included.

2.3. Step 2: Outlining Relevant Investigations

A systematic search was conducted on the following scientific databases: PubMed, Cumulative Index to Nursing and Allied Health Literature (CINAHL), Psychological Abstracts Information Services (PsycINFO), Scopus, and National Library of Medicine (MEDLINE) via EBSCO and Cochrane Library. Observational, experimental, and quasi-experimental studies on malnourished elderly inpatients (>65 years) with cognitive impairment (including dementia and Alzheimer's disease) in English, Spanish, and Italian were included. Studies involving people living at home and the adult and pediatric population (<65 years old) were excluded. No limits for country of origin or geographical context were applied. No time limits were applied.

2.4. Step 3: Study Selection

Citations were imported into Zotero® Reference Manager, and the duplicates were eliminated. Two independent researchers conducted the initial screening, from March 2022 to May 2022, by reading the titles and abstracts of the publications. Unrelated studies were removed. If the publication's relevance was undefined based on the title or abstract reading, the reviewers read the paper in full text to determine its eligibility.

The same investigators retrieved and assessed the whole text of articles deemed eligible for inclusion criteria. Any disagreement was resolved by discussion and final consensus. When the latter was not reached, arbitration was sought from a third researcher who supervised the study.

2.5. Step 4: Data Charting

The selected articles were summarized in Table 1 [23–31], including authors and year, aim, method, and main results.

Table 1. Summary of findings.

Authors and Year	Aim	Method	Results
Lauque et al., 2004 [26]	To study the effects of OS (oral supplement) on body weight, body composition, nutritional status, and cognition in elderly patients with Alzheimer's disease.	Prospective, randomized, controlled study. A total of 46 patients (intervention group) received 3-month OS. The other 45 patients (control group) received usual care.	Between baseline and 3 months, energy and protein intake significantly improved in the intervention group, resulting in a significant increase in weight and fat-free mass. No significant changes were found for dependence, cognitive function, or biological markers. The nutritional benefit was maintained in the intervention group after discontinuation of OS at 3 months.
Wong et al., 2008 [25]	To evaluate strategies designed to improve nutrition in elderly hospitalized patients with dementia.	Interventions: Phase 1: Observation. Phase 2: Encouraging dietary, "Grazing". Phase 3: Using volunteers to feed patients. Phase 4: Improving dining room ambience by playing soothing music.	There were no differences between the groups concerning age, length of stay, gender, or baseline anthropometric scores. Simple, inexpensive and easy to implement strategies can improve nutrition in hospital in patients with dementia.
Orsitto et al., 2009 [23]	To assess the prevalence of malnutrition in older patients with mild cognitive impairment.	A total of 623 hospitalized elderly patients underwent the comprehensive geriatric assessment to evaluate medical, cognitive, affective and social aspects. Nutritional status was assessed by using the mini nutritional assessment. According to the neuropsychological evaluation cognitive function was categorized into three levels—normal cognition, mild cognitive impairment and dementia.	According to the mini nutritional assessment classification, 18% of the sample study was assessed as well-nourished, 58% at risk of malnutrition and 24% as malnourished. Patients with mild cognitive impairment and dementia had a significantly lower frequency of well-nourished and higher frequency of being at risk of malnutrition or malnourished than patients with normal cognition.

Table 1. *Cont.*

Authors and Year	Aim	Method	Results
Salva et al., 2009 [28]	To describe the study design, intervention program, recruitment, randomization, and patients' baseline characteristics.	Intervention: the NutriAlz program 1. A personalized presentation and handover of a briefcase. 2. Training for families, caregivers. 3. Support in weight monitoring. 4. Periodic information for the families. 5. Action protocols and standardized help decision.	Evaluation of the risk for malnutrition using the mini nutritional assessment resulted in 5% of malnourished patients, 37% at risk for malnutrition and 58% well-nourished subjects. The MNA score was significantly different between the two groups. Patients with dementia showed a high risk of malnutrition.
Lin et al., 2010 [29]	To investigate the risk factors of older people with dementia for developing low food intake.	Four hundred seventy-seven participants with dementia from nine dementia special care units in licensed long-term care facilities (LTCFs) in Northern and Central Taiwan. Data were collected using the Barthel index, Mini Mental State Examination, and Edinburgh Feeding Evaluation in Dementia scale.	The prevalence of low food intake at meals in patients with dementia in LTCFs was 30.7%. Eating difficulty, no feeding assistance, moderate dependence, fewer family visits, and being female and older were six independent factors associated with low food intake after controlling for all other aspects.
Allen et al., 2013 [30]	Investigate the impact of the provision of OS on protein and energy intake from food and the ability to meet protein and calorie requirements in people with dementia.	After consent by proxy was obtained, participants were enrolled in a cross-over study comparing oral intake on an intervention day to an adjacent control day.	More people achieved their energy and protein requirements with the supplement drink intervention without sufficient impact on habitual food consumption. Findings from these 26 participants with dementia indicate that supplement drinks may be beneficial in reducing the prevalence of malnutrition within the group as more people meet their nutritional requirements. As the provision of supplement drinks has an additive effect on consumption of habitual foods, these can be used alongside other measures to also improve oral intake.
Allen et al., 2014 [27]	To analyze the influence of the serving method on compliance and consumption of nutritional supplement drinks in older adults with cognitive impairment.	Participants were randomized to the serving method. Nursing and care staff were instructed to give the supplement drinks three times per day on alternate days over a week by the allocated serving method. The researcher weighed the amount of supplement drink remaining after consumption.	Participants randomized to consume nutritional drinks from a glass/beaker drank statistically significantly more than those who consumed them via a straw inserted directly into the container. However, supplements placed in a glass/beaker were more frequently omitted. Nutritional supplement drinks should be given to people with dementia who are able to feed themselves in a glass or a beaker if staffing resources allow.
Avelino-Silva et al., 2014 [24]	Assess the applicability of the proposed model comprehensive geriatric assessment (CGA) for thoroughly characterizing patients with cognitive impairment and analyze the impact of this strategy on the prediction of mortality and adverse hospital outcomes.	This prospective observational study included 746 patients aged 60 years and over. The proposed CGA was applied to evaluate all patients at admission. Impairment in ten CGA components was mainly investigated: polypharmacy, activities of daily living (ADL) dependency, instrumental activities of daily living (IADL) dependency, depression, dementia, delirium, urinary incontinence, falls, malnutrition, and poor social support.	CGA was a useful tool to identify patients at higher risk of in-hospital death and adverse outcomes, of which those with malnutrition were foremost.
Baumgartner et al., 2021 [31]	This article aimed to study the effects of individualized nutritional support for patients with ageing-related vulnerability in the acute hospital setting on mortality and other clinical outcomes.	The study analyzed data of patients at nutritional risk (Nutritional Risk Screening 2002 score $\geq$ 3 points) with ageing-related vulnerability, randomized to receive protocol-guided individualized nutritional support to reach specific protein and energy goals (intervention group) or routine hospital food (control group). The primary endpoint was all-cause 30 d mortality. Trained study nurses performed structured telephone interviews with all patients 30 days after inclusion to collect outcome information.	This study found a more than 50% reduction in mortality at 30 days in hospitalized patients with ageing-related vulnerability at nutritional risk receiving protocol-guided individualized nutritional support to reach specific protein and energy goals. Significant improvements were also found for longer-term mortality at 180 days. Individualized nutritional support also improved functional outcomes and quality of life (QoL) over 30 and 180 days. These data support the early screening of hospitalized patients with aging-related vulnerability for nutritional risk, followed by the implementation of individualized nutritional interventions.

2.6. Step 5: Collating, Summarization, and Presenting the Findings

Lastly, the results were collated and summarized according to Arksey and O'Malley's framework [18], respecting the proposed search strategy.

3. Results

The search strategy yielded 638 articles; 142 duplicates were excluded. A further 421 records were excluded after applying the title and abstract eligibility criteria. The full texts of 75 articles were reviewed. Of these, nine articles met the inclusion criteria and were included in this scoping review [23–31]. Figure 1 shows the search and selection process according to the PRISMA statement [21]. (Figure 1).

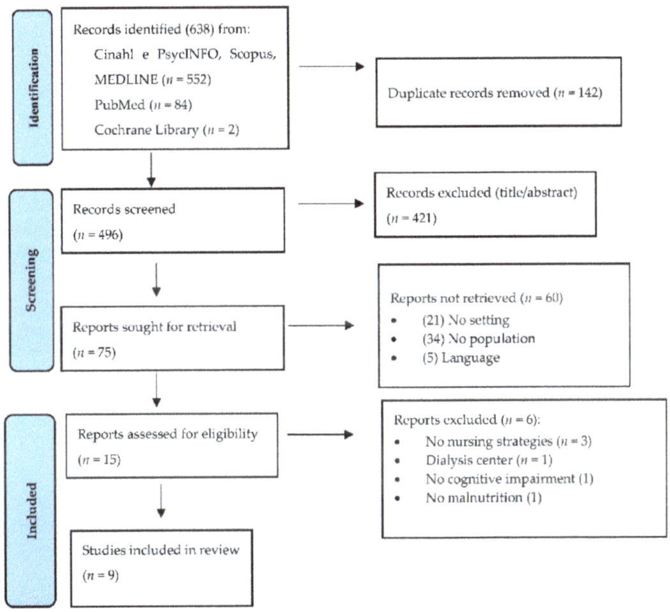

Figure 1. Flow diagram of the search and selection process, based on PRISMA flowchart.

The relationship between older people's nutritional status, hospitalization, and cognitive impairment as a result of the included studies is shown in Table 1 and in the Supplementary Material Table S1.

According to the scoping review framework, the main themes were divided into two results sections.

3.1. Prevalence and Risk Factors of Malnutrition in Older Patients with Cognitive Impairment

Orsitto et al. [23] described an extremely high prevalence of poor nutritional status in a sample of hospitalized older patients with different grades of cognitive impairment. Only 18% of the sample was well nourished, while 82% were at risk of malnutrition or malnourished. Findings showed a significantly greater malnutrition rate in hospitalized patients with severe cognitive impairment. This study showed the evidence of poor nutritional status even in patients with mild cognitive impairment who had not yet progressed to dementia [23]. Moreover, according to Salva et al. [28] patients with dementia showed a high risk of malnutrition with respect to other patients. According to Lin et al. [29], eating difficulty, no feeding assistance, moderate dependence, fewer family visits, and being female and older were six independent factors associated with low food intake after controlling for all other aspects [29].

Hospitalized frail patients develop a major risk of under-nutrition and weight loss [23,25]. However, according to the findings of another study [25] there are no differences in malnutrition among different groups of hospitalized patients concerning age, length of stay, gender, or baseline anthropometric scores.

3.2. Nursing Strategies Used to Enhance Clinical Outcomes

Simple, inexpensive, and easy-to-implement strategies, such as early dietary assessment; dietary "grazing" and staggered mealtimes, can improve nutrition in hospitalized elderly patients [25]. Nursing strategies that provide information on the clinical, functional, and cognitive aspects of the disease should be used in hospitalized patients, especially those with cognitive impairment [25]. The immediate evaluation of eating abilities, nutritional needs, and dietary preferences is a simple and inexpensive strategy that can lead to positive changes in nutritional intake in this population [25,31]. Indeed, assessing the patient's nutritional needs early is critical to reducing hospitalization [24], as it improves the patient's weight, but this does not affect cognitive impairment [25]. A positive element emerges from the study of Avelino-Silva et al.: hospitalization, by allowing more time to assess each patient, provides the opportunity for a detailed and structured nutritional clinical assessment through a CGA tool that has proven useful in reducing mortality in these patients [24]. Other tools have been used in order to evaluate nutritional status, for instance, Salva et al. used the mini nutritional assessment scale [28].

Some studies have also shown other strategies to enhance the nutritional outcome in patients with cognitive impairment. For example, the main objective of Lauque et al. [26] was to evaluate the effects of OS. Overall, 46 patients (intervention group) received 3-month OS, while the other 45 patients (control group) received standard care in geriatric wards and daycare centers in the Toulouse region. Protein and energy consumption considerably increased in the intervention group between baseline and 3 months, leading to a considerable gain in weight as well as fat-free mass. Nevertheless, no substantial changes in biological markers, cognitive function, or dependence were observed. Therefore, the authors conclude that the regular OS assumption can aid in preserving the gain in fat-free mass and enhance these individuals' nutritional status. Additionally, a study conducted by Allen et al. [30] showed that supplement drinks may be beneficial in reducing the prevalence of malnutrition within the group, as more people meet their nutritional requirements. Moreover, another study by Allien et al. [27] showed that drinking nutritional beverages with a glass makes the patient more stimulated to drink rather than using a straw. Baumgartner et al. [31] found that individualized nutritional support improves functional outcomes and quality of life (QoL) over 30- and 180-day periods of nutritional support.

Finally, if nurses take the time to assess the nutritional status and needs, implement suitable care plans and provide food and drink in ways that ensure their safe consumption, this can positively affect both patients' nutritional status and their general health condition, reducing the risk of mortality too [28–31].

4. Discussion

4.1. Nutritional Assessment and Screening

Malnutrition has a high prevalence in hospitalized elderly patients [25,26]. This review confirms that the potential associated factors are different: medical history, medicines intakes, diet, oral health, swallowing ability, physical and cognitive function, gastrointestinal, psychiatric, and neurological conditions, and also social aspects of a person's life [24]. Therefore, every hospital should establish an interdisciplinary approach to nutrition care based on formal policies and procedures, ensuring the early identification of malnourished patients or malnutrition risk and implementing comprehensive nutrition care plans [24]. In fact, the review results suggest that patients should be screened for malnutrition within the first 24 h of admission and screened regularly during their hospital stay [23,24]. Moreover, it is critically important to establish individualized nutritional support to these patients, as this improves functional outcomes and QoL, as well as reducing mortality by 50% at 30 days in hospitalized elderly patients [31].

The MNA has become a tool allowed standardized, reproducible, and reliable determination of nutritional status [26,31]. However, some studies reported that the MNA-SF (mini nutritional assessment short-form) could overestimate malnutrition risk [28].

Avoiding and solving malnutrition in elderly individuals is a crucial element of geriatric care, since healthy people can also become malnourished during hospitalization. In this context, early dietary assessment and implementation of feeding strategies are crucial not only for vulnerable patients [25].

4.2. Nutritional Strategies and Management

Breaking the vicious circle between malnutrition and cognitive impairment can help patients and reduce the impact of this degenerative condition [25,28].

The management of elderly adults who are malnourished or at malnutrition risk should be multimodal and multidisciplinary, as reported by Salva et al. in their nutritional intervention program, "The NutriAlz", aimed at preventing weight loss patients affected by dementia [28].

A routine and impersonalized hospital nutrition in elderly patients carries an increased risk of mortality compared with individualized nutritional support [31]. Especially, patients with dementia and cognitive impairment hospitalized for acute diseases often require individualized strategies to maintain adequate caloric intake.

Moreover, eating difficulties is a factor associated with low food intake [29]. Therefore, attention must be paid to oral frailty, defined as a gradual age-related loss of oral function along with decline in cognitive and physical function [32]. In addition, altered eating behavior and dysphagia are factors to be managed in these frail hospitalized patients [26]. These patients have impaired oral motor skills, particularly chewing function, oral diadochokinesis, swallowing and salivary disorders, all associated with few teeth left in the oral cavity [29,32].

4.3. Environmental Changes to Improve Mealtime Habitat and Experience

Environmental elements such as food accessibility (for example a glass door refrigerator with snack food easily visible), companions and furniture, smell, ambient sounds, lighting and temperature, size of the portion, eating location, and presentation of food play a crucial role during mealtimes, improving the patients' compliance to eat [25]. It is demonstrated that changes in mealtime habits and atmosphere, based on the personal needs of patients, could increase nutritional intake and reduce the malnutrition risk [25,29], especially among older people.

A great variety of environmental variables might have an influence on the nutritional intake of elderly inpatients as also reported in the review findings [25,30]. Wong, et al. have shown that appetite increased if those people needing more assistance were fed earlier than other patients; it is not the time of meal initiation that is important, but the longer duration [25]. To reduce the prevalence of malnutrition in these patients, it is useful to offer supplementary drinks, as they can improve oral intake and increase appetite [30]. Another environmental change to improve mealtime habitats is music. Wong et al. have shown that patients spent more time at the table when music was played, so it appears to be an effective strategy to lengthen mealtime and increase patients' appetite [25].

Family style meals, eating meals with caregivers, relaxing music during mealtime, patient education, protected mealtimes, and additional food assistance (implemented alone or in combination) were among the most promising interventions to improve mealtime experiences [25,29,30]. It is crucial that these fragile patients receive support from nurses or family members at mealtimes [25]. This relational strategy helps the patient to increase food intake during meals [29]. Receiving good mealtime assistance and increasing time spent by nurses or volunteers on feeding or helping during meals may positively affect eating behavior with a positive effect on the nutritional intake in older inpatients [27,29].

The physical presence of a caregiver helps the patient to be more focused on his or her meal [25,29]. Caregiver education is crucial to ensure proper weight maintenance for the patient who may go through weight loss even if they have a positive energy balance [26].

4.4. Limits of the Study

This scoping review focused only on the relationship between nursing activities and identification, prevention, and management of cognitive impairment in elderly hospitalized individuals, and the main limitation is the reduced availability of studies in the hospital environment. The majority of published studies recruited patients in the home setting. Hospitalization is usually due to further acute illnesses; therefore, general conditions and eventually chronic diseases associated with acute events worsen. Moreover, in this study, we could not take into consideration the intensity of care, including some invasive treatments such as intravenous or enteral support. Finally, we could not stratify nursing activities considering the cause of cognitive impairment, its worsening, its stage, or the duration of the cognitive impairment.

5. Conclusions

The findings of this scoping review suggest that malnutrition may have high frequency in hospitalized elderly patients, especially those affected by cognitive impairment. The relationship between nutritional status and cognitive impairment is complex and reciprocal. Therefore, an appropriate nutritional status evaluation along the hospitalization, followed by both healthcare and environmental managing strategies is necessary to maintain or improve the patients' nutritional status. A multidisciplinary team is essential to fulfilling the nutritional needs of these patients, and the role of nursing activities is crucial.

Supplementary Materials: The following supporting information can be downloaded at: https://www.mdpi.com/article/10.3390/nu14194036/s1, Table S1: Summary of findings.

Author Contributions: Conceptualization, M.D.M., G.B.O. and C.N.; methodology, S.D., E.D.S., V.D., and N.G.; data collection: F.R., V.D., G.L., F.F. and A.D.L.; original draft preparation, G.L., C.N. and R.M.A.; writing—review and editing, E.D.S., D.D.N., C.N. and N.G.; supervision, M.D.M., G.B.O. and C.N.; project administration, M.D.M. All authors have read and agreed to the published version of the manuscript.

Funding: This research received no funding.

Institutional Review Board Statement: Not applicable.

Informed Consent Statement: Not applicable.

Data Availability Statement: All data are available upon request.

Conflicts of Interest: The authors declare no conflict of interest.

References

1. Scotellaro, M.; Ruggeri, M.; D'inzeo, V.; Di Simone, E.; Di Muzio, M.; Dionisi, S.; Giannetta, N. Multidimensional assessment using a modified S.Va.M.A scale, a pilot study in patients aged 65 or older. *J. Gerontol. Geriatr.* **2021**, *69*, 171–176. [CrossRef]
2. Di Muzio, M.; Rosa, F.; Moriconi, A. Nutrition in dementia: A challenge for nurses. *Prog. Nutr.* **2018**, *20*, 5–11. [CrossRef]
3. Prince, M.J.; Wu, F.; Guo, Y.; Gutierrez Robledo, L.M.; O'Donnell, M.; Sullivan, R.; Yusuf, S. The burden of disease in older people and implications for health policy and practice. *Lancet* **2015**, *385*, 549–562. [CrossRef]
4. Giraldi, G.; Montesano, M.; Napoli, C.; Frati, P.; La Russa, R.; Santurro, A.; Scopetti, M.; Orsi, G.B. Healthcare-Associated Infections Due to Multidrug-Resistant Organisms: A Surveillance Study on Extra Hospital Stay and Direct Costs. *Curr. Pharm. Biotechnol.* **2019**, *20*, 643–652. [CrossRef] [PubMed]
5. Park, J.E.; Baik, H.W. Aging and Nutrition. *J. Korean Soc. Parenter. Enter. Nutr.* **2013**, *5*, 42–49. [CrossRef]
6. Zupo, R.; Castellana, F.; Guerra, V.; Donghia, R.; Bortone, I.; Griseta, C.; Lampignano, L.; Dibello, V.; Lozupone, M.; Coelho-Jùnior, H.J.; et al. Associations between nutritional frailty and 8-year all-cause mortality in older adults: The Salus in Apulia Study. *J. Intern. Med.* **2021**, *290*, 1071–1082. [CrossRef]
7. Di Simone, E.; Giannetta, N.; Gazzelloni, A.; Torre, R.; Livadiotti, J.; Dionisi, S.; Di Muzio, F.; Montesano, M.; Moriconi, A.; Fabbian, F.; et al. Nutrition in patients with unilateral neglect. *Prog. Nutr.* **2020**, *22*, e2020001. [CrossRef]
8. Zupo, R.; Castellana, F.; Bortone, I.; Griseta, C.; Sardone, R.; Lampignano, L.; Lozupone, M.; Solfrizzi, V.; Castellana, M.; Giannelli, G.; et al. Nutritional domains in frailty tools: Working towards an operational definition of nutritional frailty. *Ageing Res. Rev.* **2020**, *64*, 101148. [CrossRef]

9. Agarwal, E.; Miller, M.; Yaxley, A.; Isenring, E. Malnutrition in the elderly: A narrative review. *Maturitas* **2013**, *76*, 296–302. [CrossRef]
10. Kirkland, L.L.; Kashiwagi, D.T.; Brantley, S.; Scheurer, D.; Varkey, P. Nutrition in the hospitalised patient. *J. Hosp. Med.* **2013**, *8*, 52–58. [CrossRef]
11. Correia, M.; Hegazi, R.; Higashiguchi, T.; Michel, J.; Reddy, B.; Tappenden, K.; Uyar, M.; Muscaritoli, M. Evidence-based recommendations for addressing malnutrition in health care: An updated strategy from feed ME. *Glob. Study Group JAMDA* **2014**, *15*, 544–550.
12. Di Stefano, V.; Rispoli, M.G.; Pellegrino, N.; Graziosi, A.; Rotondo, E.; Napoli, C.; Pietrobon, D.; Brighina, F.; Parisi, P. Diagnostic and therapeutic aspects of hemiplegic migraine. *J. Neurol. Neurosurg. Psychiatry* **2020**, *91*, 764–771. [CrossRef] [PubMed]
13. Shlisky, J.; Bloom, D.E.; Beaudreault, A.R.; Tucker, K.L.; Keller, H.H.; Freund-Levi, Y.; Fielding, R.A.; Cheng, F.W.; Jensen, G.L.; Wu, D.; et al. Nutritional Considerations for Healthy Aging and Reduction in Age-Related Chronic Disease. *Adv. Nutr.* **2017**, *8*, 17–26. [CrossRef] [PubMed]
14. Peng, L.N.; Cheng, Y.; Chen, L.K.; Tung, H.H.; Chu, K.H.; Liang, S.Y. Cognition and social-physiological factors associated with malnutrition in hospitalised older adults in Taiwan. *J. Nurs. Res.* **2015**, *23*, 1–5. [CrossRef] [PubMed]
15. Mudge, A.M.; Ross, L.J.; Young, A.M.; Isenring, E.A.; Banks, M.D. Helping understand nutritional gaps in the elderly (HUNGER): A prospective study of patient factors associated with inadequate nutritional intake in older medical inpatients. *Clin. Nutr.* **2011**, *30*, 320–325. [CrossRef] [PubMed]
16. Del Parigi, A.; Panza, F.; Capurso, C.; Solfrizzi, V. Nutritional factors, cognitive decline, and dementia. *Brain Res. Bull.* **2006**, *69*, 1–19. [CrossRef]
17. De Giorgi, A.; Di Simone, E.; Cappadona, R.; Boari, B.; Savriè, C.; López-Soto, P.J.; Rodríguez-Borrego, M.A.; Gallerani, M.; Manfredini, R.; Fabbian, F. Validation and Comparison of a Modified Elixhauser Index for Predicting In-Hospital Mortality in Italian Internal Medicine Wards. *Risk Manag. Healthc. Policy* **2020**, *13*, 443–451. [CrossRef]
18. Arksey, H.; & O'Malley, L. Scoping studies: Towards a methodological framework. *Int. J. Soc. Res. Methodol.* **2005**, *8*, 19–32. [CrossRef]
19. Levac, D.; Colquhoun, H.; O'Brien, K.K. Scoping studies: Advancing the methodology. *Implement Sci.* **2010**, *5*, 69. [CrossRef]
20. Grant, M.J.; Booth, A. A typology of reviews: An analysis of 14 review types and associated methodologies. *Health Inf. Libr. J.* **2009**, *26*, 91–108. [CrossRef]
21. Tricco, A.C.; Lillie, E.; Zarin, W.; O'Brien, K.K.; Colquhoun, H.; Levac, D.; Moher, D.; Peters, M.D.J.; Horsley, T.; Weeks, L.; et al. PRISMA Extension for Scoping Reviews (PRISMA-ScR): Checklist and Explanation. *Ann. Intern. Med.* **2018**, *169*, 467–473. [CrossRef] [PubMed]
22. Peters, M.D.J.; Godfrey, C.; McInerney, P.; Munn, Z.; Tricco, A.C.; Khalil, H. Chapter 11: Scoping Reviews. In *JBI Manual for Evidence Synthesis*; Aromataris, E., Munn, Z., Eds.; JBI: Adelaide, Australia, 2020; Available online: https://synthesismanual.jbi.global (accessed on 24 September 2022). [CrossRef]
23. Orsitto, G.; Fulvio, F.; Tria, D.; Turi, V.; Venezia, A.; Manca, C. Nutritional status in hospitalised elderly patients with mild cognitive impairment. *Clin. Nutr.* **2009**, *28*, 100–102. [CrossRef] [PubMed]
24. Avelino-Silva, T.J.; Farfel, J.M.; Curiati, J.A.; Amaral, J.R.; Campora, F.; Jacob-Filho, W. Comprehensive geriatric assessment predicts mortality and adverse outcomes in hospitalised older adults. *BMC Geriatr.* **2014**, *14*, 129. [CrossRef] [PubMed]
25. Wong, A.; Burford, S.; Wyles, C.L.; Mundy, H.; Sainsbury, R. Evaluation of strategies to improve nutrition in people with dementia in an assessment unit. *J. Nutr. Health Aging* **2008**, *12*, 309–312. [CrossRef]
26. Lauque, S.; Arnaud-Battandier, F.; Gillette, S.; Plaze, J.M.; Andrieu, S.; Cantet, C.; Vellas, B. Improvement of weight and fat-free mass with oral nutritional supplementation in patients with Alzheimer's disease at risk of malnutrition: A prospective randomised study. *J Am Geriatr. Soc.* **2004**, *52*, 1702–1707. [CrossRef] [PubMed]
27. Allen, V.J.; Methven, L.; Gosney, M. Impact of serving method on the consumption of nutritional supplement drinks: Randomised trial in older adults with cognitive impairment. *J. Adv. Nurs.* **2014**, *70*, 1323–1333. [CrossRef] [PubMed]
28. Salva, A.; Andrieu, S.; Fernandez, E.; Schiffrin, E.J.; Moulin, J.; Decarli, B.; Guigoz, Y.; Vellas, B.; NutriAlz Group. Health and nutritional promotion program for patients with dementia (NutriAlz Study): Design and baseline data. *J. Nutr. Health Aging* **2009**, *13*, 529–537. [CrossRef]
29. Lin, L.C.; Watson, R.; Wu, S.C. What is associated with low food intake in older people with dementia? *J. Clin. Nurs.* **2010**, *19*, 53–59. [CrossRef] [PubMed]
30. Allen, V.; Methven, L.; Gosney, M. The influence of nutritional supplement drinks on providing adequate calorie and protein intake in older adults with dementia. *J. Nutr. Health Aging* **2013**, *17*, 752–755. [CrossRef]
31. Baumgartner, A.; Pachnis, D.; Parra, L.; Hersberger, L.; Bargetzi, A.; Bargetzi, L.; Kaegi-Braun, N.; Tribolet, P.; Gomes, F.; Hoess, C.; et al. The impact of nutritional support on malnourished inpatients with aging-related vulnerability. *Nutrition* **2021**, *89*, 111279. [CrossRef]
32. Dibello, V.; Zupo, R.; Sardone, R.; Lozupone, M.; Castellana, F.; Dibello, A.; Daniele, A.; De Pergola, G.; Bortone, I.; Lampignano, L.; et al. Oral frailty and its determinants in older age: A systematic review. *Lancet Healthy Longev.* **2021**, *2*, e507–e520. [CrossRef]

nutrients

MDPI

Communication

Food Composition Data and Tools Online and Their Use in Research and Policy: EuroFIR AISBL Contribution in 2022

Alessandra Durazzo [1,*], Siân Astley [2], Maria Kapsokefalou [3], Helena Soares Costa [4], Angelika Mantur-Vierendeel [2], Loek Pijls [5], Luca Bucchini [6], Marija Glibetić [7], Karl Presser [8] and Paul Finglas [9,*]

1 CREA—Research Centre for Food and Nutrition, Via Ardeatina 546, 00178 Rome, Italy
2 EuroFIR AISBL, Rue Washington 40, 1050 Brussels, Belgium
3 Department of Food Science & Human Nutrition, Agricultural University of Athens, 11855 Athens, Greece
4 Department of Food and Nutrition, National Institute of Health, 1649-016 Lisbon, Portugal
5 Loekintofood, 3524 GG Utrecht, The Netherlands
6 Hylobates, Consulting S.R.L., 00135 Rome, Italy
7 Centre of Research Excellence in Nutrition and Metabolism, Institute for Medical Research-National Institute of the Republic of Serbia, University of Belgrade, 11129 Belgrade, Serbia
8 Premotec GmbH, 8400 Winterthur, Switzerland
9 Quadram Institute Bioscience, Norwich NR4 7UQ, UK
* Correspondence: alessandra.durazzo@crea.gov.it (A.D.); paul.finglas@quadram.ac.uk (P.F.)

Abstract: Food, nutrition, and health are linked, and detailed knowledge of nutrient compositions and bioactive characteristics is needed to understand these relationships. Additionally, increasingly these data are required by database systems and applications. This communication aims to describe the contribution to databases and nutrition fields as well as the activities of EuroFIR AISBL; this member-based, non-profit association was founded to ensure sustained advocacy for food information in Europe and facilitate improved data quality, storage, and access as well as encouraging wider exploitation of food composition data for both research and commercial purposes. In addition to the description of its role and main objectives, a snapshot of EuroFIR AISBL's activities over the years is also given using a quantitative research literature analysis approach. The focus of this communication is to provide descriptions and updates of EuroFIR's online tools, i.e., FoodEXplorer, eBASIS, and PlantLIBRA, by highlighting the main uses and applications. Integrating food-related infrastructures and databases, following standardized and harmonized approaches, and considering interoperability and metrological principles are significant challenges. Ongoing activities and future plans of EuroFIR AISBL are highlighted, including, for instance, work within the Food Nutrition Security Cloud (FNS-Cloud) to make food, nutrition, and (food) security data more findable, accessible, interoperable, and ultimately reusable.

Keywords: EuroFIR AISBL; food data banks; nutrients; bioactive compounds; standardization; harmonization; interoperability

Citation: Durazzo, A.; Astley, S.; Kapsokefalou, M.; Costa, H.S.; Mantur-Vierendeel, A.; Pijls, L.; Bucchini, L.; Glibetić, M.; Presser, K.; Finglas, P. Food Composition Data and Tools Online and Their Use in Research and Policy: EuroFIR AISBL Contribution in 2022. *Nutrients* **2022**, *14*, 4788. https://doi.org/10.3390/nu14224788

Academic Editor: Laura Di Renzo

Received: 30 September 2022
Accepted: 4 November 2022
Published: 12 November 2022

1. Introduction

1.1. Food Databases and Nutrition: The Current Context

Research exploring relationships between diet and health have, in recent decades, garnered increasing interest in biologically active components in foods alongside nutrients. In addition to nutritional function, bioactive components of the diet have potentially beneficial health properties, which has led to greater perception of foods as functional ingredients or nutraceuticals. Moreover, new properties attributed to nutrients, and the interactions between nutrients and bioactive compounds, are also being explored. Food, nutrition, and health are linked, but detailed knowledge of nutrient compositions and bioactive characteristics is needed to understand these connections, and data characterizing bioactive compounds are required.

In this context, the development of specialized databases for components with nutritional and nutraceutical properties, as well as updating food composition databases (FCDBs) and publishing other specialized datasets (e.g., plant botanicals), at national and European levels, to supply knowledge that can help reduce the burden of chronic diseases and adopt sustainable nutrition patterns, is a challenge [1–4].

Food composition compilers aim to produce, collect, and present data in standardized formats to "speak a common language": this allows comparisons across national databases and fosters exchange and collaboration among countries [5,6]. Simultaneously, researchers are publishing databases compiling information about metabolites in humans and identifying novel dietary biomarkers.

Databases comprised of nutrients, bioactive compounds, metabolites, or food supplements are essential tools for understanding human nutrition and public health and are vital resources for nutritionists, dietitians, food developers, and researchers, with a range of different applications, e.g., dietary assessment, exposure studies, food labeling, epidemiological studies, clinical research, nutritional education, and support for food industries and SMEs for nutrient labeling and health claims. These databases are exploited in epidemiology, food production and nutraceutical, pharmaceutical, and therapeutic interventions, and research trends are frequently redefined.

Initial construction of a dataset for nutrients, bioactive compounds, or compounds classes, and their inclusion in a specialized database, should be monitored to ensure approaches are standardized and database functionalities harmonized with existing resources. Moreover, updating and expanding existing databases, as more comprehensive resources, should be encouraged, perhaps through certification. Databases dedicated to particular or characteristic categories of foods are also valuable (e.g., traditional and ethnic foods, and recipe databases). Traditional and ethnic foods should also be included in national FCDBs and recipe collections. These foods constitute an important part of culture, history, identity, heritage, and local economy of a region or country and are key elements in the dietary patterns of each country [7,8].

Databases dedicated to bioactive compounds, particular individual classes of compounds, as reported by Scalbert et al. [9], can fail to reflect numbers and diversity of chemical features, range of dietary sources, variability from one source to another, and different procedures used to extract compounds as well as analytical techniques used. Additional factors that should be considered are that (i) only a few compounds within a class are investigated, and (ii) there is a lack of appropriate well-documented analytical methods [9] for application in food research.

Technological advances that allow management of "big data", management of distributed and secured data using blockchain or process data using natural language processing, algorithms, or artificial intelligence are relatively new in the exploitation of food composition data. Nevertheless, technologies, tools, and infrastructures are now emerging with properly orchestrated processes leading to delivery of more findable, accessible, interoperable, and reusable (FAIR) big data ecosystems [10–15].

In this context, this communication aims to describe the contribution of the international, member-based, non-profit association EuroFIR AISBL to the status of FCDBs and related information being published in Europe and beyond.

1.2. EuroFIR AISBL: Role, Organization, and Main Features

The mission of EuroFIR AISBL is to promote harmonization and exploitation of high-quality food composition data and foster cooperation and participation in development with national compiler organizations. EuroFIR AISBL coordinates activities with experts and national compilers, contributing to worldwide efforts to produce and maintain high-quality food information, datasets and tools.

EuroFIR AISBL was formed in 2009, arising from the European Food Information Resource (EuroFIR) Network of Excellence (Grant agreement ID: 513944) and NEXUS project (2005–2013, Grant agreement ID: 265967) [16], to ensure sustained advocacy for food information in Europe and beyond in partnership with FAO INFOODS, and facilitate improved data quality, storage and access, and reuse for research and commercial purposes.

To provide a brief snapshot of the research history and status related to the contribution of EuroFIR to food information databases and nutrition fields, a quantitative literature analysis was carried out on 6 June 2022 using Scopus (https://www.scopus.com/home.uri, accessed on 6 June 2022). The search string "EuroFIR" was used, and bibliographic data (i.e., year, count, document type, origin, institutions, etc.) were recorded. Scopus functions "analyze" and "create citation report" were utilized for basic analyses. The search returned 101 documents covering the period 2005–2022, and the main subject areas were *Agricultural and Biological Sciences*, *Nursing*, *Medicine*, and *Chemistry*.

The oldest work was published by McKevith, B. in the journal *Nutrition Bulletin* during 2005 and describes working towards a European food information resource—EuroFIR, but also more specifically FCDBs or tables to be used by dietitians and health professionals, food manufacturers and producers, and other researchers; keywords for this paper were database; European Commission; food composition; and food information resource [17]. Further works, published in 2006, were "EuroFIR update—One pagers and web features" [18], as well as a paper published by the network describing development of a comprehensive, coherent, and validated food composition databank in Europe for nutrients [19]. The most cited work was by Trichopoulou et al. [20], where the importance of including traditional foods in current national FCDBs was highlighted. Papers identified in the search, distributed by typology mainly included, "articles" (74.3%), "reviews" (9.9%), "conference papers" (6.9%), and "book chapters" (3%) (data from Scopus database). Two documents belonging to "editorial" category were also reported, one dedicated to the Second International EuroFIR Congress 2007 [21], and the other to the 3rd International EuroFIR Congress 2009 [22].

Limiting the search to documents including "EuroFIR" as a keyword identified 35 publications, the most recent of which was published by Westenbrink et al. [23] and focused on EuroFIR activities to improve harmonization of documentation for aggregated/compiled values in FCDBs. Kapsokefalou et al. [5] described challenges related to quality of food composition data with a particular emphasis on needs in the Mediterranean area. Machackova et al. [24] published guidelines for calculating nutrient contents of foods by calculation for food business operators. Some works published in 2016 addressed (i) EuroFIR quality approaches for managing food composition data [25]; (ii) implementation of EuroFIR document and data repositories as accessible resources of food composition information [26]; and (iii) GAMA-EuroFIR guidelines for the assessment of methods of analysis [27].

The "full records and cited references" (document title, citation counts, abstract, author, and index keywords) were exported and processed using VOSviewer software (version 1.6.16, 2020; www.vosviewer.com, accessed on 6 June 2021) [28–30]. In total, 58 terms were identified and are visualized as a term map in Figure 1. Figure 1 allowed for the identification of terms correlated with research related to EuroFIR activities, and existing research focused on these topics. Among recurring keywords, food composition, food analysis, Europe, food composition database/s, database (factual)/factual database, human/s, data base, food quality, nutrition, nutrition value, information processing, food, quality control, data quality, nutrient content, nutritional assessment, reference database, food composition data, documentation, food intake, food packaging, food industry, diet, information storage, database system, software, and dietary intake appeared most.

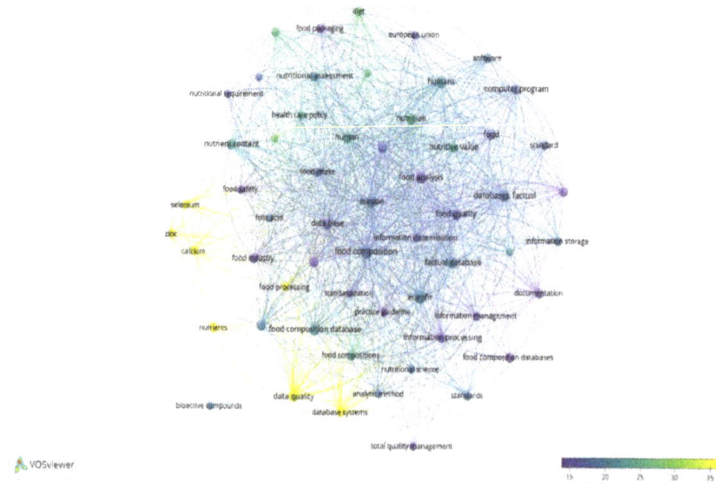

Figure 1. Term map for EuroFIR activities. Bubble size represents numbers of publications. Bubble color represents citations per publication (CPP). Bubbles are closer to one another if terms co-appeared more frequently (bibliometric data were extracted from Scopus and elaborated using VOSviewer software).

2. Updates and Results of EuroFIR AISBL Activities on Implementation of EuroFIR AISBL's Food Data Banks

EuroFIR AISBL provides a resource at the European level for compilers and user communities through online tools, e.g., FoodEXplorer, eBASIS, PlantaLIBRA, FoodWaste-EXplorer [1] (https://www.eurofir.org/our-tools/, accessed on 26 October 2022) (Figure 2).

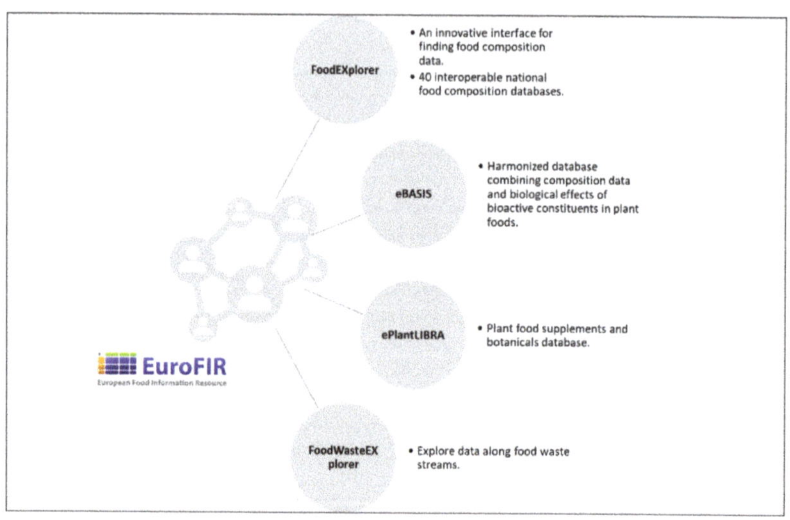

Figure 2. Representation of main EuroFIR AISBL tools.

eBASIS, ePlantLIBRA and FoodWasteExplorer are based on data from peer-reviewed literature evaluated critically by experts. National FCDBs, which form part of FoodExplorer, are based mostly on experimental data and follow EuroFIR compilation guidelines. All FCDBs included in FoodEXplorer are based on a quality assessment system. EuroFIR also set up technical working groups that continue to progress underpinning topics, such as documentation, branded food datasets, recipe calculation, laboratory analysis, and use of FoodCASE for managing food composition data (https://www.eurofir.org/discussion-groups/, accessed on 26 October 2022).

Description of EuroFIR AISBL's Food Data Banks is organized in two subsections: the subsection *EuroFIR's Approaches* gives an overview of: (i) quality management system and standard operating procedures; and (ii) food description and classification systems, while the subsection *EuroFIR AISBL's Food Data Banks: Main features and updates* describes functionalities, updates and use of FoodEXplorer, eBASIS, ePlantLIBRA, and FoodWasteEXplorer.

2.1. EuroFIR AISBL's Approaches

Many international projects and research networks have tried to standardize methods for collection, management, and publication of food data. Efforts in the development of procedures to define and establish standardized collections of food composition data, specifically nutrient content, have also been carried out (e.g., description, selection, preparation, references, analytical or computational approach, compilation) [5,31,32]. EuroFIR AISBL, along with national compilers, have put considerable effort, now recognized globally, in establishing standardized and harmonized food datasets to assure the quality of both compilation processes and their presentation [1,2,5,23–27,33–44].

2.1.1. Quality Management System and Standard Operating Procedures

To ensure the conformity (interoperability) of FCDBs, datasets must demonstrate transparency in aggregation, validation, and compilation based on standardized documentation and quality evaluation. EuroFIR AISBL has established a quality-data management system and harmonized and standardized processes.

EuroFIR AISBL's standard operating procedures (SOPs) are identified in various publications such as: (i) documentation of foods, nutrients, and background information (i.e., metadata); (ii) attribution of quality indices to original data; (iii) coding of original data before data entry; (iv) quality check on original data coding and data entry; (v) physical storage of original data; (vi) selection of original data for aggregation; (vii) selection and application of algorithms to produce aggregated and compiled datasets; (viii) validation of aggregated and compiled datasets; and (ix) selection of aggregated and compiled datasets for dissemination as a published database or tables as well as guidelines for quality data evaluation [23,45,46].

Documentation of information concerning foods, components, values, and references is essential in maintaining a FCDB. One working groups developed guidelines for default value documentation of aggregated/compiled values using the EuroFIR AISBL's standard and thesauri. Options for aggregation/compilation in the FoodCASE data management system were taken as the starting point [23].

2.1.2. Food Description and Classification Systems

There is a consensus on the importance of nomenclature, (food) descriptions, and classification of foods. In this context, and with a view to the exchange of data, design, and development of a database primarily requires exact identification of a food. FoodEx2 is a standardized food classification and description system developed by EFSA, and supported by FAO INFOODS, to describe characteristics of foods and food supplements in exposure assessment studies. This system comprises flexible combinations of classifications and descriptions based on a hierarchical system for food safety-related domains (i.e., food consumption, contaminants, pesticide residues, veterinary drug residues, zoonoses–biological and microbiological aspects, botanicals, and food composition) [47–51].

LanguaL^{TM} or "Langua aLimentaria" or "Language of food" (LanguaL^{TM}) is generally recognized as a method for describing foods, facilitating the capture and exchange of food data. More specifically, LanguaL^{TM} has a controlled vocabulary for systematic food descriptions that can be used with thesauri for faceted classification [52]. As described by Møller and Ireland [53], any food (or food product) can be described systematically using a combination of characteristics. In turn, these characteristics can be categorized and coded for computer processing, and resulting viewpoint/characteristic codes can be used to retrieve data about foods from similarly coded external sources. Each food is described using a set of standard, controlled terms taken from facets characteristic of the nutritional and/or quality aspects of a food, such as: food source, i.e., ANIMAL USED AS FOOD SOURCE [B1297], PLANT USED AS FOOD SOURCE [B1347]; cooking, i.e., TOASTED [G0010], BOILED [G0014], STEAMED WITH PRESSURE [G0022], DEEP-FRIED [G0029]; preservation methods, i.e., PASTEURIZED BY IRRADIATION [J0119], PRESERVED BY FREEZING [J0136], PRESERVED BY STORAGE IN CONTROLLED ATMOSPHERE [J0176]; and treatment applied, i.e., BLEACHED [H0197], PUFFED [H0268], EXTRUDED [H0352]. Several applications of simple foods, food preparations, recipes, food supplements, and agro-food wastes have been carried out [54–57].

LanguaL™ was developed principally to support data exchange, whereas FoodEx2 was developed as a food classification and description system for exposure and risk assessment studies, i.e., exposure to contaminants. LanguaL™ codes are assigned following a facet scheme set in advance, which defines and describes foods (i.e., source, origin, physical state, heat treatment, cooking method, treatment, preservation, packaging, etc.), and this scheme must be applied and maintained for all food items. FoodEx2 coding aggregates food products according to need without following a pre-agreed scheme. For instance, POACHED EGGS are codified by FoodEX2, revision 2 [47–51] using a single base term [A032D], whereas LanguaL™ codifies them using terms string: 02 EGG AND EGG PRODUCTS (EUROCODE2) [A0725], HEN [B1713], WHOLE EGG WITHOUT SHELL [C0225], WHOLE, SHAPE ACHIEVED BY FORMING [E0147], FULLY HEAT-TREATED [F0014], SIMMERED, POACHED OR STEWED [G0020], HUMAN CONSUMER, NO AGE SPECIFICATION [P0024]. Recently, examples of applications using both systems on food preparations and recipes were given by Durazzo et al. [54]. FoodEx2 uses implicit descriptors to reduce code length, whereas LanguaL™ descriptors are fully explicit and structured. Both LanguaL™ and FoodEx2 are updated regularly based on feedback from users. User training courses are run for both LanguaL™ and FoodEx2.

LanguaL™ and FoodEx2 are the main food description and classification systems, and both are well developed, widely used, and recognized at European and International levels [42]. Their use also represents the likely direction of future work, specifically the automation of matching, mapping, and data quality checking. Consequently, maintenance and updating of both systems must be carried out regularly through exchanges between users and developers, considering evolution of the food market and new food classification needs in different applications. Subsequently, the correct application of classification and description systems relies on standard operating procedures (SOPs), regular updates, and multi-disciplinary cooperation [42].

These schemes are, however, not the only coding approaches, and their use can be supplemented with other systems such as ontologies. FoodOn is an open source, harmonized, and comprehensive food ontology that supports global food traceability, quality control, and data integration [58,59]. It is composed of term hierarchy facets that cover basic raw food source ingredients; process terms for packaging, cooking, and preservation; and an upper-level variety of product type [58,59]. For nutrient composition, and likely also bioactives and botanicals, however, EuroFIR AISBL recommends ongoing activities use of LanguaL™ and FoodEx2 [42].

2.2. *EuroFIR AISBL's Food Data Banks: Main Features and Updates*

2.2.1. FoodEXplorer

FoodEXplorer [1] is an innovative interface for searching simultaneously food composition data in most publicly available national FCDBs in the European Union (EU) Member States as well as Canada, the United States, New Zealand, and Japan. Currently, FoodEXplorer [60] host 40 interoperable national FCDBs (EuroFIR AISBL FoodExplorer, https://www.eurofir.org/foodexplorer/foodgroups.php, accessed on 6 June 2022). Food and nutrient data are linked throughout LanguaL^{TM}.

For the search, "access African and EMR data" (https://www.eurofir.org/FoodEXplorer/foodgroups.php?data=D2, accessed on 6 June 2022) was also created, in addition to "access on FoodEXplorer". Open (publicly available) datasets have been developed and published for Australia and New Zealand, Iran, Iraq, Kuwait, Morocco, Pakistan, South Africa, and Tunisia, supported by projects including EMR (Eastern Mediterranean Regional data, funded by UK Medical Research Council Global Challenges Research Fund in collaboration with the World Health Organization's Eastern Mediterranean Regional Office), African data (funded by the UK Biotechnology and Biological Sciences Research Council Global Challenges Research Fund in collaboration with the FAO INFOODS AFROFOODS network), and Food Standards Australia New Zealand (funded by the Commonwealth of Australia and Food Standards Australia New Zealand, 2018).

In this regard, it is worth mentioning the work of Ene-Obong et al. [61], which describes the importance and use of reliable food composition data by nutrition/dietetic professionals in solving Africa's nutrition problems and focuses on constraints and the roles of FAO INFOODS and AFROFOODS as well as other stakeholders in future initiatives. The authors noted how AFROFOODS recommended that compilation, dissemination, and use of food composition tables (FCTs)/FCDBs should be given priority and included in country and regional development and investment plans. Similarly, AFROFOODS has called on governments to incorporate food composition into curricula for higher education, particularly nutrition and dietetics professional learning, but also health and agriculture [61]. More recently, EuroFIR AISBL and Quadram Institute Bioscience (QIB, UK) have assisted AFROFOODS in capacity building and development of a website with help from Premotec GmbH (PMT, CH)—a Swiss company experienced into implementation of software solutions for food data, i.e., food composition, food consumption and total diet studies—to increase visibility and enhance networking, and development of a road map for future activities.

In 2019, analysis of harmonized EuroFIR documentation for macronutrient values in 26 European FCDBs was carried out by Westenbrink et al. [41] to evaluate the impact of harmonized documentation and its usefulness for research and/or policy; documentation of most properties describing nutrient values was complete, even if the percentage coded as unknown varied from 14% to 49% for value and method types, method indicator, and acquisition type. The same authors reported some inconsistencies and incomplete information (about 65% missing) in coding and documentation [41]. Additionally, they noted how easy data exchange was supported by harmonized procedures for data documentation according to EuroFIR guidelines, even if comparability of carbohydrate, dietary fiber, protein, and energy values remained difficult due to multiple definitions and formulae, particularly lack of details about analytical and calculation methods [41].

A potential solution to improve harmonization was defined and published in 2020 in EuroFIR FoodEXplorer Standard [42], providing updated guidelines for collecting, compiling, and updating food composition data. In particular, the following actions were proposed for datasets being uploaded to FoodEXplorer: (i) before uploading, EuroFIR will standardize data units; (ii) energy will be re-calculated using European labelling legislation EU Regulation No. 1169/2011 (https://bit.ly/3g5yegE, accessed on 26 October 2022) recommendations and presented as both kcal and kJ; and data on less common components, such as polyols, organic acids, and salatrims, should be provided and included in recalculation; (iii) vitamin A will be presented as retinol activity equivalents (RAE); (iv) for calculated components, only one value per component ID will be shown; and (v) the use of both LanguaL^{TM} food description coding and FoodEx2 classification and description coding is recommended but not mandatory [42]. In 2020, following user feedback, functionalities of FoodEXplorer were updated, specifically: (i) advanced search functionalities; (ii) formatting of downloads for Excel; (iii) options for sorting components;

(iv) presentation of component values and documentation; and (v) selection of foods for comparison.

Elaborations and applications using data from FoodExplorer were carried out among users and compilers. An example of a FoodEXplorer application for creating specialized food composition datasets, in this case for vitamin D in foods based on European standards for dietary intake assessment, was described by Milešević et al. [62] while Gurinović et al. [63] elaborated development, functionalities, and application of DIET ASSESS & PLAN (DAP) software, a platform for standardized and harmonized food consumption collection, comprehensive dietary intake assessment, and nutrition planning to support public health nutrition research in Central Eastern European Countries (CEEC). DAP enabled exploitation of national FCDBs from FoodEXplorer and their exploration using other online tools [63].

Another example of the utilization of data from FoodExplorer was given by FishChoice 2.0 (www.fishchoice.eu, accessed on 26 October 2022) [64]. FishChoice 2.0 is a tool, relaunched by Marquès et al. [64] as a tool for consumers and nutrition professionals, which delivers information about health benefits/risks as well as some sustainability information for fish and seafood on an individual basis, based on calculation of nutrients and contaminant intakes [64]; FoodEXplorer was used to collect nutrient data for fish and seafood species typically consumed in Europe for inclusion in FishChoice 2.0 [64].

2.2.2. eBASIS—Bioactive Substances in Food Information System

Demand for easily accessible information on composition, intakes, and activities of bioactive compounds is significant among researchers. Bioactive Substances in Food Information System (eBASIS) [65] is a web-based database containing scientifically validated information describing the composition of bioactive compounds in major European plant foods. eBASIS was launched in 2006 [66,67] as a user-friendly, efficient, and flexible interface for the scientific community and food industry. It was the first EU harmonized database combining composition data and biological effects for compound classes, including polyphenols, isoflavones, glucosinolates, phytosterols, glycoalkaloids, and xanthine alkaloids, in 15 languages [68,69].

EuroFIR eBASIS was compiled using data from the peer-reviewed literature evaluated critically by experts. Tutorials for users are available online (https://www.eurofir.org/our-tools/ebasis/, accessed on 26 October 2022) as well as via a short video demonstrating how eBASIS can be used (*Introduction to eBASIS*, https://ebasis.eurofir.org/Default.asp, accessible on 6 June 2022). Currently, eBASIS contains 44,664 datapoints for bioactive compounds for 276 plant-based foods, distributed in main classes, e.g., 677 datapoints for phenols, 3945 datapoints for flavonols, 4581 datapoints for anthocyanins, 881 datapoints for carotenoids, 2695 datapoints for lignans, and 2654 datapoints for glucosinolates (https://ebasis.eurofir.org/Default.asp, accessed on 6 jube 2022).

Information included in eBASIS was described by Pilegaard et al. [70] and, in 2011, the utility of eBASIS tested in a phytosterols case study [71]. In 2017, a new interface linking the eBASIS bioactives database and the Creme Nutrition® model was developed for the BACCHUS project (http://bacchus.cremeglobal.com/bacchus/, accessed on 26 October 2022) [72]. The eBASIS-Creme Global exposure tool enables users to assess compound intakes from various foods across populations to determine whether compounds required to obtain a claimed effect can be reasonably consumed within a balanced diet [72]. In 2018, an update on extractable and non-extractable antioxidants was completed [73] with the addition of 437 quality-evaluated datapoints. This update was the first example of building a resource dedicated to antioxidant properties within the existing resource. An updated eBASIS user guide was published at the same time, covering data concerning antioxidant properties and extractable and non-extractable compounds (https://ebasis.eurofir.org/files/basis_antiox.pdf, accessed on 6 June 2022).

The input form for data includes bibliographic references, food information (i.e., plant, part, subspecies/cultivar, maturity, season, growing conditions, etc.), processing (i.e., shape, state or form, heat treatment, cooking method, treatment applied, preservation method), sampling information (i.e., primary sample unit size, analytical sample size, sample plan, sample handling, etc.), compositional information (i.e., compound class, analytical method, concentration, extraction, and preparation, identification, etc.), and quality assessment. For each eBASIS section (plant/food description, processing defined, sampling plan, sample handling, compound identification, analytical method, analytical performance), transparent quality systems are included, ensuring eBASIS as a reliable resource for research with up-to-date information about plant food phytochemicals.

eBASIS was developed to present raw rather than aggregated data, reflecting variations in bioactive compositions related to cultivar, plant part, growing conditions, processing, and country of origin; there are multiple datapoints for each compound/food combination. To better meet requirements for aggregated bioactive composition data in dietary intake assessment, eBASIS data structures are being organized to link plant food data and bioactives with dietary intake assessment outputs and coding systems. At the same time, the architecture permits future inclusion of food data from animal origins and/or addition of new data on other plant foods/products or classes of compounds, emphasizing the need to envisage potential needs and gaps during development.

2.2.3. ePlantLIBRA

In the area of dietary supplements (FDA definition)/food supplements (EFSA definition) [74], ePlantLIBRA [75,76] presents comprehensive and searchable data describing bioactive compounds specific to plant-based food supplements and botanicals, reporting health benefits, adverse effects, contaminants, and residues. ePlantLIBRA was developed by the PlantLIBRA project (PLANT food supplements: Levels of Intake, Benefit and Risk Assessment, Grant Agreement ID: ID: 245199) [77], which addressed development, validation, and dissemination of data and methodologies for risk and benefit assessment of plant food supplements and botanicals, and sustainable international cooperation in this domain [77].

ePlantLIBRA has the same structure as eBASIS; it is based on a user-friendly, efficient, and flexible interface for searching, extracting, and exporting data including links to the original references [76]. The architecture is based on eBASIS, MoniQA contaminant (FP6 Monitoring and Quality Assurance in the total food supply chain, Grant Agreement ID: 36337), and FERA's HorizonScan databases (https://www.eurofir.org/our-tools/eplantlibra/, accessed on 26 October 2022). A webinar is available (https://www.eurofir.org/our-tools/eplantlibra/, accessed on 26 October 2022) with short videos covering the functionality of ePlantLIBRA (https://eplantlibra.eurofir.org/Default.asp, accessed on 6 June 2022).

Currently, 45,168 and 117 datapoints are available for composition and beneficial data, respectively, and 55 are specifically addressed to plant-based food supplements or botanicals, e.g., aloe vera extract, borage oil, pomegranate supplement, boswellia products, cinnamon products, dandelion products, and so on (https://eplantlibra.eurofir.org/Default.asp, accessed on 6 June 2022).

2.2.4. FoodWasteExplorer

Advances in food research are increasingly directed towards sustainability of food chains, including exploitation of unconventional foods/waste for biologically active compounds, and reuse or recycling to achieve a circular economy. FoodWasteEXplorer [78] brings together the compositions of some of the most common products and their associated side streams and was developed within the EU-founded project REFRESH (REFRESH: Resource Efficient Food and dRink for the Entire Supply cHain, Grant Agreement ID: 641933, https://eu-refresh.org/, accessed on 6 June 2022). Currently, FoodWasteEXplorer contains 27,069 datapoints, including 587 nutrients, 698 bioactives, and 49 toxicants, gathered from peer-reviewed papers, grey literature (e.g., manufacturers' data), and other sources (https://ws.eurofir.org/foodwasteexplorer/about, accessed on 6 June 2022). Food and side streams in FoodWasteEXplorer are searchable and grouped under areas of interest such as wine and beer, spirits, cider, cereals, chocolate, (fruit and vegetable) juices, cheese, animal products, sugar, vegetable oil, and coffee production. They are also grouped into food categories, e.g., cereals; milk and dairy; eggs; fats and oils, nuts and seeds; fish and seafood; fruits and vegetables; beverages; and other (i.e., algae, frog, snail, etc.). Finally, specific searchable functions—by foods, side streams, components—are available, e.g., by searching for foods, coffee, related side stream* information about the compositions of coffee grounds, coffee husks, coffee hulls, coffee leaves, coffee pulp (dried), coffee oil meal, malt coffee marc, instant coffee by product, and coffee parchment are described.

2.2.5. Other Developing/Ongoing Resources: FoodCASE

FoodCASE was developed by Premotec GmbH (CH) in partnership with EuroFIR AISBL to manage food composition, food consumption, total diet study (TDS), laboratory food analysis, and branded food data, assembling food information in one system to promote re-use by linking food lists to other datasets and resources [79,80]. This data management system has wizards to support advanced data operations such as data import and export, recipe calculations, dataset linkage, nutrient estimation, data issue, and data quality analysis. It also supports different processes involved in the acquisition, management, and processing of data and uses European and international standards for the different datasets [80].

3. Ongoing Work and Future Directions

To ensure that EuroFIR AISBL resources remain valuable to user communities, it is important not only to update, expand, and enhance databases, but also to do these in standardized and harmonized ways among organizations and countries, considering existing and emerging food sources, and adding new descriptors and markers as necessary. To this end, engagement with networks and research infrastructures is a priority, creating synergies necessary to generate high-quality data and develop tools for the production, management, and exploitation of food data. In line with the European Strategy Forum on Research Infrastructures (ESFRI), the research infrastructure METROFOOD-RI and the European Open Science Cloud (EOSC), strategies leading to reliable and comparable analytical measurements in foods along food chains, from primary producers to consumers and beyond (food waste) and increasingly FAIR data [81] are valuable for researchers, food producers, and consumers. However, continued cooperation and sharing of data between compilers and users, within an integrated approach for agro-food, nutrition, and health, are key to success. Management of data at agro-food, nutrition, and health interfaces is a priority, but integrating FCDBs and infrastructures (interoperability) can only be achieved if approaches are applied based on metrological principles [81–86].

In this context, EuroFIR AISBL is involved in Member and Client activities and EU or otherwise-funded projects considering a range of relevant topics. The Food Nutrition Security Cloud (FNS-Cloud, Grant Agreement ID: 863059, www.fns-cloud.eu, accessed on 26 October 2022) aims to support integration of existing and emerging food research data and tools to address diet and health research questions across agro-food, nutrition and lifestyle, and non-communicable disease and healthy diet domains [87].

EuroFIR AISBL is also active in the proposed Food Nutrition Health Research Infrastructure (FNH-RI), which aims to link food production (agriculture and food technology) and food consumption (food determinants, intake, nutrition, and health) domains. To this end, a prototype *Determinants and Intake Platform*, harmonizing and linking consumer food behaviors, was formulated based on EuroDISH (Study on the need for food and health research infrastructures in Europe, Grant Agreement ID: 311788) and RICHFIELDS (Research infrastructure on consumer health and food intake using e-science with linked data sharing, Grant Agreement ID: 654280) outputs [88].

With the food environment undergoing vast changes, the need to study the nutritional variation in processed foods has driven an international move for branded food composition databases (BFCDBs). EuroFIR AISBL is working with its members to create a platform for collaboration and advocacy around BFCDBs, addressing user needs and gaps surveyed in 2020–2021. During the EuroFIR Food Forum 2021, a workshop was dedicated to BFCDBs, discussing advances at the European level and open access issues.

Author Contributions: Conceptualization, A.D. and P.F.; writing—original draft preparation, A.D., S.A., M.K., H.S.C., A.M.-V., L.P., L.B., M.G., K.P. and P.F. All authors have made substantial, direct, and intellectual contributions to the work and approved it for publication. All authors have read and agreed to the published version of the manuscript.

Funding: This research received no external funding.

Institutional Review Board Statement: Not applicable.

Informed Consent Statement: Not applicable.

Data Availability Statement: Not applicable.

Conflicts of Interest: The authors declare no conflict of interest.

References

1. Finglas, P.M.; Berry, R.; Astley, S. Assessing and improving the quality of food composition databases for nutrition and health applications in Europe: The contribution of EuroFIR. *Adv. Nutr.* **2014**, *5*, 608S–614S. [CrossRef] [PubMed]
2. Finglas, P.; Roe, M.; Pinchen, H.; Astley, S. The contribution of food composition resources to nutrition science methodology. *Nutr. Bull.* **2017**, *42*, 198–206. [CrossRef]
3. Durazzo, A.; Lucarini, M. Food Composition and Dedicated Databases: Key Tools for Human Health and Public Nutrition. *Nutrients* **2021**, *13*, 4003. [CrossRef] [PubMed]
4. Durazzo, A.; Lucarini, M. Editorial: Databases and Nutrition. *Front. Nutr.* **2022**, *9*, 853600. [CrossRef] [PubMed]
5. Kapsokefalou, M.; Roe, M.; Turrini, A.; Costa, H.S.; Martinez-Victoria, E.; Marletta, L.; Berry, R.; Finglas, P. Food Composition at Present: New Challenges. *Nutrients* **2019**, *11*, 1714. [CrossRef]
6. Traka, M.H.; Plumb, J.; Berry, R.; Pinchen, H.; Finglas, P.M. Maintaining and updating food composition datasets for multiple users and novel technologies: Current challenges from a UK perspective. *Nutr. Bull.* **2020**, *45*, 230–240. [CrossRef]
7. Costa, H.S.; Vasilopoulou, E.; Trichopoulou, A.; Finglas, P. New nutritional data on traditional foods for European food composition databases. *Eur. J. Clin. Nutr.* **2010**, *64*, S73–S81. [CrossRef] [PubMed]
8. Costa, H.S.; Albuquerque, T.G.; Sanches-Silva, A.; Vasilopoulou, E.; Trichopoulou, A.; D'Antuono, F.; Alexieva, I.; Boyko, N.; Costea, C.; Fedosov, S.; et al. New nutritional composition data on selected traditional foods consumed in Black Sea Area countries. *J. Sci. Food Agric.* **2013**, *93*, 3524–3534. [CrossRef]
9. Scalbert, A.; Andres-Lacueva, C.; Arita, M.; Kroon, P.; Manach, C.; Urpi-Sarda, M.; Wishart, D. Databases on food phytochemicals and their health-promoting effects. *J. Agric. Food Chem.* **2011**, *59*, 4331–4348. [CrossRef]
10. Cade, J.E. Measuring Diet in the 21st century: Use of New Technologies. *Proc. Nutr. Soc.* **2017**, *76*, 276–282. [CrossRef]
11. Koroušić Seljak, B.; Korošec, P.; Eftimov, T.; Ocke, M.; van der Laan, J.; Roe, M.; Berry, R.; Crispim, S.P.; Turrini, A.; Krems, C.; et al. Identification of Requirements for Computer-Supported Matching of Food Consumption Data with Food Composition Data. *Nutrients* **2018**, *10*, 433. [CrossRef] [PubMed]
12. Pehrsson, P.R.; Mitchell, D.C. Advancing Food and Nutrient Databases through Partnerships and Technology. *J. Food Comp. Anal.* **2020**, *92*, 103527. [CrossRef]
13. Mansouri, Y.; Prokhorenko, V.; Ali Babar, M. An Automated Implementation of Hybrid Cloud. *J. Netw. Comp. Appl.* **2020**, *167*, 102740. [CrossRef]
14. Balech, B.; Brennan, L.; de Santa Pau, E.C.; Cavalieri, D.; Coort, S.; D'Elia, D.; Dragsted, L.O.; Eftimov, T.; Evelo, C.T.; Ferk, P.; et al. The future of food and nutrition in ELIXIR [version 1; peer review: 1 approved with reservations]. *F1000Research* **2022**, *11*, 978. [CrossRef]
15. Durazzo, A.; Sorkin, B.C.; Lucarini, M.; Gusev, P.A.; Kuszak, A.J.; Crawford, C.; Boyd, C.; Deuster, P.A.; Saldanha, L.G.; Gurley, B.J.; et al. Analytical Challenges and Metrological Approaches to Ensuring Dietary Supplement Quality: International Perspectives. *Front. Pharmacol.* **2022**, *12*, 714434. [CrossRef] [PubMed]
16. EuroFIR Network of Excellence and NEXUS Projects. Available online: http://www.eurofir.org/our-resources/noe-and-nexus-projects/ (accessed on 20 May 2019).
17. McKevith, B. Working towards a European food information resource–EuroFIR. *Nutr. Bull.* **2005**, *30*, 290–293. [CrossRef]
18. McKevith, B.J. EuroFIR update-One pagers and web features. *Nutr. Bull.* **2006**, *31*, 69–70. [CrossRef]
19. European Food Information Resources (EuroFIR) Network. *Nutr. Food Sci.* **2006**, *36*, 103–107. [CrossRef]
20. Trichopoulou, A.; Soukara, S.; Vasilopoulou, E. Traditional foods: A science and society perspective. *Trends Food Sci. Technol.* **2007**, *18*, 420–427. [CrossRef]
21. Denny, A.R.; Buttriss, J.L.; Finglas, P.M. Second International EuroFIR Congress 2007: The role of food composition data in improving quality, healthiness and safety of European diets. *Food Chem.* **2009**, *113*, 733–735. [CrossRef]

22. Finglas, P.; Weichselbaum, E.; Buttriss, J.L. The 3rd international EuroFIR congress 2009: European food composition data for better diet, nutrition and food quality. *Eur. J. Clin. Nutr.* **2010**, *64*, S1–S3. [CrossRef] [PubMed]
23. Westenbrink, S.; Presser, K.; Roe, M.; Ireland, J.; Finglas, P. Documentation of aggregated/compiled values in food composition databases; EuroFIR default to improve harmonization. *J. Food Comp. Anal.* **2021**, *101*, 103968. [CrossRef]
24. Machackova, M.; Giertlova, A.; Porubska, J.; Roe, M.; Ramos, C.; Finglas, P. EuroFIR Guideline on calculation of nutrient content of foods for food business operators. *Food Chem.* **2018**, *238*, 35–41. [CrossRef] [PubMed]
25. Westenbrink, S.; Roe, M.; Oseredczuk, M.; Castanheira, I.; Finglas, P. EuroFIR quality approach for managing food composition data; where are we in 2014? *Food Chem.* **2016**, *193*, 69–74. [CrossRef] [PubMed]
26. Unwin, I.; Jansen-van der Vliet, M.; Westenbrink, S.; Presser, K.; Infanger, E.; Porubska, J.; Roe, M.; Finglas, P. Implementing the EuroFIR Document and Data Repositories as accessible resources of food composition information. *Food Chem.* **2016**, *193*, 90–96. [CrossRef] [PubMed]
27. Castanheira, I.; Saraiva, M.; Rego, A.; Ollilainen, V. EuroFIR guidelines for assessment of methods of analysis: GAMA. *Food Chem.* **2016**, *193*, 82–89. [CrossRef] [PubMed]
28. Van Eck, N.J.; Waltman, L. Software survey: VOSviewer, a computer program for bibliometric mapping. *Scientometrics* **2009**, *84*, 523–538. [CrossRef]
29. Waltman, L.; van Eck, N.J.; Noyons, E.C.M. A unified approach to mapping and clustering of bibliometric networks. *J. Inf.* **2010**, *4*, 629–635. [CrossRef]
30. Van Eck, N.J.; Waltman, L. Text mining and visualization using VOSviewer. *ISSI Newslett.* **2011**, *7*, 50–54.
31. Slimani, N.; Deharveng, G.; Unwin, I.; Vignat, J.; Skeie, G.; Salvini, S.; Møller, A.; Ireland, J.; Becker, W.; Southgate, D. Standardisation of an European end-user nutrient database for nutritional epidemiology: What can we learn from the EPIC Nutrient Database (ENDB) Project? *Trends Food Sci. Technol.* **2007**, *18*, 407–419. [CrossRef]
32. International Network of Food Data Systems (INFOODS). Available online: http://www.fao.org/infoods/infoods/en/ (accessed on 20 May 2019).
33. Castanheira, I.; André, C.; Oseredczuk, M.; Ireland, J.; Owen, L.; Robb, P.; Earnshaw, A.; Calhau, M.A. Improving data quality in food composition databanks: A EuroFIR contribution. *Accred Qual Assur* **2007**, *12*, 117–125. [CrossRef]
34. Castanheira, I.; Roe, M.; Westenbrink, S.; Ireland, J.; Møller, A.; Salvini, S.; Beernaert, H.; Oseredczuk, M.; Calhau, M.A. Establishing quality management systems for European food composition databases. *Food Chem.* **2009**, *113*, 776–780. [CrossRef]
35. Møller, A.; Unwin, I.D.; Becker, W.; Ireland, J. EuroFIR's food databank systems for nutrients and bioactives. *Trend. Food Sci. Technol.* **2007**, *18*, 428–433. [CrossRef]
36. Khokhar, S.; Gilbert, P.A.; Moyle, C.W.A.; Carnovale, E.; Shahar, D.R.; Ngo, J.; Saxholt, E.; Ireland, J.; Jansen-van der Vliet, M.; Bellemans, M. Harmonised procedures for producing new data on the nutritional composition of ethnic foods. *Food Chem.* **2009**, *113*, 816–824. [CrossRef]
37. Hollman, P.C.H.; Witthöft, C.M.; Busstra, M.C.; Elburg, L.; Hulshof, P. Training aspects in the use and production of food composition databases. The EuroFIR experience. *Food Chem.* **2009**, *113*, 842–845. [CrossRef]
38. Bell, S.; Colombani, P.C.; Pakkala, H.; Christensen, T.; Møller, A.; Finglas, P.M. Food composition data: Identifying new uses, approaching new users. *J. Food. Comp. Anal.* **2011**, *24*, 727–731. [CrossRef]
39. Bell, S.; Pakkala, H.; Finglas, P.M. Towards a European food composition data interchange platform. *Int. J. Vitam. Nutr. Res.* **2012**, *82*, 209–215. [CrossRef]
40. Westenbrink, S.; Oseredczuk, M.; Castanheira, I.; Roe, M. Food composition databases: The EuroFIR approach to develop tools to assure the quality of the data compilation process. *Food Chem.* **2009**, *113*, 759–767. [CrossRef]
41. Westenbrink, S.; Kadvan, A.; Roe, M.; Koroušić, B.S.; Mantur-Vierendeel, A.; Finglas, P. 12th IFDC 2017 Special Issue–Evaluation of harmonized EuroFIR documentation for macronutrient values in 26 European food composition databases. *J. Food Comp. Anal.* **2019**, *80*, 40–50. [CrossRef]
42. Westenbrink, S.; Roe, M.; Kadvan, A.; Koroušić, B.S.; Mantu-Vierendeel, A.; Finglas, P.M. *EuroFIR AISBL FoodEXplorer Standard Version 2020–01*; EuroFIR AISBL: Brussels, Belgium, 2020; *(Numero d'entreprise: 811.804.183)*; Available online: https://www.eurofir.org/wp-content/uploads/2020/08/2020-08-19-FoodEXplorer-standard-edn2020-01.pdf (accessed on 26 October 2022).
43. Reinivuo, H.; Bell, S.; Ovaskainen, M.L. Harmonisation of recipe calculation procedures in European food composition databases. *J. Food Comp. Anal.* **2009**, *22*, 410–413. [CrossRef]
44. Gibson-Moore, H. EuroFIR: Where we are now? *Nutr. Bull.* **2013**, *38*, 358–362. [CrossRef]
45. Astley, S.; Bell, S.; Beernaert, H.; Black, L.; Borgejardet, A.; Cavi, E.; Colombani, P.; Finglas, P.; Ireland, J.; Gnagnarella, P.; et al. Standard Operating Procedures-Technical Manual 2019. Available online: https://www.eurofir.org/wp-content/uploads/2019/03/2019-03-.pdf (accessed on 26 October 2022).
46. Eurofir Workpackage 1.3, Task Group 4 Guidelines For Quality Index Attribution To Original Data From Scientific Literature Or Reports For Eurofir Data Interchange. Available online: https://www.eurofir.org/wp-admin/wp-content/uploads/Deliverables/EuroFIR_Quality_Index_Guidelines.pdf (accessed on 26 October 2022).
47. European Food Safety Authority. The food classification and description system FoodEx2 (revision 2). *EFSA Support. Publ.* **2015**, *12*, 804E.
48. EFSA (European Food Safety Authority); Ioannidou, S.; Nikolic, M.; Gibin, D. FoodEx2 maintenance 2016–2018. *EFSA Support. Publ.* **2019**, *16*, 1584E. [CrossRef]

49. EFSA (European Food Safety Authority); Nikolic, M.; Ioannidou, S. FoodEx2 maintenance 2019. *EFSA Support. Publ.* **2020**, *17*, 1810E. [CrossRef]
50. EFSA (European Food Safety Authority); Nikolic, M.; Ioannidou, S. FoodEx2 maintenance 2020. *EFSA Support. Publ.* **2021**, *18*, 6507E. [CrossRef]
51. EFSA (European Food Safety Authority); Nikolic, M.; Ioannidou, S. FoodEx2 maintenance 2021. *EFSA Support. Publ.* **2022**, *19*, 7220E. [CrossRef]
52. Møller, A.; Ireland, J. *LanguaL™ 2017—The LanguaL™thesaurus Technical Report*; Danish Food Informatics: Roskilde, Denmark, 2018; p. 2724. ISBN 978-87-92125-26-2. Available online: https://www.langual.org/download/LanguaL2017/LanguaL%202017%20Thesaurus.pdf (accessed on 26 October 2022).
53. Møller, A.; Ireland, J. *LanguaL™ 2014-The LanguaL™ Thesaurus. EuroFIR Technical Report*; Danish Food Informatics: Roskilde, Denmark, 2015; p. 1486. ISBN 978-87-92125-24-8. Available online: http://www.langual.org/download/LanguaL2014/LanguaL%202014%20Thesaurus%20-%20Final.pdf (accessed on 26 October 2022).
54. Durazzo, A.; Camilli, E.; D'Addezio, L.; Sette, S.; Marconi, S.; Piccinelli, R.; Le Donne, C.; Turrini, A.; Marletta, L. Italian composite dishes: Description and classification by LanguaL™ and FoodEx2. *Eur. Food Res. Technol.* **2019**, *246*, 287–295. [CrossRef]
55. Durazzo, A.; Camilli, E.; D'Addezio, L.; Piccinelli, R.; Mantur-Vierendeel, A.; Marletta, L.; Finglas, P.; Turrini, A.; Sette, S. Development of Dietary Supplement Label Database in Italy: Focus of FoodEx2 Coding. *Nutrients* **2020**, *12*, 89. [CrossRef]
56. Di Stefano, V.; Bongiorno, D.; Buzzanca, C.; Indelicato, S.; Santini, A.; Lucarini, M.; Fabbrizio, A.; Mauro, M.; Vazzana, M.; Arizza, V.; et al. Fatty Acids and Triacylglycerols Profiles from Sicilian (Cold Pressed vs. Soxhlet) Grape Seed Oils. *Sustainability* **2021**, *13*, 13038. [CrossRef]
57. Durazzo, A.; Lucarini, M.; Plutino, M.; Lucini, L.; Aromolo, R.; Martinelli, E.; Souto, E.B.; Santini, A.; Pignatti, G. Bee Products: A Representation of Biodiversity, Sustainability, and Health. *Life* **2021**, *11*, 970. [CrossRef]
58. Dooley, D.M.; Griffiths, E.J.; Gosal, G.S.; Buttigieg, P.L.; Hoehndorf, R.; Lange, M.C.; Schriml, L.M.; Brinkman, F.S.L.; Hsiao, W.W.L. FoodOn: A harmonized food ontology to increase global food traceability, quality control and data integration. *NPJ Sci. Food.* **2018**, *2*, 23. [CrossRef] [PubMed]
59. FoodOn. Available online: http://foodon.org (accessed on 23 October 2022).
60. EuroFIR–FoodEXplorer-Website. Available online: https://www.eurofir.org/foodexplorer/login1.php2020 (accessed on 6 June 2022).
61. Ene-Obong, H.; Schönfeldt, H.C.; Campaore, E.; Kimani, A.; Mwaisaka, R.; Vincent, A.; El Ati, J.; Kouebou, P.; Presser, K.; Finglas, P.; et al. Importance and use of reliable food composition data generation by nutrition/dietetic professionals towards solving Africa's nutrition problem: Constraints and the role of FAO/INFOODS/AFROFOODS and other stakeholders in future initiatives. *Proc. Nutr. Soc.* **2019**, *78*, 496–505. [CrossRef] [PubMed]
62. Milešević, J.; Samaniego, L.; Kiely, M.; Glibetić, M.; Roe, M.; Finglas, P. Specialized food composition dataset for vitamin D content in foods based on European standards: Application to dietary intake assessment. *Food Chem.* **2018**, *240*, 544–549. [CrossRef] [PubMed]
63. Gurinović, M.; Milešević, J.; Kadvan, A.; Nikolić, M.; Zeković, M.; Djekić-Ivanković, M.; Dupouy, E.; Finglas, P.; Glibetić, M. Development, features and application of DIET ASSESS & PLAN (DAP) software in supporting public health nutrition research in Central Eastern European Countries (CEEC). *Food Chem.* **2018**, *238*, 186–194. [CrossRef] [PubMed]
64. Marquès, M.; Torres, C.M.; García-Fernández, F.; Mantur-Vierendeel, A.; Roe, M.; Wilson, A.M.; Reuver, M.; Nadal, M.; Domingo, J.L. FishChoice 2.0: Information on health benefits / risks and sustainability for seafood consumers. *Food Chem Toxicol.* **2021**, *155*, 112387. [CrossRef] [PubMed]
65. eBASIS—Bioactive Substances in Food Information System. Available online: http://ebasis.eurofir.org/Default.asp (accessed on 6 June 2022).
66. Gry, J.; Black, L.; Eriksen, F.D.; Pilegaard, K.; Plumb, J.; Rhodes, M.; Sheehan, D.; Kiely, M.; Kroon, P.A. EuroFIR-BASIS—A combined composition and biological activity database for bioactive compounds in plant-based foods. *Trends Food Sci. Technol.* **2007**, *18*, 434–444. [CrossRef]
67. Black, L.; Kiely, M.; Kroon, P.; Plumb, J.; Gry, J. Development of EuroFIR-BASIS–A composition and biological effects database for plant-based bioactive compounds. *Nutr. Bull.* **2008**, *33*, 58–61. [CrossRef]
68. Kiely, M.; Black, L.J.; Plumb, J.; Kroon, P.A.; Hollman, P.C.; Larsen, J.C.; Speijers, G.J.; Kapsokefalou, M.; Sheehan, D.; Gry, J.; et al. EuroFIR eBASIS: Application for health claims submissions and evaluations. *Eur. J. Clin. Nutr.* **2010**, *64*, S101–S107. [CrossRef] [PubMed]
69. Plumb, J.; Pigat, S.; Bompola, F.; Cushen, M.; Pinchen, H.; Nørby, E.; Astley, S.; Lyons, J.; Kiely, M.; Finglas, P. eBASIS (Bioactive Substances in Food Information Systems) and bioactive intakes: Major updates of the bioactive compound composition and beneficial bio effects database and the development of a probabilistic model to assess intakes in Europe. *Nutrients* **2017**, *9*, 320. [CrossRef]
70. Pilegaard, K.; Eriksen, F.D.; Soerensen, M.; Gry, J. Information on plant foods in eBASIS: What is in a correct botanical scientific name? *Eur. J. Clin. Nutr.* **2010**, *64*, S108–S111. [CrossRef]
71. Plumb, J.A.; Rhodes, M.J.C.; Lampi, A.M.; Buchgraber, M.; Kroon, P.A. Phytosterols in plant foods: Exploring contents, data distribution and aggregated values using an online bioactives database. *J. Food Comp. Anal.* **2011**, *24*, 1024–1031. [CrossRef]

72. Pigat, S.; Bompola, F.; Cushen, M.; Plumb, J.; Finglas, P. Combining National Food Consumption Survey Data with a Bioactive Database in a Probabilistic Intake Model to Assess Bioactive Intakes in Europe-A Tool Resulting from the European BACCHUS Project. *FASEB J.* **2017**, *31*, 648-11.
73. Plumb, J.; Durazzo, A.; Lucarini, M.; Camilli, E.; Turrini, A.; Marletta, L.; Finglas, P. Extractable and Non-Extractable Antioxidants Composition in the eBASIS Database: A Key Tool for Dietary Assessment in Human Health and Disease Research. *Nutrients* **2020**, *12*, 3405. [CrossRef] [PubMed]
74. Durazzo, A.; Lucarini, M.; Heinrich, M. Editorial: Dietary Supplements, Botanicals and Herbs at The Interface of Food and Medicine. *Front. Pharmacol.* **2022**, *13*, 899499. [CrossRef] [PubMed]
75. ePlantLIBRA database. Available online: https://eplantlibra.eurofir.org/Default.asp (accessed on 6 June 2022).
76. Plumb, J.; Lyons, J.; Nørby, K.; Thomas, M.; Nørby, E.; Poms, R.; Bucchini, L.; Restani, P.; Kiely, M.; Finglas, P.; et al. PlantLIBRA Consortia. ePlantLIBRA: A composition and biological activity database for bioactive compounds in plant food supplements. *Food Chem.* **2016**, *193*, 121–127. [CrossRef] [PubMed]
77. Bucchini, L.; Rodarte, A.; Restani, P. The PlantLIBRA project: How we intend to innovate the science of botanicals. *Food Funct.* **2011**, *2*, 769–773. [CrossRef] [PubMed]
78. FoodWaste Explorer. Available online: https://ws.eurofir.org/foodwasteexplorer/ (accessed on 6 June 2022).
79. Presser, K.; Hinterberger, H.; Weber, D.; Norrie, M. A scope classification of data quality requirements for food composition data. *Food Chem.* **2016**, *193*, 166–172. [CrossRef]
80. Presser, K.; Weber, D.; Norrie, M. FoodCASE: A system to manage food composition, consumption and TDS data. *Food Chem.* **2018**, *238*, 166–172. [CrossRef]
81. Rychlik, M.; Zappa, G.; Añorga, L.; Belc, N.; Castanheira, I.; Donard, O.F.; Kouřimská, L.; Ogrinc, N.; Ocké, M.C.; Presser, K.; et al. Ensuring Food Integrity by Metrology and FAIR Data Principles. *Front Chem.* **2018**, *6*, 49. [CrossRef]
82. Dwyer, J.T.; Saldanha, L.; Bailen, R.; Durazzo, A.; Le Donne, C.; Piccinelli, R.; Andrews, K.; Pehrsson, P.; Gusev, P.; Calvillo, A.; et al. Commentary: An Impossible Dream? Integrating Dietary Supplement Label Databases Needs, Challenges, Next Steps. *J. Food Comp. Anal.* **2021**, *102*, 103882. [CrossRef]
83. Durazzo, A.; Souto, E.B.; Lombardi-Boccia, G.; Santini, A.; Lucarini, M. Metrology, Agriculture and Food: Literature Quantitative Analysis. *Agriculture* **2021**, *11*, 889. [CrossRef]
84. Zeb, A.; Soininen, J.-P.; Sozer, N. Data harmonisation as a key to enable digitalisation of the food sector: A review. *Food Bioprod. Proces.* **2021**, *127*, 360–370. [CrossRef]
85. Tsimidou, M.Z.; Ordoudi, S.A.; Mantzouridou, F.T.; Nenadis, N.; Stelzl, T.; Rychlik, M.; Belc, N.; Zoani, C. Strategic Priorities of the Scientific Plan of the European Research Infrastructure METROFOOD-RI for Promoting Metrology in Food and Nutrition. *Foods* **2022**, *11*, 599. [CrossRef] [PubMed]
86. Andrés-Hernández, L.; Blumberg, K.; Walls, R.L.; Dooley, D.; Mauleon, R.; Lange, M.; Weber, M.; Chan, L.; Malik, A.; Møller, A.; et al. Establishing a Common Nutritional Vocabulary—From Food Production to Diet. *Front. Nutr.* **2022**, *9*, 928837. [CrossRef] [PubMed]
87. Finglas, P.; Astley, S.; Korousic-Seljak, B.; Gibney, E.; Presser, K.; de la Cueva Gonzalez-Cotera, J.; Fillery-Travis, A.; Webb, S.; Frankte, P.; on behalf of the FNS Consortium. *The Food Nutrition Security Cloud in Europe. Book of Abstracts. 5th Imeko Foods, Metrology for Sustainable Food Production*; European Commission: Prague, Czech Republic, 2020.
88. Timotijevic, L.; Astley, S.; Bogaardt, M.; Bucher, T.; Carr, I.; Copani, G.; de la Cueva, J.; Eftimov, T.; Finglas, P.; Hieke, S.; et al. Designing a research infrastructure (RI) on food behaviour and health: Balancing user needs, business model, governance mechanisms and technology. *Trend. Food Sci. Technol.* **2021**, *116*, 405–414. [CrossRef]

MDPI
St. Alban-Anlage 66
4052 Basel
Switzerland
Tel. +41 61 683 77 34
Fax +41 61 302 89 18
www.mdpi.com

Nutrients Editorial Office
E-mail: nutrients@mdpi.com
www.mdpi.com/journal/nutrients

www.ingramcontent.com/pod-product-compliance
Lightning Source LLC
Chambersburg PA
CBHW061500210926
51850CB00008B/1770